Current Technologies in Plant Molecular Breeding

Hee-Jong Koh • Suk-Yoon Kwon
Michael Thomson
Editors

Current Technologies in Plant Molecular Breeding

A Guide Book of Plant Molecular Breeding for Researchers

Springer

Editors
Hee-Jong Koh
College of Agriculture and Life Sciences
Crop Science and Biotechnology
Seoul National University
Seoul, Korea, Republic of (South Korea)

Suk-Yoon Kwon
Plant Systems Engineering
Research Center
KRIBB, Daejeon, Korea, Republic of
(South Korea)

Michael Thomson
Plant Breeding, Genetics,
and Biotechnology
International Rice Research Institute
Laguna, Philippines

ISBN 978-94-017-9995-9 ISBN 978-94-017-9996-6 (eBook)
DOI 10.1007/978-94-017-9996-6

Library of Congress Control Number: 2015946612

Springer Dordrecht Heidelberg New York London

Printed on acid-free paper

Springer Science+Business Media B.V. Dordrecht is part of Springer Science+Business Media (www.springer.com)

Preface

Plant breeding, the science for plant genetic improvement, made great progress in the twentieth century with the rediscovery of Mendelian genetic principles in 1900. Most of the traditional breeding methods were established before the 1960s leading to the development of high-yielding varieties in cereal crops which brought the "Green Revolution" during the 1960s–1980s. Recent progress in biotechnology and genomics has expanded the breeders' horizon providing a molecular platform on the traditional plant breeding, which is now known as "plant molecular breeding." Under a new paradigm of plant breeding in the twenty-first century, breeders try to create new variation through the direct manipulation of target genes instead of phenotype-based trait selection. Genetic resources are extended to the unrelated species because transgenic technologies break through the sexual limit for gene transfer. In addition, selection and genetic fixation in the progeny can be performed by monitoring of genes and genomics information by which breeders can develop new varieties precisely and quickly.

Although diverse technologies for molecular breeding have been developed and applied individually for plant genetic improvement, the common use in routine breeding programs seems to be limited probably due to the complexity and incomplete understanding of the technologies. This book is intended to provide a guide for researchers or graduate students involved in plant molecular breeding by describing principles and application of recently developed technologies with actual case studies for practical use.

This book is organized in nine chapters. In Chap. 1, a brief history and perspectives of plant breeding are presented, including the directions of future development of breeding methods. In Chap. 2, the basics on genetic analysis of agronomic traits are described, including how to construct molecular maps and how to develop DNA markers. In Chap. 3, methods of detecting QTLs are illustrated, while in Chap. 4, the application of molecular markers in actual plant breeding is described in detail with case studies. In Chap. 5, genome sequencing and how to analyze the association between sequencing data and phenotype are introduced, including the epigenome and its possible application to plant breeding. In Chap. 6, genome-wide association studies are explained so that researchers can analyze the data following the manual including the introduction of software for analysis of population structure. In Chap. 7, methods for mutation screening and targeted mutagenesis are described. In Chap. 8, how to isolate the genes of interest and how to analyze the gene function are presented with case studies. In Chap. 9, the basics of gene transfer in major crops and the procedures for commercialization of GM crops are explained.

We attempted to cover most of the molecular tools applicable in plant breeding; however, due to the limitation of the book volume, we had to skip some skills that are still under development. Therefore, in this book, only key technologies which are currently used in plant breeding are mentioned. Since technologies per se are being advanced, we may add newly emerging ones with a chance given later. We hope this book would be a valuable reference for plant molecular breeders and, in addition, will become a cornerstone for the development of new technologies in plant molecular breeding for the future.

We are indebted to all the authors for their dedicated efforts and their time in writing the chapters despite the busy schedule. We are greatly thankful to Springer Publishing Co., Editorial Team, and particularly to Ms. Sophie Lim of Springer Korea for her support during the process of preparation and editing of the manuscripts. Our thanks extend to Dr. Mi-ok Woo for her clerical assistance.

Seoul, Republic of Korea Hee-Jong Koh
Daejeon, Republic of Korea Suk-Yoon Kwon
Metro Manila, Philippines Michael Thomson

Contents

Contributors

Sang Nag Ahn Department of Crop Science, Chungnam National University, Daejeon, Republic of Korea

Joong Hyoun Chin International Rice Research Institute (IRRI), Manila, Philippines

Ho-Jun Joh Department of Plant Science, Seoul National University, Seoul, Republic of Korea

Hong-Il Choi Advanced Radiation Technology Institute, Korea Atomic Energy Research Institute, Jeongeup, Republic of Korea

Ik-Young Choi Green-Bio Institute of Science and Technology, Seoul National University, Seoul, Republic of Korea

Pil-Son Choi Oriental Pharmaceutical Development, Nambu University, Gwangju, Republic of Korea

Sang-Bong Choi Division of Bioscience and Bioinformatics, Myongji University, Yongin, Republic of Korea

Tae-Ho Ham Department of Agricultural Sciences, Korea National Open University, Seoul, Republic of Korea

Chee Hark Harn Nongwoo Bio Co., Yeoju, Republic of Korea

Sherry Lou Hechanova Plant Breeding, Genetics and Biotechnology Division, International Rice Research Institute, Metro Manila, Philippines

Jin Hoe Huh Department of Plant Science and Plant Genomics and Breeding Institute, Seoul National University, Seoul, Republic of Korea

Sun-Goo Hwang Department of Applied Plant Sciences, Kangwon National University, Chuncheon, Republic of Korea

---

All of the co-authors equally contributed to this chapter.

Cheol Seong Jang Department of Applied Plant Sciences, Kangwon National University, Chuncheon, Republic of Korea

Kshirod K. Jena Plant Breeding, Genetics and Biotechnology Division, International Rice Research Institute, Metro Manila, Philippines

Soon-Chun Jeong A Bio-Evaluation Center, Korea Research Institute of Bioscience and Biotechnology, Cheongwon, Republic of Korea

Young Hee Joung School of Biological Sciences & Technology, Chonnam National University, Gwangju, Republic of Korea

Byoung-Cheorl Kang Department of Plant Science, Seoul National University, Seoul, Republic of Korea

Won-Hee Kang Department of Plant Science, Seoul National University, Seoul, Republic of Korea

Dong-Gwan Kim Department of Plant Science, Seoul National University, Seoul, Republic of Korea

Dong Sub Kim Advanced Radiation Technology Institute, Korea Atomic Energy Research Institute, Jeongeup, Republic of Korea

Sunggil Kim Department of Plant Biotechnology, Biotechnology Research Institute, Chonnam National University, Gwangju, Republic of Korea

Tae-Sung Kim Department of Plant Resources, Kongju National University, Yesan, Republic of Korea

Hee-Jong Koh Department of Plant Science, Seoul National University, Seoul, Republic of Korea

Soon-Jae Kwon Advanced Radiation Technology Institute, Korea Atomic Energy Research Institute, Jeongeup, Republic of Korea

Soon-Wook Kwon Department of Plant Biosciences, Pusan National University, Miryang, Republic of Korea

Suk-Yoon Kwon Korea Research Institute of Bioscience and Biotechnology, Daejeon, Republic of Korea

Chang-Yong Lee Department of Industrial and Systems Engineering, Kongju National University, Cheonan, Republic of Korea

Geung-Joo Lee Department of Horticultural Science, Chungnam National University, Daejeon, Republic of Korea

Hyun Sook Lee Department of Crop Science, Chungnam National University, Daejeon, Republic of Korea

Joohyun Lee Department of Applied Biosciences, Konkuk University, Seoul, Republic of Korea

Sanghyeob Lee Department of Bio Resource Engineering & Plant Engineering Research Institute, Sejong University, Seoul, Republic of Korea

Kesavan Markkandan Department of Plant Science, Seoul National University, Seoul, Republic of Korea

Seok-Hyeon Nahm NongWoo Bio Co., Ltd, Yeoju, Republic of Korea

Nam-Chon Paek Department of Plant Science, Seoul National University, Seoul, Republic of Korea

Yong-Jin Park Department of Plant Resources, Kongju National University, Yesan, Republic of Korea

Younghoon Park Department of Horticultural Biosciences, Pusan National University, Miryang, Republic of Korea

Hak Soo Seo Department of Plant Science, Seoul National University, Seoul, Republic of Korea

Michael Thomson Plant Breeding, Genetics and Biotechnology Division, International Rice Research Institute, Metro Manila, Philippines

Mi-Ok Woo Department of Plant Science, Seoul National University, Seoul, Republic of Korea

Hee-Bum Yang Department of Plant Science, Seoul National University, Seoul, Republic of Korea

Tae-Jin Yang Department of Plant Science, Seoul National University, Seoul, Republic of Korea

Gibum Yi Department of Plant Science and Plant Genomics and Breeding Institute, Seoul National University, Seoul, Republic of Korea

Jae Bok Yoon R&D Unit, Pepper & Breeding Institute, Suwon, Republic of Korea

Chapter 1
Brief History and Perspectives on Plant Breeding

Joohyun Lee, Joong Hyoun Chin, Sang Nag Ahn, and Hee-Jong Koh

Abstract Following the rediscovery of Mendel's principles of heredity in 1900, plant breeding has made tremendous progress in developing diverse methodologies to create and select variation by using genetic principles. Since the beginning of the twenty-first century, plant breeding has been systematized with state-of-the-art technologies aided by transgenic and genomics approaches. In the future, breeders will be able to assemble desirable alleles or genes into promising varieties with optimized performance using an approach that integrates scientific fields. Recent concerns about global warming, abnormal weather patterns, and unfavorable environments have pushed breeders to speed up the breeding process. In this chapter, the history of plant breeding, methods for creating variation, selection and generation advance strategies, and challenges and perspectives are briefly reviewed and discussed.

1.1 Brief History of Plant Breeding

Humans began managing wild plants in fields about 12,000 years ago; since then, plants have undergone a series of adaptive changes in production and food-associated traits, called domestication or adaptation syndromes. Early human farmers acted as

---

Author contributed equally with all other contributors.

J. Lee
Department of Applied Biosciences, Konkuk University, Seoul, Republic of Korea
e-mail: edmund@konkuk.ac.kr

J.H. Chin
International Rice Research Institute (IRRI), Manila, Philippines
e-mail: j.chin@cgiar.org

S.N. Ahn
Department of Crop Science, Chungnam National University, Daejeon, Republic of Korea
e-mail: ahnsn@cnu.ac.kr

H.-J. Koh (✉)
Department of Plant Science, Seoul National University, Seoul, Republic of Korea
e-mail: heejkoh@snu.ac.kr

H.-J. Koh et al. (eds.), *Current Technologies in Plant Molecular Breeding*,
DOI 10.1007/978-94-017-9996-6_1

breeders by cultivating and selecting better plants or seeds in anticipation of better performance in the next season (Hancock 2004).

Systematic breeding was not performed until 200 years ago. The existence of sex in plants was first recognized by Rudolf Camerarius in 1694. The first manual crossing was performed in 1717 by Thomas Fairchild, who developed the first artificial hybrid by crossing carnation (*Dianthus caryophyllus*) and sweet William (*Dianthus barbatus*). In the early nineteenth century, Patrick Shirreff developed new varieties of oat and wheat via selection and crossbreeding and are now regarded as the first cereal breeder. Knowledge began to accumulate on plant biology, such as cells, sexual reproduction, and chromosomes. However, because the gene concept was not formulated at that time, plant breeding was performed empirically, without a theoretical background. In 1865, Gregor Mendel published 'Experiments on plant hybridization', his genetic experiment with garden pea that is now the foundation for modern genetics and breeding. However, his hypothesis on plant genetics was not widely accepted scientifically for 40 years (Stoskopf 1993).

After Mendelian genetic law was confirmed in 1900, breeders began to develop new varieties based on these genetic principles. Despite the short history of scientific plant breeding, conventional breeding methods have dramatically improved crop yields in corn, rice, wheat, and other crops. The Food and Agriculture Organization of the United Nations (FAO) reported that in the two decades from 1965 to 1985, crop yield increased 56 % worldwide, whereas from 1985 to 2005, only a 28 % increase was recorded. The rapid yield improvements from 1965 to 1985, called the "Green Revolution", resulted from the introduction of genetically-improved varieties, treatment with fertilizers and pesticides, improved irrigation systems, and mechanization of agriculture. We are now facing new challenges to a stable food supply because of global warming, abnormal weather patterns, water shortages, increased demands on crops for bio-fuel, reduced arable land, and mounting population pressure. The global human population is expected to increase by 1 billion people every 14 years and to reach 10 billion within 25–30 years; stable food supply will require 70–100 % more crop production by then. Moreover, this goal must be achieved under unfavorable environmental conditions (Foley et al. 2011). To overcome these challenges, breeders should use all possible technologies to improve yield. Thus advanced biotechnology which can create new genetic variation, and the molecular technology for selecting superior genotypes will be essential in breeding programs to increase crop yield and provide a stable and sustainable food supply.

Plant breeding comprises two main steps: creating or expanding new variation and selecting and fixing desirable genotypes in the progeny (Fig. 1.1). Variation that meets breeders' goals should primarily exist in the germplasm. In the history of plant breeding, methods for creating useful variation, such as artificial crossing, induced mutation, and polyploidization (chromosome manipulation), were used relatively early. Once tissue-culture techniques were established, cell fusion, tissue culture, and inter-specific hybridization were added to the repertoire of methods. Recently, transgenic technology for introducing foreign genes into crops has

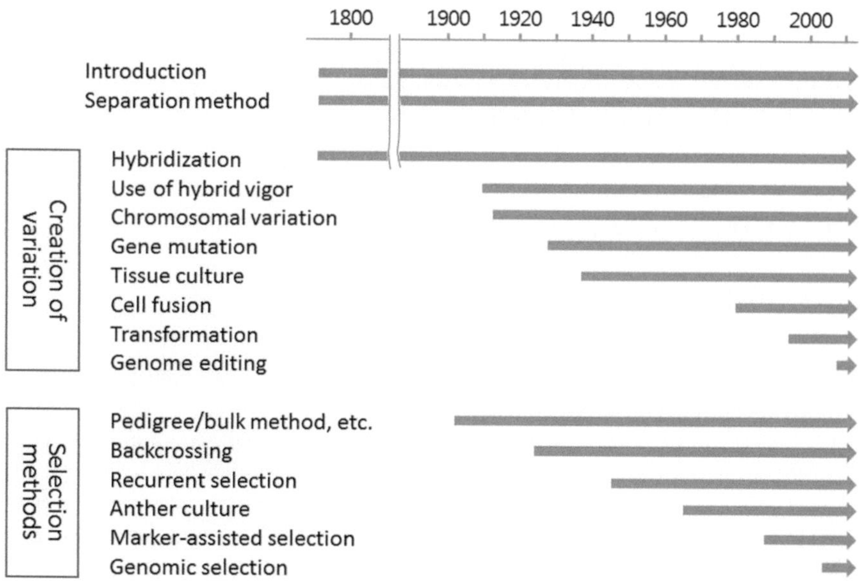

Fig. 1.1 Brief history of plant breeding methods

become available, and targeted mutation and genome editing technologies are under development for crop breeding. Lusser et al. (2011) listed the new plant breeding techniques with a focus on the creation of novel variation.

Even when useful variation exists, without an appropriate method to detect it and to select progeny that contain the target genotypes, breeding goals cannot be accomplished. Thus, the selection method is a critical determinant for successful plant breeding. Pedigree or bulk selection, backcross breeding, recurrent selection, and anther culture are methods for selecting and propagating progeny to fix a desirable trait for the next generation. Several types of molecular marker techniques are available to evaluate selection efficiency and eventually improve breeding efficiency. With accumulated genome sequence information and next-generation sequencing (NGS) techniques for high throughput sequencing, pioneering efforts for sequence-based genomics election have been initiated.

The most important milestone in plant breeding was the Green Revolution, the drastic increase in crop productivity through the development of high-yielding semi-dwarf varieties of wheat and rice. Norman Borlaug, Nobel laureate and father of the Green Revolution indicated that the main reasons for the success of these semi-dwarf varieties were wide adaptation, short plant height, high responsiveness to fertilizer, and disease resistance (1971). The International Rice Research Institute (IRRI) team developed a semi-dwarf rice variety, IR8, in 1962. IR8 had stiff straw to resist lodging and was insensitive to photoperiod, making it widely adaptable. The increase in food production led to more crops being grown per unit of land and with similar effort to that before the Green Revolution. Thus, production costs were reduced, eventually resulting in cheaper food prices at market. Also, this high productivity benefitted the environment because fewer natural areas were needed to cultivate crops. From 1961 to 2008, as the human population increased by 100 % and food production rose by 150 %, the amount of forests and natural land converted to farmland increased by only 10 % (FAO 2014).

1.2 Methods for Creating Variation

1.2.1 Hybridization

Artificial hybridization has long been a main method for creating new varieties through recombination between parents. In general, hybridization achieves breeding objectives in two ways. One is combination breeding, which combines beneficial traits from parents into desirable genotypes. For example, when one parent has disease resistance and another has insect resistance, a breeder can combine both traits in a line by crossing and progeny selection. The other method is transgression breeding: the selection of transgressive segregants in the progeny that outperform parents. This phenomenon usually occurs with quantitative traits, because several alleles at different loci for a certain trait accumulate through gene recombination. For example, when breeders attempt to develop extremely early-flowering varieties, they cross early varieties to generate transgressive segregants that flower extremely early as a result of the accumulation of alleles for earliness. Transgressive segregation is detected more frequently in progeny from wide crosses than crosses between parents of similar genetic makeup.

The principle underlying artificial hybridization is the recombination or reassembly of genes in which additive effects and epistasis among the reassembled genes improve traits in the progeny, as compared with their parents. However, using a wide range of germplasm in hybridization breeding remains difficult because artificial hybridization can only occur between cross-compatible germplasm.

1.2.2 Induced Mutation

After the first discovery by Muller and Dippel (1926) that X-rays could cause mutations in *Drosophila*, induced mutants in tobacco, *Datura*, and corn were quickly reported (1927–1928). X-rays and chemical mutagens were used not only in plant science for gene identification but also in plant breeding to develop new varieties. In the 1940s, German scientists introduced *Arabidopsis thaliana* as a model plant in mutation research. Although many plants required screening to select mutant phenotypes, and a fairly long time was needed to develop a new variety from the selected mutants, mutation breeding initiated in the 1950s was quite successful.

Most mutant varieties have been developed in China, India, and developing countries. Mutant varieties of fruits, such as apple, grape, peach, pear, and papaya, were mostly developed by private companies. According to the mutant variety database of the Joint FAO/IAEA Program (http://mvgs.iaea.org/Search.aspx), more than 3,000 mutant varieties have been developed worldwide. Mutation breeding by physical or chemical mutagens is not as popular as the recently developed method of transformation, but it has been widely used worldwide. Recently, new mutation methods have been introduced, such as space mutation and ion-beam mutation.

1.2.3 Chromosome Manipulation

Autopolyploids, which are developed by duplicating a genome, have been successfully adopted in vegetatively propagated and ornamental crops. Among grain crops, only tetraploid rye in Eastern Europe has been commercialized. Amphidiploids, which have pairs of more than two different genome sets and thus exhibit normal fertility despite their polyploidy, are expected to be useful, even in grain crops. The classic example of an artificial amphidiploid crop is triticale, which is hexaploid or octaploid and considered the first manmade crop. Triticale was developed by the doubling chromosomes of an F_1 hybrid between tetraploid or hexaploid wheat (*Triticum* spp.) and rye (*Secale cereale*). In addition, chromosomal aneuploidy and structural variations cause abnormal morphology and thus are highly applicable to breeding new varieties of vegetatively propagated ornamental plants.

1.2.4 F_1 Hybrids

F_1 hybrids with hybrid vigor have been a main source of new varieties in cross-pollinated crops, such as maize, since the 1940s, and they are widely used even in self-pollinated crops today. Currently, most commercial seeds for maize, cabbage, radish, and pepper sold by commercial seed companies are F_1 hybrid varieties. Hybrid vigor is the phenomenon that the F_1 hybrids perform better than both parents. Although the mechanism of hybrid vigor is not fully understood, hypothesized explanations include dominance and overdominance, epistatic interactions, and epigenetic control (Chen 2013).

To develop superior F_1 hybrid varieties, breeding lines with high general combining ability and parental combinations having high specific combining ability should be chosen. Then, a method for large-scale F_1 seed production should be established using genetic tools such as cytoplasmic or genic male sterility, self-incompatibility, monoecy, dioecy, or gametocide (chemical pollen suppressor). Because of the costs of hybrid seed production, hybrid vigor must be high enough to compensate for the expenses. In the United States, maize yield was increased from 1.8 to 7.8 t/ha (≈430 %) by cultivation of F_1 hybrids. F_1 hybrid varieties account for 65 % of the worldwide cultivated area for maize, 60 % for sunflower, 48 % for sorghum, and 12 % for rice. Most commercial varieties of Brassicaceae and Cucurbitaceae are F_1 hybrids.

1.2.5 Transgenic Approach

Transgenic crops can be developed by artificially introducing genes. This technology has been the most rapidly applied biotechnology in agronomic history. From 1996 to 2013, the area cultivated with genetically modified (GM) organisms increased from 1.7 to 175 Mha, roughly a 103-fold increase. Recently, biotech crops with stacked traits, involving the introduction of more than two genes, are gaining popularity and were planted in 47.1 Mha in 2013 (27 % of total GM crop area). The market share for GM crops in 2013 was 79 % for soybean, 70 % for cotton, 32 % for maize, and 24 % for rape (James 2013).

From the viewpoint of plant breeding, transformation can introduce novel variations originated from other species. Today, most cultivated transgenic crops have herbicide or insect resistance obtained owing to a few modified genes. Although resistances to weeds and insects are not directly associated with yield, weeds or insects can reduce yield. In addition, biotic stresses, such as viruses, bacteria, fungi, and nematodes, can cause more serious yield losses. A few transgenic crops with virus resistance are available in potato and papaya, but transgenic virus resistance in cereal crops is still years from the market. Moreover, resistance to bacteria, fungi, and nematode is much more difficult to develop transgenically than virus resistance. In particular, for broad-spectrum resistance to fungi, several genes must be introduced, a very challenging task. In this case, conventional breeding strategies may be more efficient.

Abiotic stresses, such as salt, drought, cold, and high temperatures, threaten stable production, and therefore a large amount of transgenic crop research has been conducted around the world; however, progress has been slow. A main reason is that the precise mechanisms of these abiotic stresses are not known. Another problem is that crops generally experience varying environmental conditions, so they can be affected by multiple abiotic stresses at the same time, which harm crops more seriously than individual abiotic stresses. Recent research has shown that when several abiotic stresses are applied, plants respond quite differently than to the individual stresses. Therefore, to develop abiotic stress-resistant transgenic crops, various stress responses must be considered together.

Developing salt-tolerant crops by conventional breeding has progressed slowly, because many quantitative genes are involved and the mechanism is quite complex. Thus, salt-tolerant varieties are unlikely to be developed by transgenic approaches alone. Drought resistance, in which responses to various abiotic stresses are associated, is also important. GM maize with a 6–10 % yield increase under drought conditions was developed by Monsanto in 2013 using a cold shock gene (*cspB*) from *Bacillus subtilis* (http//Monsanto.mediaroom.com).

The next target for GM crops will be improving quality. Attempts have begun to modify the composition of fatty acids in rape seed to produce bio-diesel, and nutrient oils to prevent heart disease. Golden rice, which biosynthesizes beta-carotene, a precursor of vitamin A, was developed to prevent deficiencies in dietary vitamin A. Additionally, GM crops, such as GM banana, tomato, or carrot that express antigens for medical vaccines are under development.

1.3 Selection and Generation Advancement

1.3.1 Simple Phenotypic Selection

In the twentieth century, selection was based mainly on the phenotypic evaluation of target traits. Semi-dwarfism for high yields in rice and wheat is a typical success story. However, because of a paucity of analytical tools for phenotyping and the complicated nature of traits, particularly quantitative ones, simple phenotypic selection has seemingly stagnated in most breeding programs.

Target phenotypes can be divided into qualitative and quantitative traits. Qualitative traits, which are controlled by one or a few genes with genotypes that are easily distinguishable by phenotype, respond readily to simple phenotypic selection. In contrast, quantitative traits, which involve multiple genes, have low heritabilities and are difficult to select phenotypically. Moreover, the effect of selection should be lower in early breeding generations because of dominance effects in the population. Although index selection may be a good alternative, it is rarely applied to field selection on a large scale because of the high cost and effort. Therefore, breeders have empirically selected desirable plants in segregating populations or lines using a truncation selection method, which resulted in limited genetic gain after selection.

1.3.2 Recurrent Selection

Recurrent selection is a method to develop promising populations or lines by pyramiding genes involved in target traits. In the beginning, it was applied to cross-pollinated crops for population improvement and breeding inbred lines. Recently, it has been widely used even in self-pollinated crops to develop useful variations using genic-male sterility in every generation (Zhao et al. 2007). In rice, a MAGIC (Multi-parent Advanced Generation Inter-Cross populations) population, developed by recurrent hybridization among progeny derived from crosses among diverse germplasm, was used to accumulate genes that improved target traits (Bandillo et al. 2013). However, a limitation of phenotypic recurrent selection is that only dominant genes are chosen for the next round of hybridization, because recessive genes are hidden and may be lost in segregating populations. Marker-assisted selection (MAS) should be incorporated for greater success.

1.3.3 Marker-Assisted Selection (MAS)

MAS applies molecular-marker technology to conventional breeding. DNA markers are commonly used in MAS. In general breeding, phenotype evaluation and selection of desirable phenotypes in the progeny are conducted repeatedly from F_2. Molecular markers can be used to reduce the time, cost, and labor of the process. In addition, molecular markers can be used in various breeding fields: genetic diversity analysis, genotype identification via DNA fingerprinting, genetic mapping of qualitative and quantitative traits, and MAS.

The most representative application of DNA markers in plant breeding is MAS, which is superior to phenotypic selection. MAS can be performed in the early seedling stage and provides high confidence in selection. For traits that are difficult to evaluate such as disease resistance, which requires a special facility to grow pathogens and an inoculating system, MAS will be very effective. MAS can also select recessive traits hidden in heterozygous genotypes. MABC (marker-assisted backcrossing) is a simple application of MAS that is widely used in breeding programs. One or a few targeted quantitative trait loci (QTLs) can be easily introgressed into elite lines through MABC. In conventional breeding, during backcrossing, the progeny are selected based on the target phenotype, whereas in MABC, the selection is based on the DNA marker genotypes. In general, both foreground and background selection are conducted at the same time. Foreground selection confirms the presence of the desired allele from the donor parents until the final backcross, whereas background selection eliminates unwanted genomic introgression from the donor parent. Thus, in background selection, markers distributed across the whole genome are required. A well-known example is IRRI's efforts to develop rice varieties through backcross breeding using MABC (Xu et al. 2006).

Many DNA markers (gene-based or simply linked to the trait) have been developed, mainly for qualitative traits. Genome sequences for most major crops and great progress in NGS technologies have accelerated the systematic development of useful markers on a large scale. However, DNA markers have limited applications for selecting polygene traits of low heritability.

1.3.4 Genomic Selection

With rapid progress in NGS technology, we are able to obtain sequence information for large numbers of germplasm lines at a reasonable cost whenever needed. NGS technology revolutionized genomics and related studies. The genome-wide association study (GWAS), a statistical examination of the association of an array of single-nucleotide polymorphisms with a trait of interest, opened a new horizon for dissecting polygenic traits into genomic information. Until GWAS became available, detailed QTL studies were the only way to interpret quantitative traits and develop appropriate markers. Similar to GWAS, where all of the major- and minor-effect QTLs can be identified, genomic selection uses high-density marker sets for simultaneous selection of trait-enhancing loci across the genome (Heffner et al. 2009). Genomic-estimated breeding values (GEBVs) for the target traits can be estimated for each breeding line through genome selection, allowing breeders to select promising genotypes with higher accuracy. Instead of full-genome sequencing by NGS, genotyping-by-sequencing (GBS) is preferred for GWAS and genomic selection because of its relatively low cost and analytical simplicity (Elshire et al.

2011). Syngenta developed a drought-tolerant maize variety 'Syngenta-Artesian Corn Hybrids' using genomic selection.

1.3.5 Heritability and Genetic Gain After Selection

Genetic gain after selection of polygenic traits, which represent most agronomically important characteristics, such as yield, is greatly affected by heritability, the barometer of selection effect. To improve selection efficiency, populations or lines for selection should be cultivated under uniform/controlled environments and/or generation-advanced to increase heritability as much as possible. Even though genomic selection for polygenic traits is performed using GEBVs, genetic gain will be greatly influenced by heritability if the analysis used for GEBV calculation was based on phenotyping in the field. Therefore, GEBV calculations should be performed using the training populations grown in a uniform environment, where heritability estimates can be maximized by minimizing the environmental effects on phenotype.

1.4 Challenges and Perspectives

Breeders are facing global weather changes, increased food demand, limited natural resources such as water and energy, and new demand for health crops. Thus, breeding remains an essential task. Most modern elite varieties were developed by conventional breeding, which is still highly effective. However, conventional breeding products may not be able to meet current demands, so molecular breeding tools need to be actively utilized in current breeding programs. The methods and strategies discussed in previous sections are focused on creating new variation and selecting progeny with superior genotypes or phenotypes. Here, we will discuss the current challenges in breeding, focusing mainly on creating variation and selection methods by both conventional and molecular methods.

1.4.1 Creation of Variation

1.4.1.1 Recombination

Meiotic recombination is a basic mechanism for creating new variation by reshuffling homologous chromosome segments. Breeders are eager to bring desired alleles together into new combinations and to maximize reshuffling of alleles to create new variability. The frequency of meiotic recombination is determined by the number of chromosomes (via independent assortment) and the number crossover events along a chromosome. Thus, variation can be created naturally by recombination events though the outcomes are quite variable and cannot be estimated.

Recently, IRRI celebrated their 100,000th crossing of rice, a distinguished achievement in hybridization breeding by a single institute. Given the number of rice crossings worldwide, we may be approaching the maximum variation possible through recombination. Although no direct evidence addresses this concern, it may explain why grain yield in rice has stagnated for decades. Similarly, the total number of crossings conducted per crop worldwide, including public and private institutes and universities, must be tremendous. Breeders may have already experienced most of the variation among frequently used germplasm, which can be induced by conventional breeding methods.

We must not overlook numerous crop germplasm resources, including wild relatives. Given that million of accessions exist, a tremendous amount of novel variation still remains to be exploited. Efforts are ongoing to use various germplasm accessions, such as land races or wild relatives, in which linkage drag would be a distinct barrier. With advanced DNA markers, MABC would increase the use of wild relatives or land races to enhance the types and range of genetic variation. Nonetheless, optimal genotype combinations might not be created without random recombination events or controlling meiotic recombination to modify the gametic allele compositions. Recently, the possibility of controlling meiotic recombination by increasing crossover incidence, altering crossover positions on chromosomes, and silencing crossover formation was reported (Wijnker and de Jong 2008; Osman et al. 2011). However, no variety or new phenotype variation has yet been developed this way. However, controlling meiotic recombination will be an important technology for creating novel variation that overcomes linkage drag and linkage blocks in the near future.

Epistasis, interactions among genes, has long been recognized as fundamentally important to understanding both the structure and function of genetic pathways and the evolutionary dynamics of complex genetic systems (Phillips 2009). Breeders make crosses or transgenic plants to introduce useful genes/alleles and to harmonize gene/allele combinations for variation and better performance in the progeny. However, epistasis is not well investigated as a genetic reservoir, even in intervarietal breeding. If recombination-controlling techniques are available, breeders will have more opportunity to produce desirable phenotypic variations and, thus, new epistasis interactions for yet more variation. Recent studies on meiosis and recombination may shed light on how to manipulate crossovers during meiosis (Osman et al. 2011; Crismani et al. 2013).

1.4.1.2 Mutation and Genome Editing

Mutation is an effective way to induce variability into plants. Along with cross-breeding, mutation breeding is a main method in conventional breeding. Many varieties that were developed by induced mutation have been released so far. However, the occurrence of novel variations from this method has gradually decreased. The success of mutation breeding largely depends on a high mutation rate and technologies for detecting mutants. Even though various chemical mutagens and radiation types are available for mutation breeding, the mutation spectrum is limited for a few reasons. (1) Most induced mutants result in loss-of-function mutations, so gain-of-function mutants are rare. (2) Methods for screening mutants are limited. Physical appearance is the primary criterion for detecting mutants and only some biotic or abiotic stresses can be used for screening. Thus, without new large-scale screening methods, detecting mutant phenotypes will be limited. Recently, TILLING (Targeting Induced Local Lesions in Genomes), which can screen the mutated genes of known function and sequence in a high-throughput manner, became available as a new strategy. If breeders intend to knockout a specific gene or gene groups, targeted mutagenesis is a feasible approach that can be applied in plants via homologous recombination (Terada et al. 2002). However, this technique is being replaced by genome-editing techniques because of its low efficiency.(3) Finally, even without considering transposable elements, there may be mutable loci responsive to specific physical or chemical mutagens. In this case, other types of mutagens can be used to induce novel mutants. For example, ion beam mutation generated many novel mutants (Yu 2006).

Genome editing of a target area is expected to be a powerful tool to create desirable variation through insertion, replacement, or removal of genes from a genome using molecular scissors, which are artificially engineered nucleases. Three families of engineered nucleases are being used in plants: zinc finger nucleases (ZFNs), transcription activator-like effector nucleases (TALENs), and the CRISPR/Cas system (Puchta and Fauser 2013).

1.4.1.3 Genetically-Modified Organisms (GMOs)

Transgenic breeding is appealing to breeders for creating totally new variation that cannot be obtained with conventional breeding. However, the issue of conventional versus transgenic breeding is still debated in both civilian and scientific communities. Debates about food/feed and environmental safety continue, despite scientific efforts toward providing an objective view on the issues (Nicolia et al. 2013). Some GM crops have been highly successful in assuring crop yield via protection against diseases, insect pests, and weeds by using alien genes. Increasing crop yield potential is the ultimate target for the future. For this, GM crops should incorporate genes to increase the efficiencies of nutrient and water uptake, nutrient use, photosynthesis, and translocation of photosynthates into storage organs. In addition, more effective and safer transgenic technologies should be developed to provide more new variants. Much as agrochemicals gained acceptance decades ago, the hope is that GM crops will be accepted in the near future because few alternative technologies exist for enhancing crop productivity.

1.4.2 Selection Methods

Breeders select improved genotypes based on phenotypes under specific or diverse environments, followed by regional performance tests before variety registration. During field surveys of early breeding generations, breeders observe and evaluate more than 100,000 lines, and genotype-by-environment interactions must be considered, so selecting the best individuals is difficult. Recently, DNA markers have enabled breeders to indirectly select desirable genotypes prior to field evaluations of traits. In most crops, however, only a small portion of genes in the genome have been associated with agronomic traits, so few markers, including linked markers, are available for MAS. Most genes important for agronomic traits remain unidentified.

In gene isolation studies, accurate phenotyping is critical because phenotypic values vary along environmental gradients in breeding fields. Therefore, stable, reproducible, and large-scale controlled environments that minimize environmental effects are crucial for determining genotype from phenotype. Moreover, phenotyping should be performed in a high-throughput system aided by remote-sensing and computer-based technologies. In reality, however, such facilities are difficult to establish. Currently, phenotyping is a limiting factor in isolating genes and subsequently developing DNA markers.

Genomic selection for quantitative traits is promising because even minor loci influencing a trait of interest can be identified and selected through whole-genome scanning. However, despite the feasibility of genomic selection, its effectiveness is limited for low-heritability traits (Nakaya and Isobe 2012). Therefore, genomic selection models should be developed using methods to eliminate environmental noise. Such markers may improve genetic gain by selection to the full extent of heritability. In addition, understanding gene-by-environment interactions would help in adopting genomic selection methods in diverse breeding fields. Recently, the epigenetic nature of agronomic traits is receiving much attention. However, whether epigenetic variations are heritable remains under study (Springer 2013).

When both genomic and epigenomic data on quantitative traits are thoroughly understood, breeders may overcome the limitation in selection gain caused by heritability.

1.5 Conclusion

Transgenic-based technologies to create novel variation, such as meiotic manipulation, targeted mutation, and genome editing, have the potential to replace traditional methods that mainly depended on the hybridization of germplasm and random mutation. Despite controversies about GM crops, transgenic technologies promise to be the basic tools for plant breeding in the future. Because GM technology has been successful for traits controlled by major-effect genes, it should be developed further so that multiple genes or poly genes can be harmoniously incorporated into target plants. Genome editing seems promising in this regard. In association with an understanding of epigenetic control of trait expression, the technology will initiate a new era in designing useful polygenic variation. Creation of harmonious and desirable gene assemblies and gene networks will provide breeders with more opportunities to select better genotypes in their breeding fields.

Genomic selection is a state-of-the-art technology that enables breeders to select quantitative traits based on genotypes. To improve the accuracy of genome selection, phenotyping technologies should be systematized and automated in controlled environments, a field known as 'phenomics'. In addition, understanding epigenomic control and gene-by-environment interactions of target traits should accompany improved selection efficiency. Nonetheless, phenotypic validation, which includes stability and adaptability tests across locations and years, will remain the most important step in developing new varieties.

During the past century, plant breeding objectives have gradually changed to address global food security, industrial needs, and human preferences. Recent trends in global warming and frequent unfavorable climatic conditions compel breeders to adjust their targets and speed the development of new varieties for sustainability. The global seed market has been growing rapidly, with an annual increase of more than 10 %. A worldwide patent system for plant variety was consolidated by The International Union for the Protection of New Varieties of Plants (UPOV), stimulating private breeding companies to create competitive varieties. Future breeding programs should be more systematized and armed with new technological tools. Much as a car production line is automated, we anticipate that a series of technologies for breeding crop cultivars will be assembled and automated to produce designer varieties in the future (Fig. 1.2).

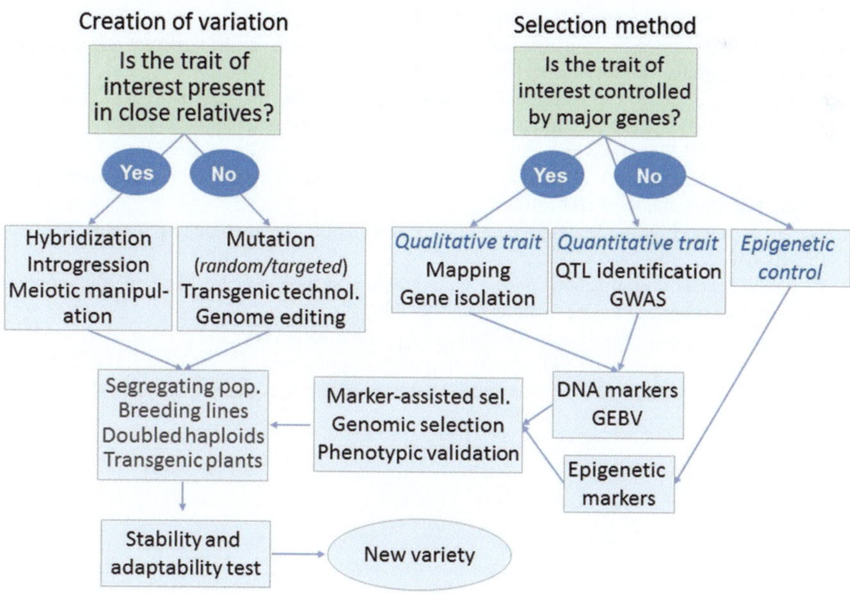

Fig. 1.2 Plant breeding in the twenty-first century (QTL, quantitative trait locus, GWAS, genome-wide association study; GEBV, genomic estimated breeding value)

References

Bandillo N, Raghavan C, Muyco PA et al (2013) Multi-parent advanced generation inter-cross (MAGIC) populations in rice: progress and potential for genetics research and breeding. Rice 6:11

Chen ZJ (2013) Genomic and epigenetic insights into the molecular bases of heterosis. Nat Rev Genet 14:471–482

Crismani W, Girard C, Mercier R (2013) Tinkering with meiosis. J Exp Bot 64:55–65

Elshire RJ, Glaubitz JC, Sun Q et al (2011) A robust, simple Genotyping-by-Sequencing (GBS) approach for high diversity species. PLoS One 6:e19379

FAO (2014) Statistics DB. http://www.fao.org/statistics/databases/en/

Foley JA, Ramankutty N, Brauman KA et al (2011) Solutions for a cultivated planet. Nature 478:337–342

Hancock JF (2004) Plant evolution and the origin of crop species, 2nd edn. CABI Publishing, Wallingford, 313 p

Heffner EL, Sorrells ME, Jannink JL (2009) Genomic selection for crop improvement. Crop Sci 49:1–12

James C (2013) Global status of commercialized biotech/GM crops 2013, ISAAA Briefs No. 46

Lusser M, Parisi C, Plan D et al (2011) New plant breeding techniques: state-of-the-art and prospects for commercial development. JRC technical report EUR 24760 – 2011, European Commission Joint Research Centre

Muller HJ, Dippel AL (1926) Chromosome breakage by x-rays and the productions of eggs from genetically male tissue in drosophila. J Exp Biol 3(2):85–122

Nakaya A, Isobe SN (2012) Will genomic selection be a practical method for plant breeding? Ann Bot 110:1303–1316

Nicolia A, Manzo A, Veronesi F et al (2013) An overview of the last 10 years of genetically engineered crop safety research. Crit Rev Biotechnol, Early online, 1–12. doi:10.3109/07388551.2013.823595

Osman K, Higgins JD, Sanchez-Moran E et al (2011) Pathways to meiotic recombination in Arabidopsis thaliana. New Phytol 190:523–544

Phillips PC (2009) Epistasis – the essential role of gene interactions in the structure and evolution of genetic systems. Nat Rev Genet 9(11):855–867

Puchta H, Fauser F (2013) Gene targeting in plants: 25 years later. Int J Dev Biol 57:629–637

Springer NM (2013) Epigenetics and crop improvement. Trends Genet 29(4):241–247

Stoskopf NC (1993) Plant breeding – theory and practice. Westview Press, Boulder, 531 p

Terada R, Urawa H, Inagaki Y et al (2002) Efficient gene targeting by homologous recombination in rice. Nat Biotechnol 20:1030–1034

Wijnker E, de Jong H (2008) Managing meiotic recombination in plant breeding. Trends Plant Sci 13:640–646

Xu K, Xu X, Fukao T et al (2006) *Sub1A* is an ethylene-response-factor-like gene that confers submergence tolerance to rice. Nature 442:705–708

Yu Z (2006) Introduction to ion beam technology. Springer, New York, 302 p

Zhao S, Zhang M, Jiang C et al (2007) Study on quality improvement effect and separate character of soybean Male Sterile (MS1) recurrent selection population. Agric Sci China 6:545–551

Chapter 2
Methods for Developing Molecular Markers

Hee-Bum Yang, Won-Hee Kang, Seok-Hyeon Nahm, and Byoung-Cheorl Kang

Abstract Molecular markers are essential for breeding major crops today and many molecular marker techniques have been developed. DNA markers are now the most commonly used. This chapter describes the principles of DNA marker techniques and methods to map major genes. DNA markers can be classified into two categories: (1) DNA hybridization-based techniques, including restriction fragment polymorphism and DNA chips, and (2) polymerase chain reaction techniques, including simple sequence repeats, random amplified polymorphic DNA, amplified fragment length polymorphism, and single nucleotide polymorphism. To develop trait-linked markers, segregating populations for the target traits and reliable phenotyping methods are indispensable. With these tools, two approaches can be used to develop trait-linked markers: (1) when there is no biological information for the trait, and (2) when biological information is available. Finally, we describe several case studies for trait-linked marker development.

2.1 Definition of Technology and Related Terminology

When a specific phenotype, such as disease resistance or crop quality in plants, is difficult to determine, a different method must be used to investigate the trait. Genetic markers are a viable alternative. Because they are located close to the target gene and are inherited with it, selecting plants with useful traits using genetic markers is relatively easy. Genetic markers can be classified into morphological markers, including plant shape and/or color; protein markers, such as isozymes; and DNA markers based on sequence differences.

Author contributed equally with all other contributors.

H.-B. Yang • W.-H. Kang • B.-C. Kang (✉)
Department of Plant Science, Seoul National University, Seoul, Republic of Korea
e-mail: yhbk0130@snu.ac.kr; hui81@snu.ac.kr; bk54@snu.ac.kr

S.-H. Nahm
NongWoo Bio Co., LTD, Yeoju, Republic of Korea
e-mail: shnahm1@nongwoobio.co.kr

H.-J. Koh et al. (eds.), *Current Technologies in Plant Molecular Breeding*
DOI 10.1007/978-94-017-9996-6_2

Morphological markers were used first for genetic analysis by geneticists like Gregor Mendel and Thomas Hunt Morgan. However, their potential numbers are low, so few examples of their practical use exist. Protein markers such as isozymes, which were developed later, can distinguish individual plants. Thanks to this method, many samples could be analyzed with low cost. However, the relatively small number of isozyme variants limits their utility. DNA-based restriction fragment length polymorphism (RFLP) markers were developed and used in the 1980s, and in the following decade, polymerase chain reaction (PCR) technology gave rise to various types of DNA markers. The strengths and weaknesses of each type of genetic marker can be seen in Table 2.1.

Typically, the term 'molecular markers' indicates technology that uses phenotype-determining or closely related genes to find similarities or differences among individual plants, cultivars, or breeding lines. By analyzing molecular markers, individual plants with useful phenotypes can be selected at the seedling stage. Molecular markers for plant breeding are based on genetic differences among individuals in alleles at a certain locus. A molecular linkage map can be constructed by investigating the marker genotypes of each individual plant and calculating the genetic distances between marker pairs based on the recombinant frequency.

Table 2.1 Comparison of different types of genetic markers

Type	Benefit	Drawback	Example
Morphological markers	Easy to assay	Highly dependent on environmental factors	Color, shape, etc.
	Low cost	Difficult to analyze for quantitative traits	
		Difficult to determine heterozygosity	
		Limited availability	
Protein markers	Low cost	Assay samples must be in good condition	Isozymes
	Co-dominant	Limited availability	
	Less dependent on environmental factors	Unstable materials (protein)	
DNA markers based on hybridization	Do not require sequence information of the target	Costly and time consuming	RFLP
	Co-dominant	Use isotopes	
	Unaffected by environmental factors	Require large quantities of high molecular weight DNA	
		Difficult to automate	
DNA markers based on PCR	Require low quantities of DNA	Require expensive equipment	SSR
	Quick and easy to assay	Require sequence information	AFLP
	High accuracy		RAPD
	Unaffected by environmental factors		SNP

2.1.1 *Types of Molecular Markers*

2.1.1.1 Protein Markers

Proteins are the products of gene expression. Different alleles encode different amino acid sequences, giving proteins different sizes or biochemical characteristics that can be observed by electrophoresis and thus used as molecular markers. Isozymes are an example. Isozymes are useful molecular markers because they can be distinguished from each other based on differences in charge or size, despite having the same enzymatic activity. Although proteins cannot usually be seen by the naked eye, isozymes can be easily detected by separating via gel electrophoresis then adding the substrate of the enzyme. The isozyme produces color from the substrate, producing a band on the gel. Isozyme markers have a few drawbacks that greatly reduce their utility. In addition to having a very limited number of possible markers (only a few dozen have been developed), they are not distributed evenly on the chromosome, and often the enzyme activity depends on the plant's age or tissue type. Even so, isozyme analysis is very cheap and simple and was used for studies of maize, wheat, and barley decades before DNA markers were developed.

2.1.1.2 DNA Markers

DNA marker techniques use sequence differences among species or individuals within a species. Genetic differences among individuals in a group are usually due to abnormal pairing of sister chromosomes or recombination that rearranges the chromosomes, for example, insertions, deletions, inversions, translocation events, or reduplication. Chromosomal rearrangements can vary in size, from just a few base pairs to millions. DNA mutations in the form of base pair substitutions also occur. To develop genetic markers using DNA variants, DNA hybridization or PCR techniques are often used. In DNA hybridization, a short DNA fragment that is homologous to the target DNA is used as a probe. The probe is tagged with a radioisotope and hybridized with the DNA being analyzed. DNA variations can be detected based on the target–probe hybridization or the size of the hybridized DNA fragment. RFLP is an example. PCR techniques require only a small amount of DNA and are relatively simple and inexpensive. They include using minisatellites or microsatellites, sequence-specific primers such as sequence tagged sites (STSs) or expressed sequence tags (ESTs), and random primers such as random amplified polymorphic DNA (RAPD) or amplified fragment length polymorphism (AFLP) to amplify the DNA fragment and analyze its variants.

Molecular Markers Using Hybridization Methods: RFLP The classic example of molecular markers using DNA hybridization, RLFPs are a first-generation technique and the basis of many DNA marker methods that are used today. To better understand RFLPs, restriction enzymes and Southern blotting must be clear (Fig. 2.1).

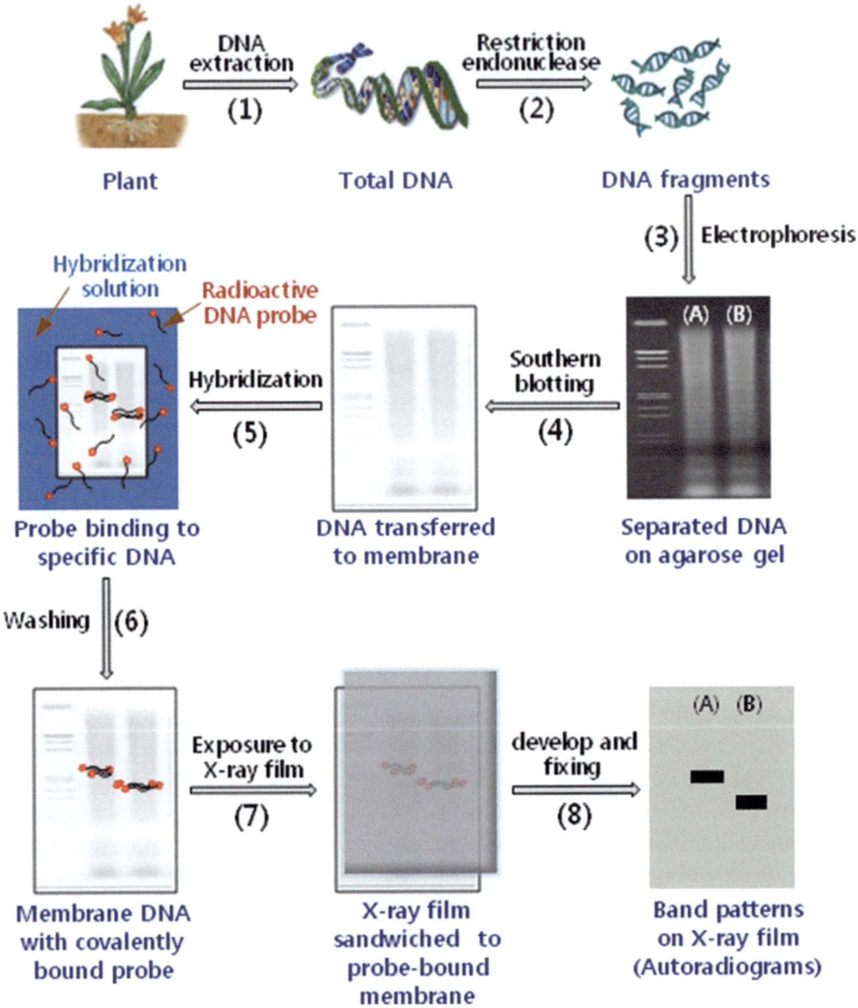

Fig. 2.1 Restriction fragment length polymorphism (RFLP) procedure. (*1*) Extract DNA from individuals A and B. (*2*) Use restriction enzymes to cut DNA. (*3*) Electrophorese DNA fragments on agarose gel to separate them by size. (*4*) Transfer the DNA in the gel to a nylon membrane by Southern blot. (*5*) Use radioactively labeled DNA fragments as probes to hybridize to the DNA. (*6*) Remove non-specifically bound or unbound probes by washing the nylon membrane. (*7*) Expose the washed membrane to X-ray film. (*8*) Develop the X-ray film to observe DNA polymorphisms

Restriction Endonucleases Restriction endonucleases (or restriction enzymes) are enzymes that recognize a specific DNA sequence (restriction sequence) and cut the DNA there. The recognized sequence can be four, six, or eight base pairs in length, depending on the enzyme. If a given DNA sequence has an equal numbers of A, T, G, and C, a restriction enzyme that recognizes six base pairs will cut every 4,096(4^6) positions on average, and the DNA of an organism with a genome size of 10^9 bp would be cut by such a restriction enzyme into about 250,000 fragments of different sizes. If a mutation happens at the restriction sequence, the restriction enzyme will no longer cut there. RFLP technology capitalizes on point mutations, which are common and widespread. Thus, individual genotypes, even within a species, result in different patterns of fragmentation. However, when too many fragments are generated, they cannot be differentiated by typical electrophoresis. Even when they can be visualized, identifying homologous fragments in different individuals is nearly impossible. Southern blotting analysis was therefore developed to overcome these problems.

Southern Blot Analysis Southern blot analysis uses 0.5–3.0 kb DNA fragments as probes to detect other fragments with the same sequence among the enzyme-digested products. The probe is selected from a genomic DNA library, a cDNA library, or DNA fragments related to the gene(s) of interest. Southern blot analysis proceeds as follows:

1. DNA is extracted from the plant materials to be surveyed and appropriate restriction enzymes are selected.
2. DNA fragments are digested with restriction enzymes and separated by size using agarose gel electrophoresis. Fragment lengths will vary from a few hundred base pairs to over 20 kb.
3. After electrophoresis, the gel is treated in a strong alkali solution to denature the DNA strands. Then, the fragments are transferred to a nitrocellulose or nylon membrane and exposed to UV light or heat to fix them onto the membrane.
4. The DNA probe is labeled with radioactive isotope or fluorescent dye and hybridized to the membrane-fixed DNA fragments. During hybridization, labeled DNA probes will bond to complementary target fragments.
5. Finally, the membrane is washed to remove unbonded or weakly bonded probes. The washed membrane is exposed to X-ray film to visualize the hybridization band patterns.

If the restriction sites differ among individuals because of mutations, the probe banding pattern will reveal this variation (Fig. 2.2). RFLP markers have many advantages, such as being mostly codominantly inherited, highly reproducible, and distributed evenly on the genome. However, because the process is complicated and requires a large amount of pure DNA, RFLP markers are not often used today.

Molecular Markers Using Polymerase Chain Reaction (PCR) Several methods have been developed based on PCR and are summarized here.

Minisatellite/Microsatellite Markers Minisatellites are 9–100 bp DNA sequences that are repeated along the genome. The number of repeats varies but is usually under 1000. Polymorphism in the number of repeats is called variable number of tandem repeats (VNTR). VNTR can be measured by either probe hybridization or PCR. When a probe hybridizes with a fragment containing minisatellite repeats, length differences in the fragments indicate VNTR. Alternately, PCR using primers that flank the repeat allow size variation of amplicon, and thus VNTR, to be detected.

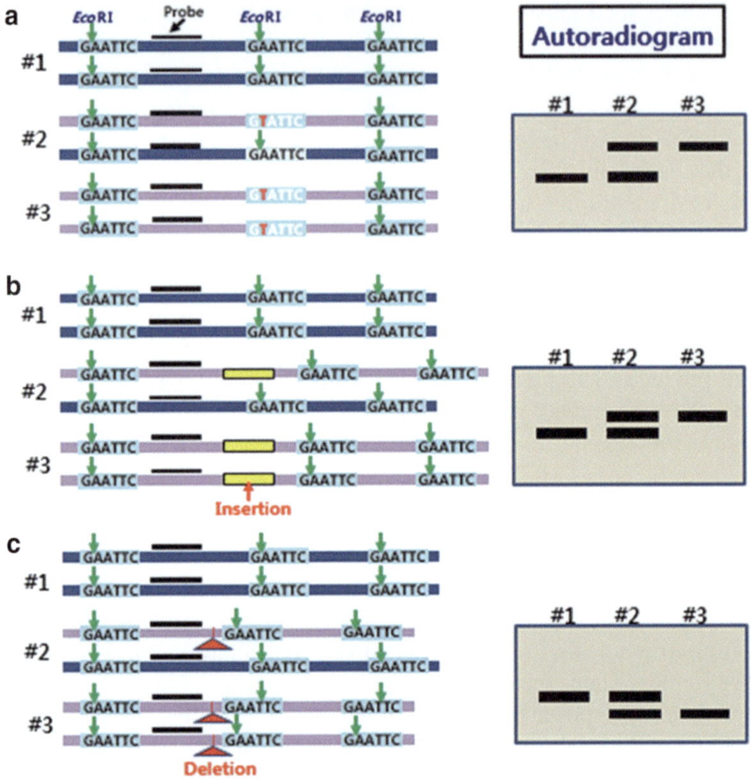

Fig. 2.2 Examples of restriction fragment length polymorphism (RFLP) analysis. (**a**) RFLP caused by substitution. A point mutation occurred at the restriction enzyme recognition site; the length of the fragment to which the probe hybridizes changes. (**b**) RFLP caused by insertion. (**c**) RFLP caused by deletion

Microsatellites, also called simple sequence repeats (SSRs), are 1–6-bp random sequences that are repeated fewer than 100 times. The most common case is two alternating bases, such as $(CA)_n$:$(GT)_n$, $(CG)_n$:$(GC)_n$, or $(AT)_n$:$(TA)_n$. Three or four base-pair repeats also exist, but are less common. In plants, the number of microsatellite repeats can differ among individuals and species. When a microsatellite repeat is discovered, the flanking sequences can be used as PCR primers to develop SSR markers (Fig. 2.3). In the past, to develop an SSR marker, a probe containing a repeat sequence was first used to identify homolog repeats in genes from a DNA library. However, because large amounts of sequence data are now easy to obtain, the need to screen a DNA library no longer exists. SSR markers can now be developed from gene bank data. After a repeat is identified, the flanking sequence is used to amplify the SSR, and length differences among the amplicons reveal SSR polymorphisms. SSR markers are usually codominant, easy to analyze, and provide reproducible results, making them ideal molecular markers.

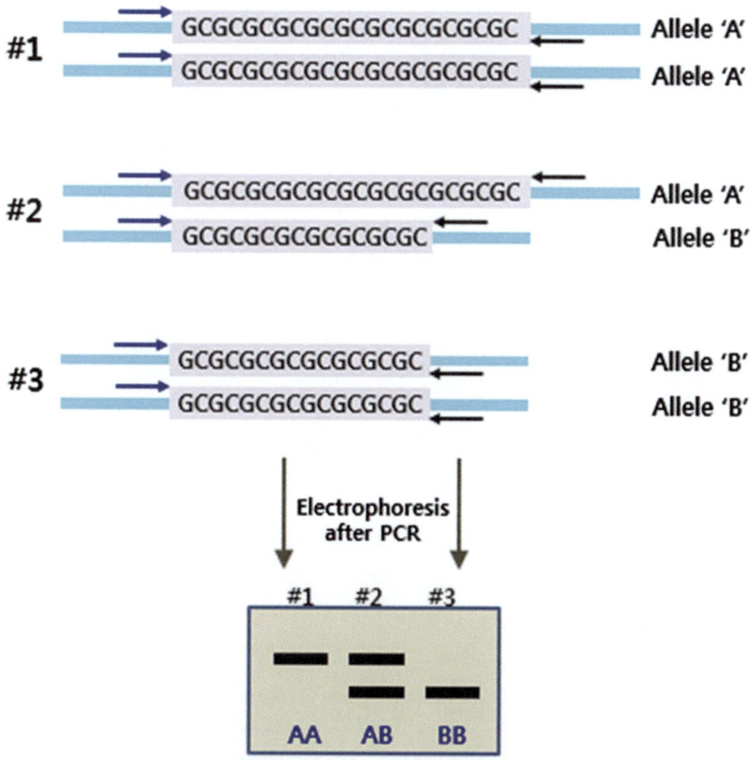

Fig. 2.3 Principle of simple sequence repeat (SSR) markers. Amplifying the SSR sequence by PCR using primers just outside the sequence reveals polymorphisms in amplicon size

Random Amplified Polymorphic DNA (RAPD) Markers RAPD markers can be developed even without DNA sequence data of the target species. These markers are generated using random PCR primers of about 10 bp in length. Because the random primers can bind with any complementary sites in the genome, when two primers are close enough to amplify the intervening sequence, the PCR product can be analyzed by electrophoresis. DNA sequence differences among individuals at the priming sites cause different band patterns, allowing polymorphisms to be observed on an agarose gel (Fig. 2.4). RAPD markers are relatively easy to develop, and their results are relatively simple to analyze. However, because RAPD markers are inherited dominantly and experimental reproducibility can vary with reaction conditions, precise genotype analysis is difficult. Despite these shortcomings, RAPD is widely used to develop markers linked to certain traits, such as disease resistance.

Amplified Fragment Length Polymorphism (AFLP) Markers To solve the reproducibility problems of RADP, the Dutch biotech company KeyGene developed the AFLP method in 1995. AFLP has the strengths of both RFLP and RAPD. In AFLP, restriction-digested DNA fragments are amplified using PCR. The basic principle of

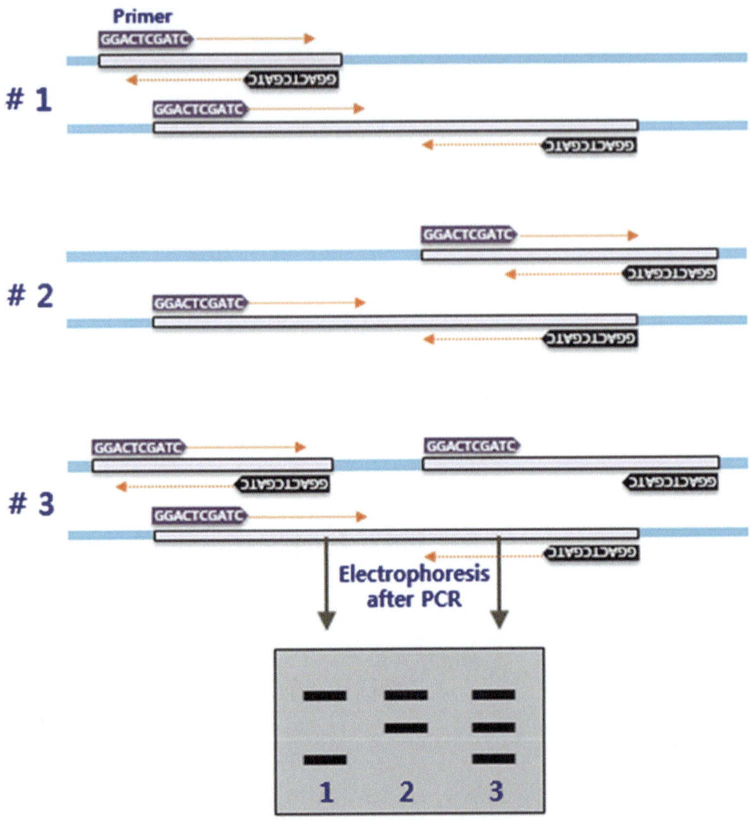

Fig. 2.4 Principle of random amplified polymorphic DNA (RAPD). A single random primer binds to complementary sites in the genome. Difference RAPD profiles on an agarose gel reveals polymorphism among individuals

AFLP is shown in Fig. 2.5. First, DNA is digested by two different restriction enzymes. An adaptor, a small double stranded oligonucleotide that complements the sticky end of the digest, is attached the end of each digested fragment. Then, the adaptor-attached DNA fragments are amplified by PCR using primers that complement the adaptors. One primer must be labeled with a radioactive isotope or fluorescence to visualize the amplified products. A few additional base pairs (selective nucleotides) can be added to the primer sequence beyond the adaptor region to decrease the number of amplified fragments. For example, if three selective nucleotides are added to each primer, the number of amplified fragments will be reduced by $1/4^6$. Changing the selective nucleotides allows different DNA fragments to be amplified.

AFLP can be used to detect differences in restriction enzyme sites, selective nucleotide sequences, and insertions/deletions among individuals. AFLP is less influenced by PCR conditions than other methods, so reproducibility is very high. Moreover, because 50–100 DNA fragments can be analyzed at once, many loci can

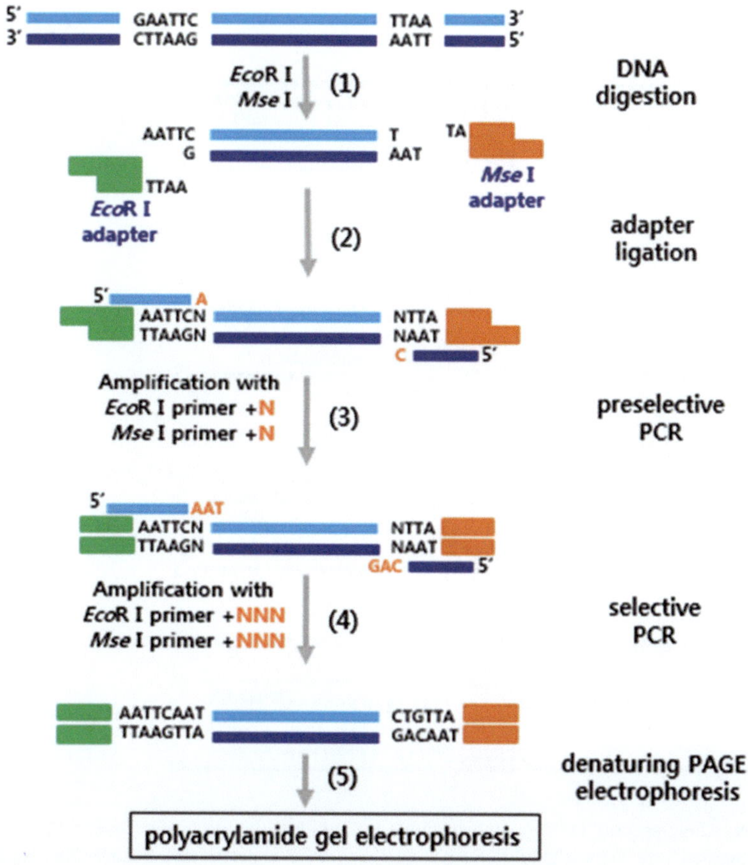

Fig. 2.5 Schematic diagram showing the amplified fragment length polymorphism (AFLP) technique. (*1*) The DNA sample is digested with two types of restriction enzymes. (*2*) Adaptors are attached to each side of DNA fragment. (*3*) Pre-amplification by PCR using primers that have the adaptor sequence and one random additional base (+1 selective nucleotide). (*4*) Selective amplification with primers that have the adaptor sequence and +2–4 selective nucleotides. (*5*) Electrophoresis of AFLP fragments on an acrylamide gel. Different individuals have different restriction recognition sites and selective nucleotides, giving different AFLP fragments

be surveyed to find polymorphisms. AFLP also has an advantage in that it does not require probes or DNA sequence information. AFLP markers are inherited dominantly, and heterozygotes cannot be distinguished from homozygotes (as the allelic relationships of AFLP bands are difficult to determine). AFLP techniques have been used in various fields, most intensively in biodiversity analyses, trait-associated marker development, and linkage map development.

Single Nucleotide Polymorphism (SNP) Markers SNPs are literally single-base differences among individuals and represent 80 % of DNA polymorphisms. On average, one SNP exists every 1 kb in the human genome and every 170 bp between the

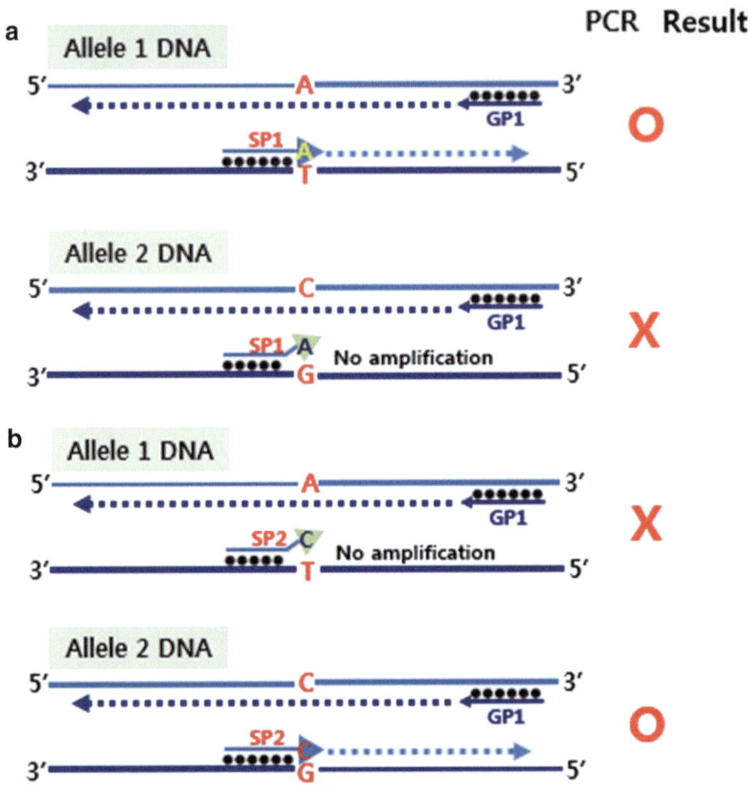

Fig. 2.6 Allele-specific PCR to analyze single nucleotide polymorphisms (SNPs). Using primer SP1, designed to have the allele "1" SNP at its 3′ end, PCR is performed. Because SP1 was based on its sequence, allele 1 amplifies well, while other alleles (e.g., allele 2) do not (**a**). However, if PCR is performed with primer SP2 with the allele 2 SNP sequence at its 3′ end, allele 2 will be amplified while allele 1 will not (**b**)

japonica and *indica* rice varieties. The most direct way of identifying SNPs in genomes is by directly comparing DNA sequences. Today, thanks to the accumulation of DNA sequence data, SNP markers can be developed *in silico* by analyzing existing databases. Once SNPs are identified, differences among individuals can be analyzed as follows.

1. Allele-specific PCR: When one of the two primers used in PCR has a 3′ end that does not complement the template DNA, DNA amplification is slower than usual. Therefore, if the SNP is set as the 3′-end of the primer sequence, one allele will amplify well, while different alleles will not, allowing easy allele distinction (Fig. 2.6).
2. 5′ Nuclease Assay: This method involves probes, such as TaqMan™, with a fluorescent label attached to the 5′ end, while the 3′ end has a chemical that inhibits

fluorescence (Fig. 2.7). The TaqMan™ probe is designed to be complementary to one allele. The probe does not cover the entire allele; instead it complements a small part that contains the SNP. If PCR is performed with the TaqMan™ probe attached, the 5′ exonuclease function of the PCR enzyme cuts the 5′ end, releasing the fluorescent compound and allowing it to light up. If PCR is attempted with different alleles, the 5′ end does not attach properly and is not cut by the enzyme, so no fluorescence is emitted. Therefore, the presence or absence of fluorescence reflects which allele the sample DNA has with respect to a given SNP. This method is expensive but can be automated.

3. DNA Chips: This method uses a DNA chip with various DNA sequences (about 25 bp-long) attached regularly on a glass or metal plate (Fig. 2.8). The DNA density of the chip is very high, and more than a million DNA fragments can be

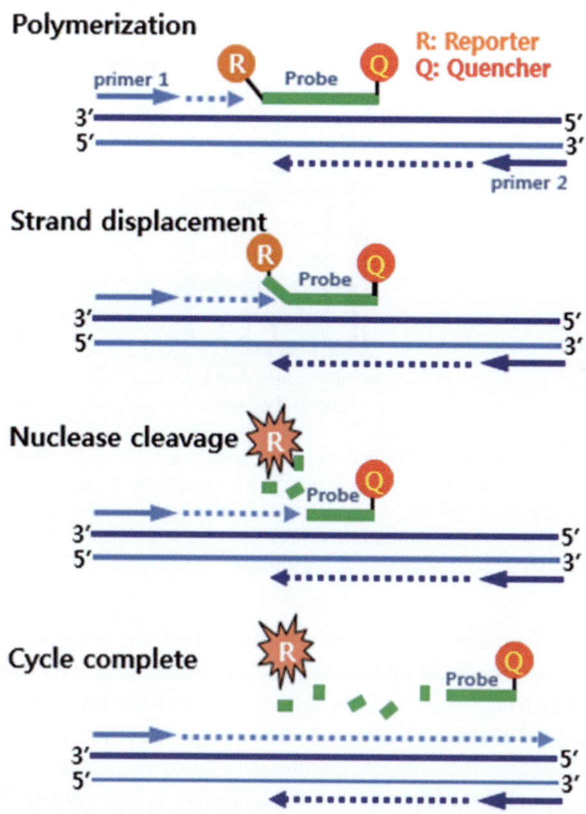

Fig. 2.7 TaqMan™ probe assay and 5′ exonuclease activity of the Taq polymerase used to analyze single nucleotide polymorphisms (SNPs). On the 5′ end of the probe, a fluorescent chemical (R) is attached. On the 3′ end, a substance that inhibits fluorescence (Q) is attached. When PCR is performed, the exonuclease activity of the DNA polymerase cleaves the 5′ fluorescent substance off, allowing fluorescence to occur. SNPs can be determined by light emission

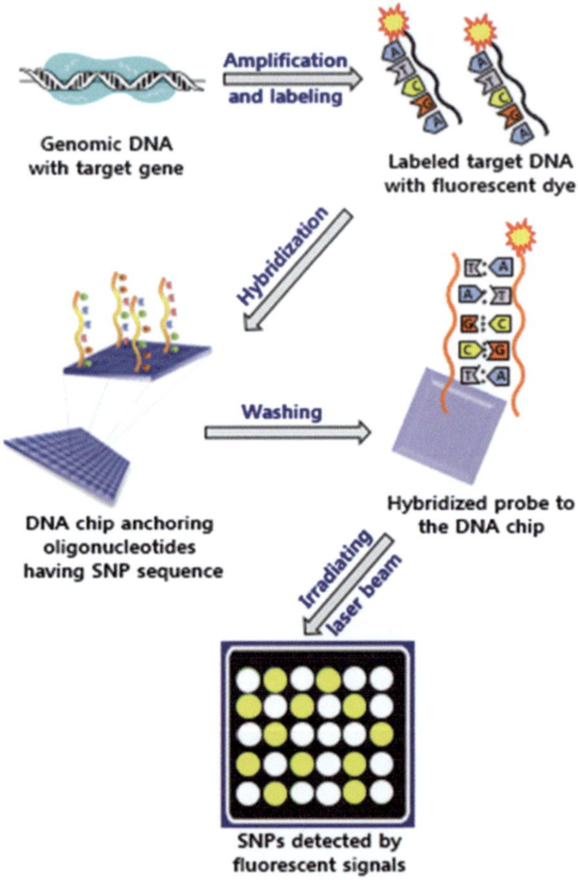

Fig. 2.8 DNA chips to analyze single nucleotide polymorphisms (SNPs). A DNA chip is prepared by attaching DNA fragments of ~25 bp in length from a species to a glass or metal plate. These fragments contain the SNP variants to be analyzed. Fluorescence-labeled probes are hybridized with the DNA fragments on the plate, and the DNA chip is washed. Finally, the probes are exposed to light of a specific wavelength to induce fluorescence

arranged per square centimeter. For the analysis, sections of the sample DNA containing SNPs must be amplified using PCR. Fluorescent dyes are attached to these fragments, which are used as probes to hybridize with the pre-prepared DNA chip. Then, the chip is washed. The probes that complement the chip sequences will attach firmly, while other allele fragments will be washed away. By analyzing the fluorescent spots, the alleles in the sample DNA can be determined. DNA chips are useful when analyzing species for which genome information is available.

Pros and Cons of Various DNA Marker Methods The ideal DNA marker would have the following characteristics: even distribution throughout the genome, high experimental reproducibility, fast and cheap analysis, little DNA requirement, and

Table 2.2 Comparison of DNA markers

Marker	Reproducibility	Technical difficulty	Inheritance	Sequence information
RFLP	High	High	Codominant	Not required
SSR	High	Medium	Codominant	Required
RAPD	Low	Low	Dominant	Not required
AFLP	High	Medium	Codominant	Not required
SNP	High	High	Codominant	Required

linkage with various phenotypes. Currently, no DNA markers meet all of these conditions; each technique has pros and cons (Table 2.2) that must be considered when planning experiments so the selected markers match the purpose of the experiment.

2.1.2 *Tools for Trait-Linked Molecular Marker Development*

2.1.2.1 Plant Material

A simple way to develop molecular markers linked to useful traits in agriculture is to find clear differences between allelic sequences of the target trait. However, because anonymous sequence differences between two individuals are not likely to be located in the target region, carefully prepared plant materials must be used to identify trait-linked molecular markers. In other words, because DNA sequence diversity often does not represent phenotype differences, plant materials must be prepared to develop trait-linked molecules, as described below.

Segregating Population This method is one of the easiest ways to prepare plant materials for developing trait-linked markers. Two plants with and without the target trait are crossed to develop an F_2 or backcross BC_1 population. The former is derived by self-pollination of F_1 individuals, whereas the latter results from crossing an F_1 hybrid with one of its parents. These populations contain individual plants segregating the target trait. Molecular markers linked to the trait can be developed by analyzing individuals that have the target trait and those do not. In other words, molecular markers linked to the target trait can be developed by first finding marker polymorphisms between two parents then analyzing the entire segregating population to find markers correlated to the phenotype.

Near Isogenic Lines (NILs) NILs are isogenic lines that have the same genomic background except at one locus that has different alleles. NILs are developed by repeatedly backcrossing a donor parent having a desired allele and a recipient (recurrent) line lacking the target allele (Fig. 2.9). By selecting individual plants having the desired allele in each generation, a line with both the recipient genome and the target trait is developed. Developing NILs is time-consuming and tedious. However, they are very useful for developing trait-linked molecular markers. For

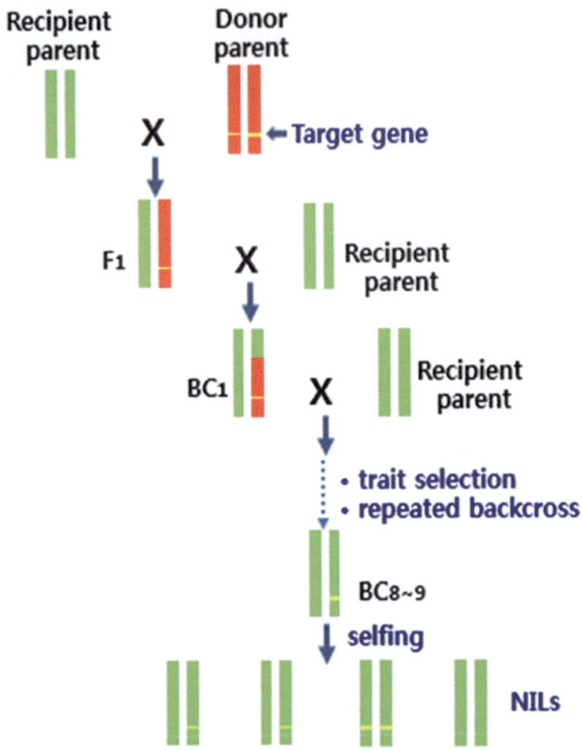

Fig. 2.9 Process of developing near isogenic lines (NILs). The F_1 is created by crossing a donor and a recipient parent. Recipient parents are repeatedly crossed with offspring that have the target allele to create NILs with the target allele in the recipient genome

example, if two individuals of a NIL set show DNA sequence difference, a DNA polymorphism between them is probably closely linked to the target trait, because their only genetic difference is near the target locus. NILs are often used for fine mapping of a specific locus and for physical mapping in map-based cloning of useful traits.

Recombinant Inbred Lines (RILs) RILs are obtained from F_2 individuals by successive self-pollination via single-seed descent (Fig. 2.10). First, the size of the F_2 generation should be determined based on the experimental goal. A single seed is chosen from a plant in each generation and propagated by self-pollination. After eight or more consecutive generations, individual plants from each family member derived from a given F_2 plant become genetically uniform. RILs are time-consuming to develop but are used frequently because their loci are genetically fixed and very useful for studying quantitative traits. Because RILs are genetically homozygous, experiments can be repeated in various labs around the world.

Double Haploid (DH) Lines DH lines are similar to RILs in that individual plants from each family derived from a given population are genetically uniform. DH lines

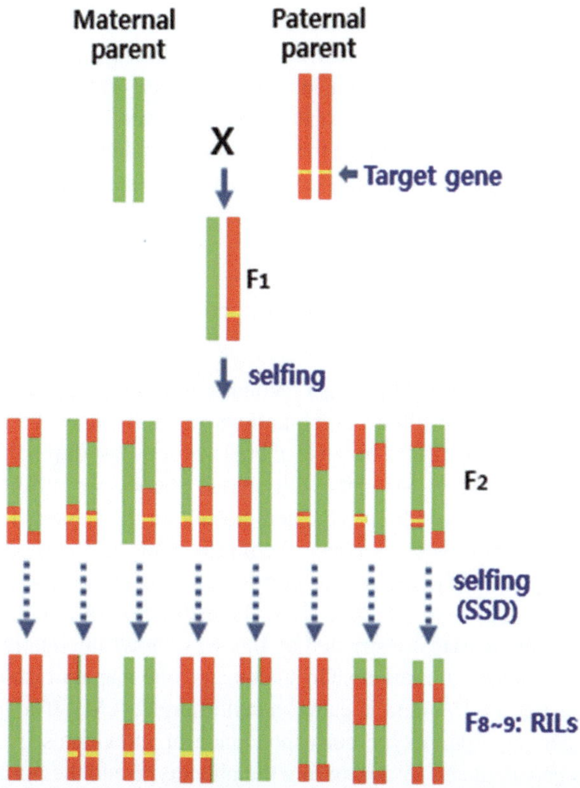

Fig. 2.10 Process of developing recombinant inbred lines (RILs). Two parents with different genotypes for the target trait are crossed to form F_1 plants, which are self-pollinated to get F_2. Single seed descent is used to obtain genetically fixed lines

are developed by culturing haploid tissue, such as the anther, microspore, or ovary. These cells become diploid either naturally or via colchicine application. DH lines have the same characteristics as RILs.

2.1.2.2 Phenotype Testing and Genetic Analysis

Agriculturally important characteristics include plant architecture, photosynthesis ability, seed size, flower-related traits, fruit color, taste, the presence of functional compounds, and abiotic resistance. Among these traits, some traits of disease resistance or color are controlled by only a few genes, and the phenotypes of these qualitative traits are relatively easy to determine, allowing accurate identification of the inheritance mode in these traits. However, traits like productivity and abiotic-stress resistance are controlled by multiple genes and show continuous phenotype variation, making the effects of the underlying genes hard to analyze and their inheritance mode are even more difficult to determine. Developing molecular markers for such quantitative traits requires an accurate phenotyping method. To investigate the inheritance of a quantitative trait, two different inbred lines are crossed to make the F_1, which is then self-pollinated or backcrossed with one of the parents to develop the F_2 or BC_1. In each population, the phenotype is analyzed to estimate the number of genes controlling the trait. Even so, the inheritance mode is hard to identify when more than three genes are involved in the trait. In these cases, quantitative trait locus (QTL) mapping can help identify the number of involved genes and their roles.

2.1.2.3 Linkage Mapping and Genome Information

Linkage Mapping (Genetic Mapping) When two parents with distinct alleles at many loci are crossed, the allelic combinations on individual chromosomes can change in subsequent generations because of cross-over events during meiosis. These novel combinations yield descendants with unique phenotypes different from their parents. However, when two loci are located closely on the same chromosome, the probability of cross-over events between them falls, and the recombinant genotype becomes relatively rare. The cross-over rate increase in proportion to the distance between genes, so cross-over rate data allow the estimation of loci distances on the chromosome. A genetic map constructed by this way shows the relative locations of morphological or molecular markers. On such maps, one map unit is defined as having a cross-over rate of 1 % and is called a centimorgan (cM). If a genetic map is available, the genotype/phenotype correspondance of individuals in a segregating population can be calculated by comparing the phenotype and the marker genotype.

Genome Information A genome is defined as all the genetic information of a species; it includes both nuclear and cytoplasmic genetic information. The unique genetic information of an organism affects its phenotype. Thanks to rapid technological developments, genome analysis has become a routine way to study gene functions and evolution of animals, plants, and microorganisms. Genome analysis can be used both within a species and among related species. Rather than focusing on the function and expression patterns of each gene, genome research surveys the entire genome sequence to analyze gene structure and find a large scale regulatory mechanisms of the genes.

2.1.3 Glossary

Allele-Specific PCR One way to test SNPs. If PCR is performed with a primer whose 3′ end complements the SNP, an allele that perfectly matches the primer will be amplified while others will not. This principle is used to distinguish SNP alleles.

Allele One of two or more variants of a gene can reside at the same locus on a chromosome. If the loci on homologous chromosomes have the same allele, the individual is said to be homozygous at that locus. If the alleles are different, the individual is heterozygous for the trait.

Amplified Fragment Length Polymorphism (AFLP) A DNA marker that combines the advantages of both RFLP and RAPD. Sample DNA is digested with two restriction enzymes, and adaptors are attached to the fragment ends. Primers specific to the adaptors are used to amplify DNA using PCR. The PCR products are separated by electrophoresis to observe individual polymorphisms.

Centimorgan (cM) The genetic distance along a chromosome with a 1 % probability of cross-over between two molecular markers or loci.

Codominant Marker Markers that can distinguish the homozygote and heterozygote states.

Codominant In heterozygotes, when the contributions of both alleles are seen in the phenotype.

Crossing Over The exchange of paternal and maternal segments of homologous chromosomes during meiosis. It causes progeny to have different allele combinations from their parents.

DNA Chip One way to test SNPs. Artificial 25 bp oligonucleotides are attached to glass or metal plates. Fluorescently-labeled probes of DNA fragments are hybridized to determine which alleles have a given SNP.

DNA Hybridization A process that combines two complementary single-stranded DNA molecules to form double-stranded molecule through base pairing.

Dominant Marker A marker that cannot distinguish the homozygous and heterozygous states.

Dominant A genetic phenomenon in which the phenotypes of dominant homozygotes (AA) and heterozygotes (Aa) are the same.

Doubled Haploid Lines (DH Lines) A plant that is created by culturing haploid cells (anther, microspore, or ovary cells) and converted diploids naturally or by adding colchicine. DH lines have the same genetic characteristics as RILs.

Expressed Sequence Tags (ESTs) 5′ and 3′ end sequences of clones in cDNA libraries. ESTs are 300–500 bp single-strand mRNA sequence reads derived from genes expressed in a given tissue and/or at a given developmental stage. The technique is used in functional genomics.

5′ Nuclease Assay A way to test SNPs that uses probes, such as TaqMan™. A fluorescent chemical is attached to the probe's 5′ end, while on the 3′ end bears a fluorescence-inhibiting chemical. During PCR, the 5′ exonuclease function of the PCR polymerase cuts the inhibiting 3′ end, allowing fluorescence to be generated to indicate SNPs.

Functional Genomics A set of methods to identify gene function using genome sequence data acquired from structural genomics research. It includes methods like high-throughput sequencing of expressed genes (ESTs), gene expression analysis using DNA chips, and two-dimensional electrophoresis to differentiate proteins.

Genetic Marker A marker that is closely linked with useful traits, such as disease resistance or abiotic stress resistance.

Genetic or Linkage Map A map that shows the relative locations of genetic or molecular markers calculated using cross-over rates. Also called gene linkage-group map.

Genome The complete set of genetic information found in an species (includes both nuclear and cytoplasmic genetic information).

Isozyme A type of protein marker. Isozymes of an enzyme have the same activities but can be distinguished by different electrophoresis speeds.

Library A gene set that contains clones of certain DNA fragments from a target crop. Various types exist, such as a genomic library of the entire genomic DNA of a sample, and a cDNA library of information about expressed genes.

Linkage The genetic proximity of two or more genes on a chromosome.

Locus The location of a gene on a chromosome or chromosome map.

Microsatellite A 1–6 bp DNA repeat that is usually repeated fewer than 100 times. Also called simple sequence repeats (SSRs).

Minisatellite A 9–100 bp random sequence that is usually repeated fewer than 1000 times.

Molecular Marker A protein or DNA marker used to easily distinguish a target trait.

Morphological Marker A marker that can distinguish the genotype by the form of the phenotype, instead of requiring biochemical or molecular biology techniques.

Near Isogenic Lines (NILs) Lines that have the same genotype except at one locus that has different alleles.

Physical Map A map that showing the positions of sequence features, including genes. Physical maps are generated using DNA sequence data.

Polymerase Chain Reaction (PCR) A technique for amplifying a specific DNA sequence. A cycle of DNA denaturation → primer annealing → polymerization (using a heat-resistant DNA polymerase) is repeated to amplify DNA within a machine.

Polymorphism The phenomenon in which more than one different allele exists at the same locus. Polymorphism can be observed at the phenotype, protein, or DNA level.

Primer A short nucleotide sequence that acts as the starting point of DNA replication. It complements the template DNA strand.

Probe A DNA fragment that complementarily bonds to a specific DNA sequence. Probes are labeled with isotopes or fluorescent dyes to easily identify homologous genes on DNA (or RNA) blots or DNA chips.

Random Amplified Polymorphic DNA (RAPD) A type of DNA marker that can be developed without prior sequence information about the target gene. A random sequence primer (usually 10 bp long) is used to perform PCR; electrophoresis of the amplicons reveals polymorphisms among individuals.

Recombinant Inbred Lines (RILs) RILs are obtained by successive self-pollination from F_2 individuals via single-seed descent. RILs are often used to find quantitative trait loci (QTL mapping).

Restriction Endonuclease An enzyme that recognizes a specific base-pair sequence (usually 4–8 bp in length) and cuts the DNA double strand.

Restriction Fragment Length Polymorphism (RFLP) DNA polymorphism revealed by digesting DNA with restriction enzymes, followed by electrophoresis. The term also denotes technology that confirms such differences using Southern blotting and DNA probes.

Sequence Tagged Site (STS) A type of molecular marker that uses the DNA sequence near a target gene as a primer for PCR to find differences among amplified regions.

Simple Sequence Repeat (SSR) (See microsatellite.)

Single Nucleotide Polymorphisms (SNPs) A single nucleotide difference between individuals of a species.

Southern Blot A method of transferring DNA fragments that were generated by restriction enzymes and electrophoresed onto a membrane; often used to test for the presence of homologous genes in a genome.

Structural Genomics A research method to understand genetic phenomena from a macroscopic perspective by studying the genome's molecular structure. Methods like high-density gene mapping and genome sequencing are included in structural genomics.

Variable Number of Tandem Repeats (VNTRs) Polymorphisms based on differences in minisatellite repeats. VNTRs can be confirmed using DNA probe hybridization or PCR.

2.2 Developing Trait-Linked Molecular Markers

When a target trait is controlled by a single gene, molecular markers linked to the trait can be developed by generating a linkage map. To check whether a molecular marker and target trait are linked and to calculate their genetic distance, co-segregation analysis must be performed. A mapping population such as F_2 or BC_1F_1 should be constructed, and then the phenotype and molecular marker genotype are analyzed. The number of recombinant individuals is counted, and the genetic distance between the molecular marker and the target gene can be calculated in cM units to generate a genetic linkage map.

However, because only a small fraction of the genome contains a given trait, making a genetic linkage map of an entire plant genome could be a waste of time and energy. Additionally, saturating a genetic linkage map is not guaranteed to lead to the development of a molecular marker closely linked to the target gene. Therefore, methods for developing molecular markers that are more efficient have been devised. Prior to the advent of sufficient biological information like genome sequences, researchers found efficient ways to develop molecular markers without a large amount of data.

2.2.1 *When Genetic Linkage Maps or Biological Data Are Not Available*

When biological data were insufficient, NILs developed during plant breeding were often used. If such plant material could not be prepared, bulked segregant analysis (BSA) techniques were used to develop molecular markers. Molecular markers such as RAPD or AFLP can be used.

2.2.1.1 Near Isogenic Lines

Theoretically, the genomes of NILs differ only in the region near the target trait. NILs can be developed using backcrossing (Fig. 2.11).

1. A 'donor' (P1) with the target gene is crossed with the 'recurrent' parent (P2) to obtain the F_1 generation.
2. The F_1 and P2 are backcrossed to make the next generation. This backcrossing process is usually repeated 4–5 times.
3. In each generation, individuals with the target trait are chosen to be backcrossed again. Thus, all genetic 'background' except the target gene is replaced with the P2's genetic information.
4. A line that is homozygous for the target gene is selected to develop an NIL through 1–2 rounds of self-fertilization.
5. To obtain NIL, plants should be advanced to the BC_4F_2–BC_5F_3 generation. NILs can be developed with fewer backcrosses when RILs or introgression lines (ILs) already exist.

Developing molecular markers linked to a target gene using NILs begins with comparing the genotype of the P1, P2, and NIL using RAPD or AFLP markers. There are three possible outcomes: (1) the P1, P2, and NIL all have the same genotype; (2) the P2 and NIL have the same genotype but not the P1; or (3) the P1 and

Tip

Making NILs requires a systematic way to select individuals with the target gene for each backcross generation. If the target gene is a male-sterility restoring or fruit-color gene, individuals can be easily selected by observing the anther or fruit, respectively. However, if the target gene is related to disease resistance, a disease-screening experiment must be performed to identify individuals that carry the gene. If the target gene makes a functional compound, the amount of product can be measured experimentally. Whether the target trait is dominant or recessive is also an important factor. If the P1's genotype is AA and the P2's genotype is aa, the genotype of the BC_1F_1 generation segregates to Aa:aa = 1:1. If the target gene is dominant, the heterozygote will show the target trait, so individuals with the target trait can be selected directly to advance to the next generation. However, if the target gene is recessive, none of the individuals will show the target trait in BC_1F_1, so progeny must be tested to find the heterozygotes.

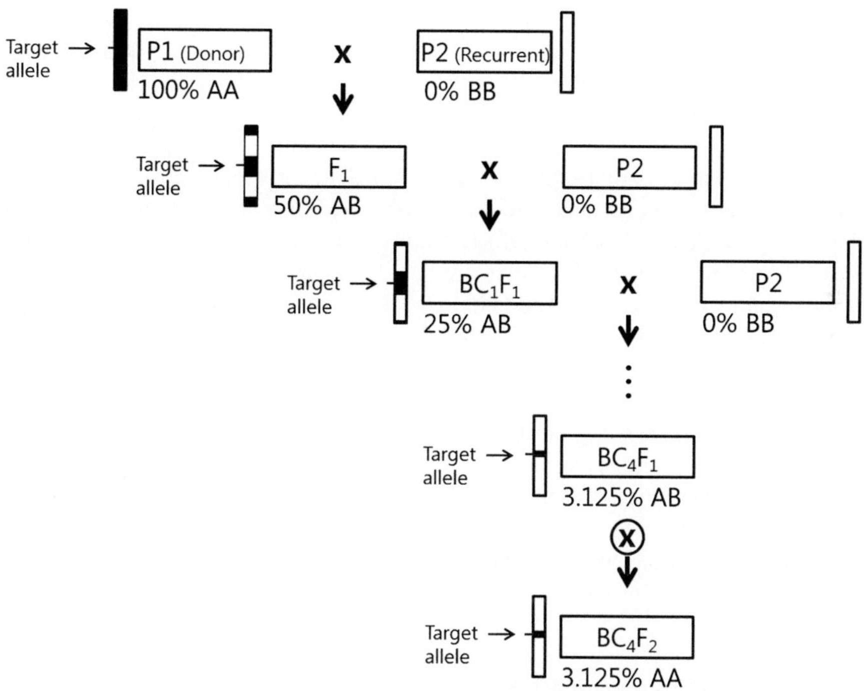

Fig. 2.11 Near isogenic line (NIL) development scheme. For each generation, the *bar* indicates a chromosome; *black* represents a genetic segment from the donor and white a segment from the recurrent. The amount of genetic data inherited from the donor at each generation is shown as a percentage at the *bottom* of each generation. The genotype of the target gene is shown as AA (homozygote) or Aa (heterozygote)

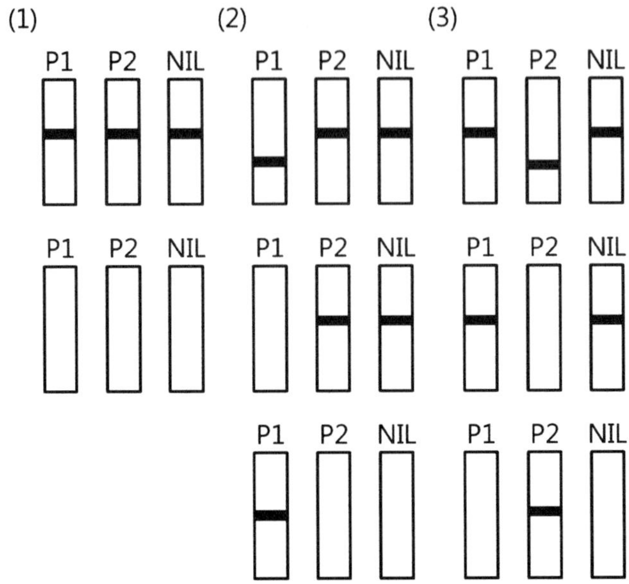

Fig. 2.12 Example of analysis results for the donor (P1, genotype AA), recurrent (P2, genotype aa), and near isogenic line (NIL, genotype AA). The picture shows each possible PCR result for molecular markers. The *boxes* represent lanes in an agarose/acrylamide gel after electrophoresis, with bands shown as *black lines*

NIL have the same genotype but not the P2 (Fig. 2.12). In the first case, because the three lines have no polymorphism, it cannot be used in co-segregation analysis. In the second case there is polymorphism, but because the genotypes of the P2, which lacks the target gene, and NIL, which has it, are the same, the marker is probably far from the target locus or not linked to it. In the third case, there is polymorphism and, because the marker genotype of the NIL is the same as that of P1, the marker is probably linked to the target locus. After choosing markers that correspond to the third category, co-segregation analysis can be performed to check the degree of linkage and, ultimately, the selected markers can be used in experiments.

NILs are very useful in genetic research, and their utility is decreased by their long development times. Additionally, in the case of a BC_4F_1 generation, 3.125 % of its genome should come from the P1 plant, but that genomic fraction can exist across several locations that are not linked to the target gene. Because, it is dispersed throughout the genome and may cause false positive errors to occur (Fig. 2.13).

2.2.1.2 Bulked Segregant Analysis (BSA)

BSA is a method that can be used when experimental material like NILs is not available. To find molecular markers linked to a target gene, genomic DNA of several plants with the same phenotype is bulked into one pool. Two bulked pools of

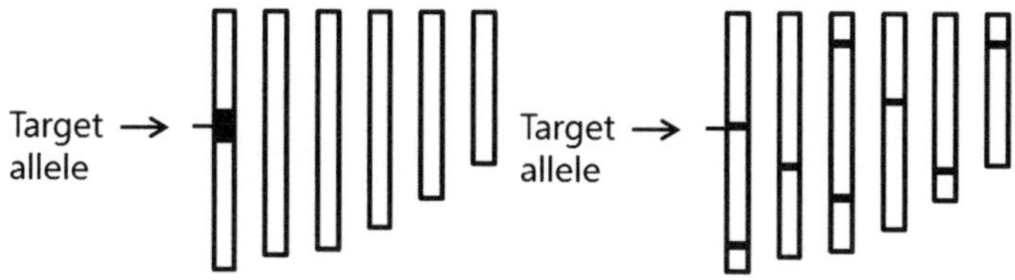

Fig. 2.13 Explanation of false positive errors in near inbred lines (NILs). The *vertical bars* represent chromosomes; the *black areas* originate from the donor. At *left*, the donor-origin areas are concentrated near the target gene, while at *right*, the donor-origin genetic material is dispersed to places unlinked with the target gene

segregants differing for one trait will differ only at the locus harboring that trait. BSA can be performed via the following steps (Fig. 2.14).

1. Cross a target gene-containing P1 line and gene-lacking P2 line to generate the F_1 generation, and self-pollinate F_1 to get the F_2 generation.
2. Determine the phenotype for the target gene in each individual and extract DNA. Dilute the DNA to the same concentration, then mix the DNA samples of the same phenotype to make a DNA 'bulk' for each phenotype.

Tip

When performing BSA, two things must be considered: the minimum individual number to comprise a DNA bulk and how to distinguish homozygotes from heterozygotes with the same phenotype. The minimum individual number for a bulk is closely related to the maximum probability that a molecular marker will actually be linked to the target locus. If the individual number in a bulk is *n*, the probability that a molecular marker will be linked with the target gene is calculated as $2\times\left(\frac{1}{4}\right)^n\times\left\{1-\left(\frac{1}{4}\right)^n\right\}$. When the individual number in a bulk is 4, the probability is 1 out of 100; for n=10 it is 2 out of a million, and for n=15 it is 2 out of a billion. DNA bulks usually consist of DNA from 10 to 15 individuals. Choosing homozygotes requires test crossing in later generations.

If one assumes that both parents have a homozygous genotype at all loci, the genotypes of all F_1 individuals will be the same. Consider two other loci, B and C, that are far from the target gene. After performing BSA and making DNA bulks, there must be an AA bulk and an aa bulk. While the allele frequency for gene A in the AA bulk is A:a=1:0, for other loci that are not linked to A, the genotype frequency will be BB:Bb:bb=CC:Cc:cc=1:2:1. Thus, the allele frequency for other loci is 1:1, showing heterozygotes as a result (Fig. 2.15).

Like NILs, when DNA bulks are used in RAPD or AFLP marker analysis, there are three possible results: (1) the AA bulk and aa bulk are both heterozygotes, (2) the AA bulk and aa bulk are both homozygotes but do not have polymorphism, and

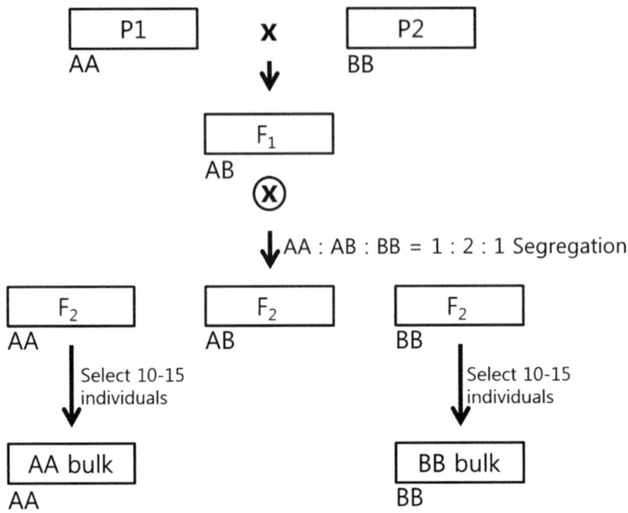

Fig. 2.14 Bulked segregant analysis (BSA) scheme, showing important steps in the progress. The genotype of each generation is written as AA (homozygous for target allele), Aa (heterozygous), and aa (homozygous for non-target allele)

(3) the AA bulk and aa bulk are both homozygotes and show polymorphism. As with NIL, molecular markers with the third outcome are selected, and co-segregation analysis is performed to develop trait-linked molecular markers (Fig. 2.16).

Because BSA can be performed with only the F_2 generation, it is one of the most efficient ways to develop molecular markers. BSA is used in molecular marker technologies like RAPD and AFLP, and applications are expanding to DNA chips or next generation sequencing (NGS) technology.

2.2.2 *When Genetic Linkage Maps or Biological Data Are Available*

Many genes have been identified and their functions were determined in various organisms. This biological information can be used to discover the functions of genes in other species. Thanks to the development of molecular genetics, the genetic linkage maps for many species have been saturated, and physical maps based on bacterial artificial chromosome (BAC) libraries have contributed greatly to elucidating the functions of previously unknown genes and developing new molecular markers. Now, NGS technologies are allowing entire genomes to be sequenced as references, and resequencing individuals of the same species has become common.

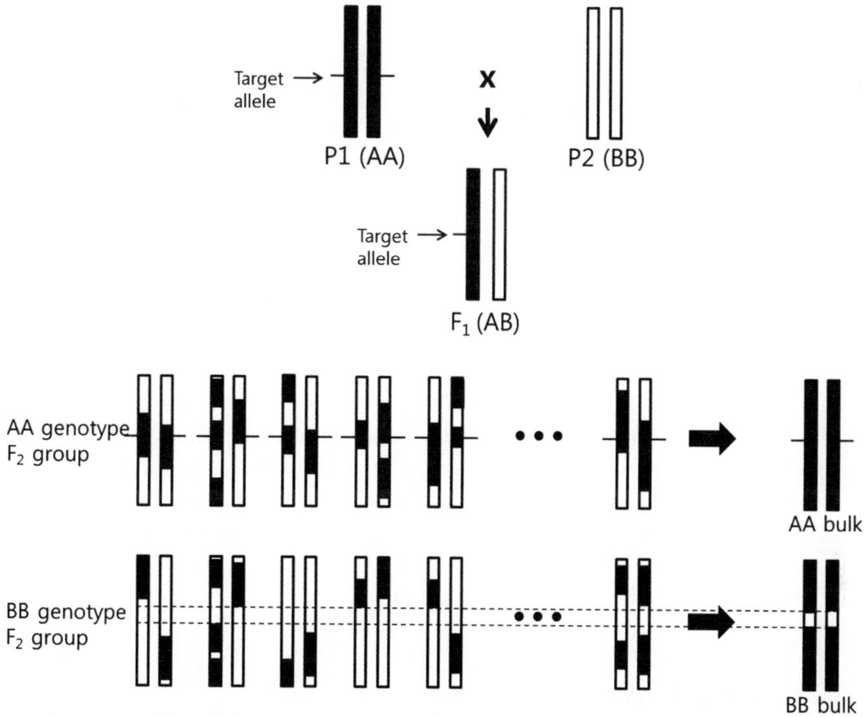

Fig. 2.15 Diagram of P1, P2, F1, and F_2 diploid group in bulked segregant analysis (BSA). The *vertical boxes* represent the chromosome on which the target locus is located; the *black area* indicates genetic material from P1 and the white area from P2. The lower picture is an example of possible genetic backgrounds for genotypes AA and aa. When these individuals are picked to make a DNA bulk, all areas of the chromosome except the target locus become similar

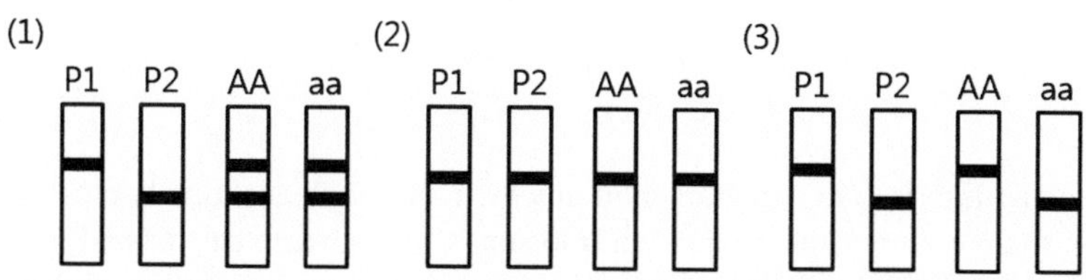

Fig. 2.16 Example of molecular genotype analysis performed on P1, P2, the AA bulk, and the aa bulk in bulked segregant analysis (BSA). The picture depicts possible genotype analysis results for PCR-based molecular markers. The *vertical boxes* represent post-electrophoresis agarose/acrylamide gel lanes, and the *black lines* signify amplified bands

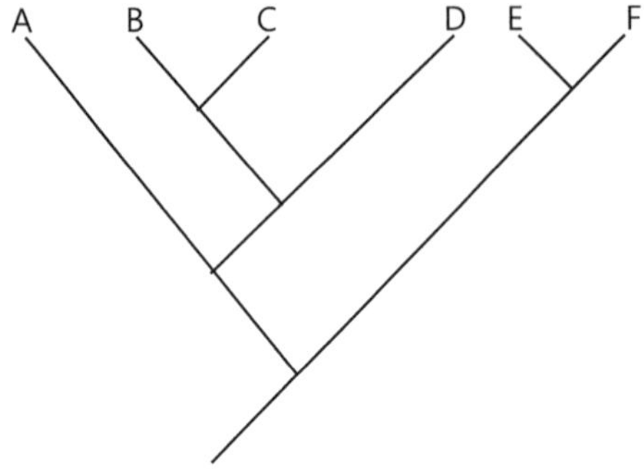

Fig. 2.17 Phylogenetic tree of six species originating from a common ancestor. Because speciation occurred early between the *A–D* and *E–F* groups, they have little in common. In contrast, species *B* and *C* are similar to one another, as are species *E* and *F*

Such abundant biological information allows new molecular markers to be developed with speed and accuracy that was only in the realm of imagination before.

Methods to develop molecular markers depend on whether existing biological information can be used or not. When there is abundant biological information about a crop plant, the data can be used to develop molecular markers via a candidate gene or comparative genetics approaches.

2.2.2.1 Using Comparative Genetics

Comparative genetics is based on evolution as a theoretical background. An important assumption of evolutionary theory is that any group of organisms will have a common ancestor (Fig. 2.17). Species that recently shared a common ancestor have similar numbers of chromosomes, and the gene loci are also similar (a feature termed synteny). The more distantly related two species are, the less commonality in their DNA, so comparative genetics is usually applied to species in the same genus or, in some cases, more broadly to species within a family.

Tip

Synteny describes the phenomenon in which two phylogenetically close species share common genes and chromosomes from their ancestor and, therefore, are similar. For example, if a gene for fruit color exists between molecular markers 1 and 2 in species B, the same gene will probably also appear between markers 1 and 2 in a closely related species C because of synteny (Fig. 2.18).

Early comparative genetics began by comparing the genetic maps of different species. This process requires the availability of a common marker that can be used in at least two genetic maps. The comparison becomes more accurate as more markers are shared by the individual genetic maps (Fig. 2.19). If we consider two linkage groups

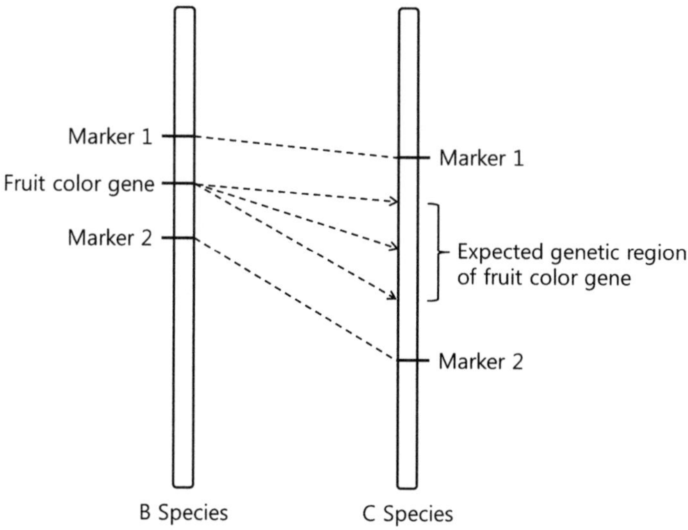

Fig. 2.18 Chromosome map used in comparative genetics showing synteny between two phylogenetically close species

with only one common molecular marker, that marker can identify the syntenic genetic region but cannot show the orientation of markers (Fig. 2.19a), which requires at least two common markers to determine (Fig. 2.19b, c). The degree of synteny can differ with genetic distance. The closer two species are phylogenetically, the greater the synteny in both local and large scales, for example at the chromosome level (Fig. 2.19d). However, if two species are distantly related, synteny is conserved only in small regions, but not in large scale ('macro synteny' is not conserved) (Fig. 2.19e). Therefore, if only micro synteny is observed, more common markers may be needed to find the genetic region containing the target gene. Also, analyses must be performed cautiously even in syntenic areas, which may contain occasional rearrangements.

If complete genome or EST sequence data exist for the species that is being used for comparison, different methods of research become possible. If the entire genome sequence is known for a model species and only EST sequence data exist for the target plant, the ESTs can be used as common markers because they originate from expressed genes. Therefore, EST sequences are highly likely to be conserved between the two species, and similar EST sequences can be assumed to be present in the genome of a model plant. When comparing species that are distantly related, the tblastx algorithm should be used, while the blastn algorithm is preferable when investigating closely related species. The tblastx algorithm finds similar amino acid sequences; therefore, it may not work accurately when the genome contains many paralogs. Because closely related species have more similar orthologs than paralogs, the blastn algorithm, which compares nucleotide sequence similarity, is better for them.

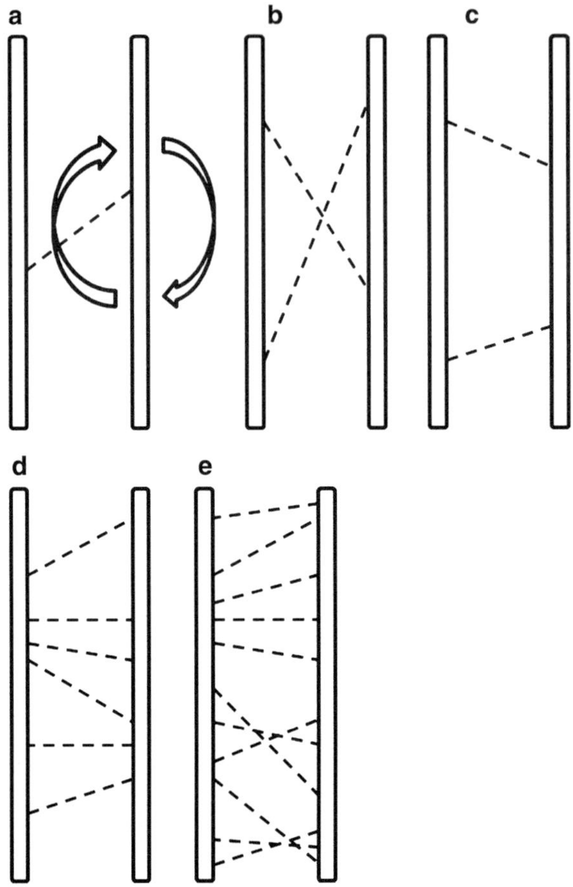

Fig. 2.19 Example of a comparative genetics analysis of two different species. The picture shows a comparative linkage map for cases when (**a**) there is one common marker, (**b**, **c**) there are two common markers, and (**d**, **e**) there are many common markers. When more than two common markers exist, two linkage groups can be compared. For example, in (**b**), the two linkage maps are reversed. In (**d**), both macrosynteny and microsynteny can be observed. In (**e**), the species evolved separately and experienced rearrangement of the chromosomal genes, leaving only microsynteny

Tip

The Basic Logical Alignment Search Tool (BLAST) is one fundamental program for analyzing nucleotide sequences. BLAST compares a submitted sequence (query) with others in a database (DB) and aligns the query with any similar sequences found. BLAST has many embedded algorithms, including blastn, blastp, blastx, tblastn, and tblastx. Their uses will vary by DB and query. Blastn is used when the query and DB are both nucleotide sequences, while blastp is used when both are amino acid sequences. The other three algorithms artificially translate nucleotide sequences to amino acid sequences to compare the query with the DB: blastx uses an expressed sequence query against an amino acid DB; tblastn uses an amino acid query against an expression sequence DB, and for tblastx, both the query and DB are amino acid sequences.

2.2.2.2 Finding Functional Candidate Genes

Candidate genes can be divided into functional and positional candidates. In plants, the sequence, structure, and function of various genes have been discovered through research using model plants such as *Arabidopsis*, rice, and tomato. Assuming that all species have a common ancestor (Fig. 2.17), a gene with a known function in one species will likely have a similar function in another. This inference justifies the development of molecular markers using functional candidate genes. The theoretical range for such comparisons can be within a kingdom, but candidate genes are usually found within lower-level taxa, such as a genus or family, because genetic similarities are more easily found.

A similar gene structure indicates that the expressed three-dimensional structures of two proteins are similar and that the amino acid sequences are identical or similar. Consider a gene that determines fruit color in species A. The following strategies can be used to find functional candidate genes in species B. If species B has its entire genome known or at least an EST database, similar genes in B can be found by querying the fruit color gene for A in the database of species B. For such a search converting the DB and query base-pair sequences to amino acid sequences, then comparing the sequence similarity instead of identity is convenient. As noted above, the tblastx algorithm is appropriate for this purpose.

Candidate genes found in EST or whole-genome DBs by BLAST should be confirmed experimentally. To find the entire sequence for the gene, primers should be designed to identify the 5′ and 3′ UTR regions. If the resulting base-pair sequence shows high similarity with the EST or whole-genome DB, the candidate gene is confirmed to be present in the genome.

Tip

If identity analysis is a process of confirming whether two sequences are the same, similarity analysis is comparing whether the two sequences are in the same 'group'. Even if the species have their origin in one species, as they each go through their own evolutionary process, many mutations accumulate on the gene. Let us assume that if gene functions are impaired, there is a disadvantage in survival. If a mutation happens to change the amino acid sequence, usually the resulting enzyme cannot function properly; therefore amino acid-changing mutations are given negative selecting pressure. In contrast, 'silent' mutations that do not change the amino acid sequence can survive to spread their genes. In such cases, the identity of the corresponding nucleotide sequences can be very low, but the similarity of the amino acid sequence can be very high.

2.2.3 General Strategies for Developing Molecular Markers

Modern comparative genetics does not simply compare genetic maps or sequences. Rather, all possible information is used to develop molecular markers. The genomes of many species are being analyzed and massive genome information is accumulating thanks to the rapidly decreasing costs of NGS. Inexpensive systems that analyze gene expression by sequencing transcriptomes with NGS technology are already available. Enormous genetic data, with functions identified, can be found at NCBI, and the rate of development of analytical instruments for the field of bioinformatics is amazing.

To develop target-gene-related molecular markers, the phenotype information and molecular marker genotype should be known for each individual in a segregating population. Methods of analyzing the marker genotypes have been in great progress, and using NGS to analyze biological information is now a necessity. Commercial bioinformatics programs provide simple interfaces for researchers with little experience in this area, but filtering and analyzing the data are time consuming. In the big-data era, analyzing a large amount of data with quick homemade scripts will become an essential skill for developing molecular markers.

2.3 Case Studies on Developing Molecular Markers

2.3.1 Using Bulked Segregant Analysis (BSA): **Phytophthora** *resistance in pepper (Liu et al. 2014)*

2.3.1.1 Preliminary Research and Background

Resistance of pepper against *Phytophthora capsici* is controlled by QTLs. Numerous sources of resistance have been reported, for example in *Capsicum annuum* CM334, PI201232, AC2258, and Perennial. The inheritance patterns of resistance vary depending on disease screening conditions and the water mold isolate. However, several studies have reported that a major QTL is located on chromosome 5 of pepper (Bonnet et al. 2007; Quirin et al. 2005; Thabuis et al. 2004).

2.3.1.2 Plant Materials and Phenotype Analysis

Many different kinds of mapping population have been used for QTL analysis, such as F_2 or BC. However, families of RILs have been the main choice, because of their homozygosity and accumulated recombination. The effects of environmental variation on quantitative traits can be much reduced by assessing several plants of the same genotype instead of a single plant. The YT population, consisting of 128 RILs at the F7–F9 generation of an intra-specific cross between *C. annuum* 'YCM334' and 'Tean' was used for inheritance analysis and marker development (Truong et al. 2012). *Capsicum annuum* 'YCM334' was the resistant parent and 'Tean' was susceptible. This study hypothesized that *Phytophthora* resistance in pepper acts as a monogenic trait with resistance being dominant over susceptibility under low disease-pressure conditions in which a low *Phytophthora* concentration (3×10^4 zoospore/mL) was used. A segregation ratio of 65 (resistant) and 78 (susceptible) plants was obtained by resistance screening of the YT RIL population.

2.3.1.3 Development of Molecular Markers

To develop molecular markers linked to the major QTL on chromosome 5, BSA-SFP (BSA-single feature polymorphism) was performed. For BSA, equal amounts of genomic DNA from 20 resistant and 20 susceptible lines selected from YT RILs were separately bulked and treated by DNase. The labeled DNA pools were hybridized on Affymetrix chips containing 30,000 ESTs, with five replicates for each DNA pool. Two candidate EST-based unigenes with different Dstat values (≥3 or ≤−3) between resistant and susceptible were identified. To develop molecular markers using these two ESTs, BLAST searches against the '*Capsicum annuum* Database' (http://peppergenome.snu.ac.kr) were performed. Several scaffold sequences were obtained and tested between YT RIL parents. Only one primer, 002466, was polymorphic in YT RILs. Co-segregation analysis of this primer showed 11 recombinant lines out of 135 lines. This marker was designated PhytoSNP5. The linkage map of chromosome 5 was constructed using developed markers in the YT RIL mapping population. As the result, the LOD score between PhytoSNP5 and IBP557 was the highest.

2.3.2 Using Candidate Genes: Markers from Tomato Linked to Cmr1 *(Kang et al. 2010)*

2.3.2.1 Preliminary Research and Background

The *Cmr1* gene, which has a single dominant phenotype, confers resistance against *Cucumber mosaic virus* (CMV) in pepper. Previously, Kim et al. (2004) reported three CAPS markers (CAPS-A, CAPS-B, and CAPS-C) linked to *Cmr1* but not the gene's map location.

2.3.2.2 Plant Materials and Phenotype Analysis

The mapping population used in this study was a *C. annuum* 'Bukang' F_2 population derived from self-pollinated F_1 'Bukang' plants. CMV screening was performed in the F_2 population using the CMV-Kor strain. Two varieties of *C. annuum*, 'Bukang' and 'Jeju', were used as negative and positive controls, respectively. The susceptible plants showed typical CMV symptoms of vein clearing mosaic and leaf distortion. Resistant plants showed no symptoms. The segregation analysis of resistance and susceptibility in the F_2 'Bukang' population showed 236 resistant plants and 73 susceptible ones, fitting the 3:1 Mendelian segregation ratio (P=0.7326).

2.3.2.3 Development of Molecular Markers

CAPS markers linked with the *Cmr1* gene were determined to be about 3 cM away in the 'Bukang' F_2 population. To locate the *Cmr1* gene in a pepper linkage map, the CAPS-A and CAPS-B markers were mapped using the AC 99 F_2 population. CAPS-A is located in the centromeric region of LG2 near TG31A (Fig. 2.20). The *Tm-1* gene, a *Tomato mosaic virus* resistance (ToMV) gene, is located in the centromeric region of tomato chromosome 2, and *Cmr1* is in a syntenic region. Therefore, the *Tm-1* sequence was used as a candidate gene for *Cmr1*. One pepper

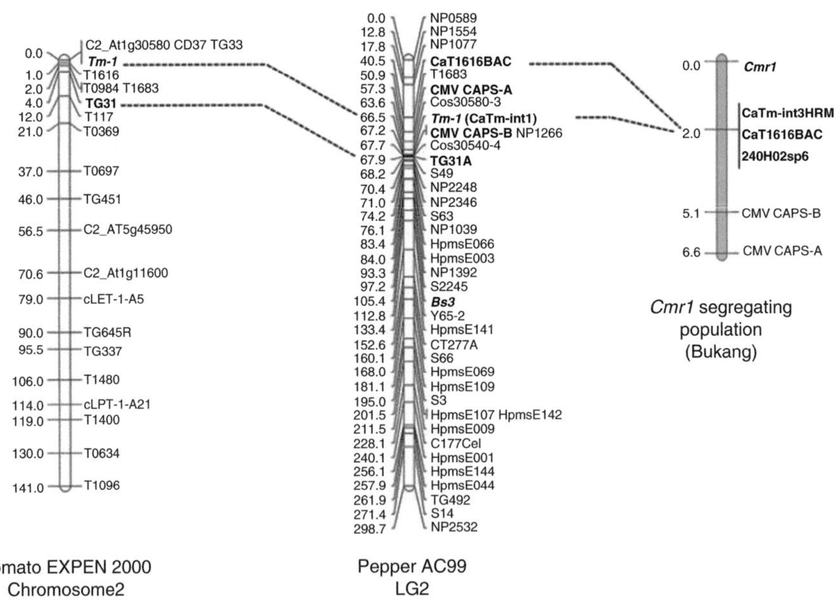

Fig. 2.20 Comparative analysis between tomato and pepper maps. The *Cmr1* gene is located in a region syntenic to the *Tm-1* gene of tomato. The molecular marker, CaTm-int1, linked to *Cmr1*, was developed using the ortholog of *Tm-1*

EST (cacn2211), which has the most nucleotide similarity with *Tm-1*, was used to develop markers. A total of five introns were predicted in this EST sequence using the Intron Finder program of SGN (http://solgenomics.net/). Sequence analysis of the first intron revealed a polymorphic endonuclease recognition site, *Hinf*I, which was used to develop the CAPS marker CaTm-int1. CaTm-int1 mapped near TG31 on chromosome 2. However, CaTm-int1 was not polymorphic in the 'Bukang' F_2 population, but further analysis showed that the third intron sequence was polymorphic. This SNP was detected by high resolution melting (HRM) analysis, and the marker was named 'CaTm-in3HRM'. Co-segregation analysis of CaTm-int3HRM and the *Cmr1* gene revealed six recombinant individuals out of 309 individuals.

2.3.3 *Using Comparative Genetics: Markers Linked to the* L *Locus (Yang et al. 2009)*

2.3.3.1 Preliminary Research and Background

Comparative genetic analysis between pepper, potato, and tomato indicates that most resistance (R) genes are not randomly distributed in the genome. Clusters of R genes in corresponding regions of the three genomes often confer resistance to unrelated pathogen types (Grube et al. 2000). A comparative genetic map revealed that the *L* locus (resistant against Tobamovirus) of pepper, the *I2* locus (resistant against *Fusarium oxysporum*) of tomato and the *R3* locus (resistant against *Phytophthora infestans*) of potato are positioned in a syntenic R gene cluster at the end of the long arm of chromosome 11.

The *I2* locus of tomato was isolated by map-based cloning (Ori et al. 1997). Two main R gene clusters, SL8D and SL8E, were found in this locus. The *I2C-1* gene, in the main cluster SL8D, had been cloned and demonstrated to confer resistance against *Fusarium*. Comparative analysis was performed using molecular markers located in the genetic region around the SL8D and SL8E clusters to isolate the *R3* gene. *R3a*, a resistance gene against *Phytophthora*, was isolated in the corresponding region of SL8E, but not SL8D in which *I2C-1* was isolated, suggesting that molecular markers linked to the *L* gene could be developed and that, furthermore, the *L* gene could be isolated using genetic information from *R3a* in potato and *I2C-1* in tomato, because the former was isolated using genetic information of the latter (Fig. 2.21).

2.3.3.2 Plant Materials and Phenotype Analysis

F_2 *L*-segregating populations were constructed, and co-segregation analysis was performed for the linkage analysis between the *L* gene and polymorphic molecular markers. F_2 populations were derived by self-pollination of the F_1 commercial varieties Cupra (L^3/L^0), Special (L^4/L^1), and Myoung-Sung (L^4/L^1). The resistant

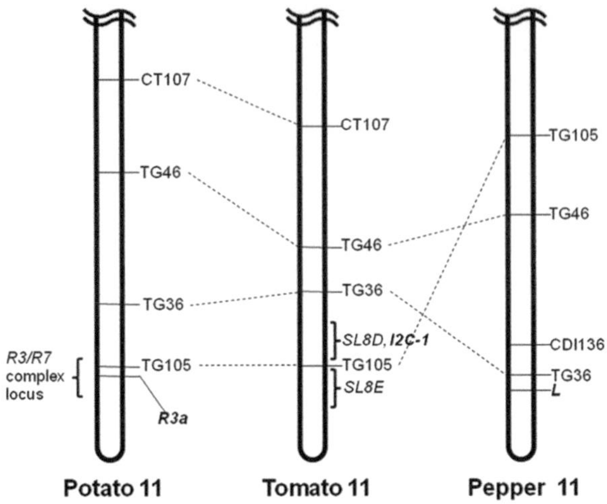

Fig. 2.21 Comparative map of the end of the long arm of chromosome 11 in potato, tomato, and pepper. The positions of *R3a* in potato, *I2C-1* in tomato, and *L* in pepper were conserved. Molecular marker analysis indicated that the syntenic relationship between potato and tomato covered most of the long arm, but genetic rearrangement occurred between tomato and pepper

phenotype was analyzed by inoculating the Tobamovirus P_0 pathotype to the Cupra F_2 population (L^3 is resistant but L^0 is susceptible to P_0 pathotype) and the $P_{1.2.3}$ pathotype to the Special and Myoung-Sung F_2 populations (L^4 is resistant but L^1 is susceptible to $P_{1.2.3}$ pathotype). Black local lesions were observed in resistant individuals, but symptoms such as mosaicism and necrosis of stem or leaves were observed in susceptible plants. The resistant-to-susceptible segregation ratio of the F_2 populations of Cupra, Special, and Myoung-Sung were 189:54, 537:109, and 504:341, respectively. The segregation ratio of the Cupra F_2 population satisfied the 3:1 Mendelian ratio of a single dominant resistance gene (P=0.3173), but the ratios of the Special and Myoung-Sung F_2 populations deviated significantly.

2.3.3.3 Development of Molecular Markers

To find the *I2* homolog in the pepper genome, hybridization of a 221,184 BAC library, derived from *Capsicum frutescens* BG2816 containing the L^2 allele, was performed using a probe based on the 3′ sequence of *I2C-1*. A total of 89 BACs with positive signal were selected and amplified by PCR using primers designed from genetic region around *R3a* to find the *R3a* homolog. Bands of 1.5 kb, 0.8–1.3 kb, and 1.2 kb in size were amplified from 37, 52, and 22 BAC clones using the R7-1, R7-2, and LRR primers, respectively. Among the 22 clones that amplified using LRR, bands were generated using both R7-1 and R7-2 from 12 BAC clones and using only one of the two primers from 10 BAC clones. These results suggested that the 22 BAC clones may contain the *L* candidate gene, because they could be amplified using primers designed based on both *I2C-1* and *R3a* and presumably included genetic information of those two genes. The PCR products of the LRR amplifications were analyzed, and the sequences were classified into nine groups. A BAC clone from each group was selected for draft sequencing. SSR type markers were developed from the BAC draft sequence and mapped to a linkage map. The pepBAC082F3-5 and pepBAC060I2-H3 markers developed based on 082F3 and 060I2 BAC clones co-segregated perfectly with the TG36 marker located 5.2 cM away from the *L* locus, and the pepBAC337L21-1 marker designed based on 337L21 BAC mapped near TG105, which is located on chromosome 11 but far from the *L* locus.

To construct a contig, PCR was performed using the pepBAC060I2-E4 marker designed based on the 060I2 BAC sequence. Among the 89 BACs, bands were clearly amplified in 087H3, 158K24, 207E3, and 290J13. Primers were designed based on the end sequences of these four BACs. A contig of 13 BACs was constructed by the presence/absence of PCR amplicons, giving an overlap order using BAC end primers (Fig. 2.22). A total of 16 primer sets were designed from the contig sequence to develop an *L*-linked molecular marker. Bands of 420 bp and 422 bp were amplified using 087H3T7 and 207E13SP6 from *Capsicum chacoense* PI260429 (L^4) and *C. annuum* 'ECW (L^0)'. Sequence analysis revealed nine and five SNPs between these two accessions in the 087H3T7 and 207E13SP6 amplicons,

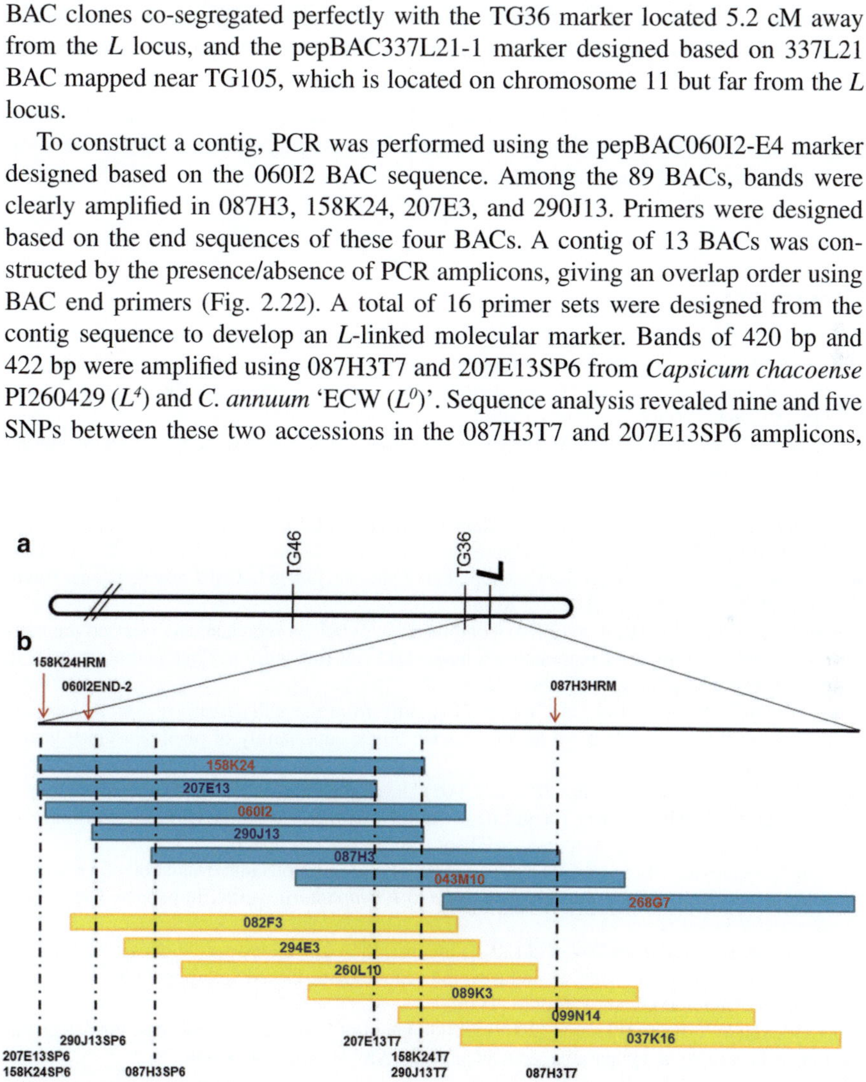

Fig. 2.22 A contig of 13 bacterial artificial chromosome (BAC) clones. The contig was constructed using a total of eight primers (listed below the *dotted lines*). Two co-dominant markers, 158K24HRM and 087G3HRM, and a dominant marker, 060I2END-2, were developed from BAC contig sequences

respectively. Primers containing one or two SNPs were designed for HRM analysis. The 087H03T7HRM marker redesigned based on the 087H03T7 BAC showed polymorphism (Fig. 2.22). A polymorphic marker, 158K24HRM, was developed in the same manner. The primer 060I2END-2, which was developed from 060I2, amplified a 700-bp band from ECW but not PI260429.

Co-segregation analysis used three polymorphic markers, 087H03T7HRM, 158K24HRM, and 060I2END-2, in three F_2 populations. The recombinant-to-total population size ratios of the three markers in Cupra, Special, and Myoung-Sung were 4/243, 5/361, and 11/858, respectively. Therefore, the genetic distances of these markers were calculated to be 0.8–1.2 cM from the *L* locus.

These case studies show different approaches to using genetic linkage mapping, bulked segregant analysis, comparative genetics, and physical mapping to develop molecular markers for key target loci.

References

Bonnet J, Danan S, Boudet C et al (2007) Are the polygenic architectures of resistance to *Phytophthora capsici* and *P. parasitica* independent in pepper? Theor Appl Genet 115:253–264

Grube RC, Radwanski ER, Jahn M (2000) Comparative genetics of disease resistance within the solanaceae. Genetics 155:873–887

Kang WH, Hoang NH, Yang HB et al (2010) Molecular mapping and characterization of a single dominant gene controlling CMV resistance in peppers (*Capsicum annuum* L.). Theor Appl Genet 120:1587–1596

Kim S, Hwang J, Kim G et al (2004) Development of markers linked to CMV resistant gene. Patent 10-2004-0086321, The Republic of Korea

Liu WY, Kang JH, Yang HB et al (2014) Combined use of bulked segregant analysis and microarrays reveals SNP markers pinpointing a major QTL for resistance to Phytophthora capsici in pepper. Theor Appl Genet 127:2503–2513

Ori N, Eshed Y, Paran I et al (1997) The I2C family from the wilt disease resistance locus *I2* belongs to the nucleotide binding, leucine-rich repeat superfamily of plant resistance genes. Plant Cell 9:521–532

Quirin EA, Ogundiwin EA, Prince JP et al (2005) Development of sequence characterized amplified region (SCAR) primers for the detection of *Phyto.5.2*, a major QTL for resistance to *Phytophthora capsici* Leon. in pepper. Theor Appl Genet 110:605–612

Thabuis A, Lefebvre V, Bernard G et al (2004) Phenotypic and molecular evaluation of a recurrent selection program for a polygenic resistance to *Phytophthora capsici* in pepper. Theor Appl Genet 109:342–351

Truong HTH, Kim KT, Kim DW et al (2012) Identification of isolate-specific resistance QTLs to phytophthora root rot using an intraspecific recombinant inbred line population of pepper (*Capsicum annuum*). Plant Pathol 61:48–56

Yang HB, Liu WY, Kang WH et al (2009) Development of SNP markers linked to the *L* locus in *Capsicum* spp. by a comparative genetic analysis. Mol Breed 24:433–446

Chapter 3
QTL Identification

Hyun Sook Lee, Sun-Goo Hwang, Cheol Seong Jang, and Sang Nag Ahn

Abstract Many important agronomic traits, such as crop yield and stress tolerance, are controlled by polygenes, each with subtle effects and influenced by the environment. Such characteristics are referred to as quantitative traits and the segregating loci as quantitative trait loci (QTLs). As quantitative traits do not segregate into discrete classes, one cannot use standard Mendelian genetic practices to study them. In the past 30 years, a new approach to understanding the genetics of quantitative traits has developed, using molecular linkage maps and populations with linkage disequilibrium. To map and tag a QTL, one needs a segregating population(s) in which linkage disequilibrium exists, a set of molecular markers, and statistical tools. Many plant species are ideally suited for quantitative trait analysis because they have short generation times and many progeny, and controlled crosses can be made in which linkage disequilibrium is known. Annual, self-pollinated species (e.g., Rice, Arabidopsis, Tomato) are best suited for QTL studies. More problematic things are outcrossing perennial species (e.g., many tree species and most animal species), in which controlled crosses are either not possible or impractical, and generation times are longer. New methods (often referred to as association genetics) are emerging that allow researchers to locate and characterize QTLs in natural populations using empirical knowledge of linkage disequilibrium and high-throughput DNA marker analysis. The importance of meta-quantitative trait loci (MQTLs) is increasing because of their role in identifying genes linked to QTL regions. They also aid marker-assisted selection (MAS) in association with statistical analyses. Recently, integrated data for use in meta-analyses of QTLs have become available in plant genome databases and have been analyzed using QTL/microarray, expression-QTL (eQTL), and MQTL methods. In this section, the requirements and methods for identifying and characterizing QTLs using the traditional approach of controlled crosses will be discussed. The information on QTL mapping will be used

---

Author contributed equally with all other contributors.

H.S. Lee • S.N. Ahn (✉)
Department of Crop Science, Chungnam National University, Daejeon, Republic of Korea
e-mail: leehs0107@gmail.com; ahnsn@cnu.ac.kr

S.-G. Hwang • C.S. Jang
Department of Applied Plant Sciences, Kangwon National University, Chuncheon, Republic of Korea
e-mail: sghwang@kangwon.ac.kr; csjang@kangwon.ac.kr

H.-J. Koh et al. (eds.), *Current Technologies in Plant Molecular Breeding*,
DOI 10.1007/978-94-017-9996-6_3

in MAS in plant breeding. We also discuss the current status of MQTL research and describe the general procedures using a case study to identify MQTL genes related to abiotic stresses in rice.

3.1 Detection and Analysis of QTLs

3.1.1 QTL Terminology

Quantitative Trait (Also Called Metric Trait or Continuous Trait) A trait that does not segregate in a discrete Mendelian manner (e.g., 3:1), but instead displays a continuous phenotypic distribution. Quantitative traits are normally conditioned by more than one gene along with substantial environmental influences.

Quantitative Trait Locus (QTL) Site on a chromosome containing a gene(s) affecting a quantitative trait. Usually detected through association with molecular marker(s) of known position in the genome.

Number of QTLs Number of significant QTLs affecting a single trait in a specific population and environment.

Magnitude of QTL Effect (Additivity) Expected phenotypic effect of substituting one allele for an alternate allele.

Degree of Dominance How two QTL alleles at the same locus interact (e.g., additive, dominant, recessive, overdominant).

Epistasis How two or more QTLs interact with each other to affect the trait of interest. Usually tested via multi-way analysis of variance (ANOVA). Non-epistatic QTLs act independently.

Pleiotropy The effect(s) of a single QTL on more than one character.

QTL × Environmental Interactions Determined by testing the same populations (containing the same segregating QTLs) in different environments. QTLs with low QTL × environment interactions are effective in different environments.

3.1.2 Methods for Mapping QTLs

QTL mapping requires a complete linkage map with neutral codominant markers (DNA markers), populations(s) in which linkage disequilibrium exists, a reliable method for measuring the trait, and statistical methods (and software) for establishing significant associations between markers and traits. QTL mapping has evolved from single-marker mapping to interval mapping through multiple marker-based approaches. The following references are recommended for a more detailed understanding of QTL mapping: Lander and Botstein (1989), Tanksley (1993), Zeng (1994), and Lynch and Walsh (1997).

3.1.2.1 Single-Marker Mapping (Point Analysis)

The simplest approach for detecting QTLs is to analyze the data one marker at a time. If an association exists between a molecular marker genotype and trait value, a trait locus is likely to be near that marker locus. The advantage of single-marker mapping is that it works for any population structure. However, it cannot determine whether the markers are associated with one or more QTLs and does not estimate the likely position of a QTL. Additionally, the effects of the QTL are likely to be confounded by recombination (Tanksley 1993).

3.1.2.2 Interval Mapping

Lander and Botstein (1989) developed a QTL mapping method, known as interval mapping, by using two flanking markers. Compared with single-marker mapping, this method allows the effect and position of the QTL to be inferred. However, if there are more than one linked QTL, a bias in the position of those QTLs may be created, and "phantom" QTLs may be inferred.

3.1.2.3 Composite Interval Mapping (CIM) (Zeng 1994. Genetics 136:1457)

This analysis conditions the search for a QTL in one interval considering effects of adjacent intervals, thus giving greater precision to the QTL's position and reducing the chances of "phantom" QTLs.

3.1.2.4 Multiple Interval Mapping (MIM) (Kao et al. 1999. Genetics 152:1203)

This QTL mapping approach can also take into account epistasis in detecting and locating QTLs.

3.1.3 Quantitative Trait Locus Mapping: An Example in Rice

Two rice varieties, Nipponbare (Variety A) and Kasalath (Variety B), are known to differ in mesocotyl length when grown in the dark. These two varieties are inbred lines (homozygous at all loci). Our goal is to answer the following:

- How many QTLs are responsible for the difference in mesocotyl length between these varieties?
- On what chromosome(s) are these QTLs located?
- What is the gene action (dominant, recessive, additive, overdominant, underdominant) of each QTL?
- What is the individual contribution (additive effect) of each QTL gene to the trait?

- Does epistasis among the QTLs play a significant role in determining the trait?
- Are these QTLs expressed across different environments?
- Do these QTLs determine differences in mesocotyl length among other rice varieties?
- For what does each of these genes code, and how do the gene products produce variation in the character?

3.1.4 *QTL Mapping Approach*

Several software programs for mapping QTLs in experimental crosses are freely available. They include QTL Cartographer (http://statgen.ncsu.edu/qtlcart/index.php), Mapmaker-QTL (http://www.broadinstitute.org/scientific-community/software), and QTL Cartographer R/qtl: A QTL mapping environment (http://www.rqtl.org/). In this example, we will demonstrate QTL mapping using QTL Cartographer.

3.1.4.1 Installation of Programs

MAPMAKER/EXP 3.0 for map construction can be freely downloaded at ftp://ftp-genome.wi.mit.edu/distribution/software/mapmaker3.QTL Cartographer V2.5 is available at http://statgen.ncsu.edu/qtlcart/WQTLCart.htm.

3.1.4.2 Data Preparation

Phenotypic Data

Numeric data can be stored in an Excel file. In this experiment, backcross inbred lines (BILs; BC_1F_7) were used. Mesocotyl length was measured in two independent experiments and the distribution of mesocotyl length is shown in Fig. 3.1 (Lee et al. 2012).

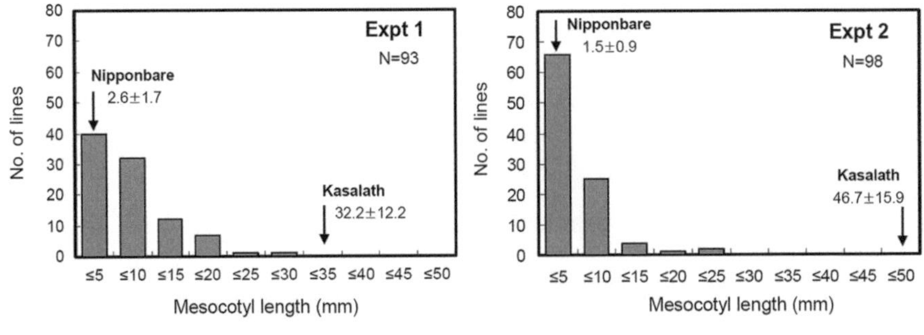

Fig. 3.1 Frequency distribution of mesocotyl length in backcross inbred lines (Lee et al. 2012)

H1

	A	B	C	D	E	F	G	H	I	J	K	L	M	N	O	P	Q	R	S
1																			
2	Chromosome	1	1	1	1	1	1	1	1	3	3	3	3	3	3	3	3	3	3
3	Marker name	580	306	8360	7303	337	346	7419	5335	8308	8309	5813	393	5443	6970	143	7000	3684	570
4	P1rent 1	1	1	1	1	1	1	1	1	1	1	1	1	1	1	1	1	1	1
5	P1rent 3	3	3	3	3	3	3	3	3	3	3	3	3	3	3	3	3	3	3
6	Line 1	3	3	-	-	-	3	-	1	1	-	3	-	1	3	3	3	3	-
7	Line 3	2	3	-	-	3	3	-	1	3	3	1	3	1	1	3	2	2	1
8	Line 3	3	1	1	1	1	1	1	1	1	1	1	1	1	1	1	1	1	3
9	Line 4	1	3	3	3	3	3	3	3	1	1	1	1	3	1	1	1	1	1
10	Line 5	-	3	3	3	3	3	3	3	3	3	3	3	3	3	3	3	3	3
11	Line 6	3	1	3	3	3	3	3	1	3	3	1	1	1	1	1	1	1	1
12	Line 7	-	2	2	2	2	2	2	1	3	3	3	3	1	3	3	-	3	3
13	Line 8	3	2	2	2	2	2	2	3	3	-	3	3	3	3	3	3	3	3
14	Line 9	-	1	1	1	1	1	1	1	3	3	3	3	1	3	3	3	3	3
15	Line 10	1	1	3	2	2	2	2	1	1	2	3	3	1	3	3	3	3	2
16	Line 11	-	1	1	1	1	1	1	3	3	1	3	3	3	2	3	2	2	2
17	Line 13	1	3	3	3	3	3	3	1	1	1	1	1	1	1	1	1	1	1
18	Line 13	3	1	1	1	1	1	1	3	1	3	3	3	3	3	3	3	3	3
19	Line 14	3	3	3	3	3	3	3	3	1	3	3	3	3	3	3	3	3	3
20	Line 15	3	3	3	3	3	3	3	1	3	3	3	3	1	3	3	3	2	2
21	Line 16	3	3	3	3	3	3	3	1	2	3	1	1	1	1	-	1	1	1
22	Line 17	3	3	3	-	3	3	3	3	2	3	3	3	3	3	3	2	2	1
23	Line 18	2	-	3	1	1	1	1	3	1	1	1	1	3	3	3	3	3	3
24	Line 19	1	3	3	3	3	3	3	3	3	3	3	3	3	3	3	3	3	3

Sheet1 Sheet2 Sheet3

E1

	A	B	C	D	E	F	G	H	I	J	K	L	M	N	O	P	Q	R	S
1																			
2	Chromosome	1	1	1	1	1	1	1	1	2	2	2	2	2	2	2	2	2	2
3	Marker name	580	306	8260	7202	237	246	7419	5335	8208	8209	5813	293	5442	6970	143	7000	3684	570
4	Parent 1	A	A	A	A	A	A	A	A	A	A	A	A	A	A	A	A	A	A
5	Parent 2	B	B	B	B	B	B	B	B	B	B	B	B	B	B	B	B	B	B
6	Line 1	B	B	-	-	-	B	-	A	A	-	B	-	A	B	B	B	B	-
7	Line 2	H	B	-	-	B	B	-	A	B	B	A	B	A	A	B	H	H	A
8	Line 3	B	A	A	A	A	A	A	A	A	A	A	A	A	A	A	A	A	B
9	Line 4	A	B	B	B	B	B	B	B	A	A	A	A	B	A	A	A	A	A
10	Line 5	-	B	B	B	B	B	B	B	B	B	B	B	B	B	B	B	B	B
11	Line 6	B	A	B	B	B	B	B	A	B	B	A	A	A	A	A	A	A	A
12	Line 7	-	H	H	H	H	H	H	A	B	B	B	B	A	B	B	-	B	B
13	Line 8	B	H	H	H	H	H	H	B	B	-	B	B	B	B	B	B	B	B
14	Line 9	-	A	A	A	A	A	A	A	B	B	B	B	A	B	B	B	B	B
15	Line 10	A	A	B	H	H	H	H	A	A	H	B	B	A	B	B	B	B	H
16	Line 11	-	A	A	A	A	A	A	B	B	A	B	B	B	H	B	H	H	H
17	Line 12	A	B	B	B	B	B	B	A	A	A	A	A	A	A	A	A	A	A
18	Line 13	B	A	A	A	A	A	A	B	A	B	B	B	B	B	B	B	B	B
19	Line 14	B	B	B	B	B	B	B	B	A	B	B	B	B	B	B	B	B	B
20	Line 15	B	B	B	B	B	B	B	A	B	B	B	B	A	B	B	B	H	H
21	Line 16	B	B	B	B	B	B	B	A	H	B	A	A	A	A	-	A	A	A
22	Line 17	B	B	B	-	B	B	B	B	H	B	B	B	B	B	B	H	H	A
23	Line 18	H	-	B	A	A	A	A	B	A	A	A	A	B	B	B	B	B	B
24	Line 19	A	B	B	B	B	B	B	B	B	B	B	B	B	B	B	B	B	B

Sheet1 Sheet2 Sheet3

Fig. 3.2 Layout of genotype data in an Excel file

Genotypic Data

The genotype of each individual is determined using various molecular markers and is expressed in numbers or letters (Fig. 3.2). In this experiment, the genotypes of 98 BILs were assayed using 245 RFLP markers (data not shown; A: Nipponbare homozygote, B: Kasalath homozygote, H: heterozygote).

```
cent func kos
make chromosome chr1 chr2 chr3 chr4 chr5 chr6 chr7 chr8 chr9 chr10 chr11 chr12
s 1-24
anchor chr1
frame chr1
s 25-41
anchor chr2
frame chr2
s 42-61
anchor chr3
frame chr3
s 62-75
anchor chr4
frame chr4
s 76-85
anchor chr5
frame chr5
s 86-98
anchor chr6
frame chr6
s 99-108
anchor chr7
frame chr7
s 109-118
anchor chr8
frame chr8
s 119-130
anchor chr9
frame chr9
s 131-141
anchor chr10
frame chr10
s 142-154
anchor chr11
frame chr11
s 155-162
anchor chr12
frame chr12
save
quit
```

Fig. 3.3 Example of a pre-file. ⓐ "cent funckos" indicates "centiMorgans function Kosambi". ⓑ List of all chromosomes. ⓒ A range of markers per chromosome. For example, 24 markers (numbered 1–24) exist on chromosome 1, and 24 markers (25–48) are mapped on chromosome 2. ⓓ and ⓔ indicate respective chromosomes. ⓕ Two commands, "save" and "quit", are required at the end

3.1.4.3 Data File for Construction of a Linkage Map

Two files, a *.pre file and a *.raw file, are needed to develop a linkage map using MAPMAKER/EXP. The format of a *.pre file is shown in Fig. 3.3. The actual *.pre file for the population is displayed in Fig. 3.4.

To prepare a *.raw file, genotype data must be changed into an alphabetic code based on the crosses used (Table 3.1). Then, the file with genotype and phenotype data is transformed into a text format file and saved as a *.raw file. An example of a raw file is shown in Fig. 3.5.

```
cent func kos
make chromosome chr1 chr2 chr3 chr4 chr5 chr6 chr7 chr8 chr9 chr10 chr11 chr12
s 1-24
anchor chr1
frame chr1
s 25-50
anchor chr2
frame chr2
s 51-72
anchor chr3
frame chr3
s 73-91
anchor chr4
frame chr4
s 92-111
anchor chr5
frame chr5
s 112-136
anchor chr6
frame chr6
s 137-155
anchor chr7
frame chr7
s 156-175
anchor chr8
frame chr8
s 176-190
anchor chr9
frame chr9
s 191-204
anchor chr10
frame chr10
s 205-229
anchor chr11
frame chr11
s 230-245
anchor chr12
frame chr12
save
quit
```

Fig. 3.4 A *.pre file for 98 backcross inbred lines

Table 3.1 Code for genotypes based on data type

General data type	Genotype code allowed	Remarks
F2 backcross	A	Recurrent parent genotype: aa
	H	Heterozygote: Aa
	–	Missing data
F2 intercross	A	Homozygote for parent 1: AA
	B	Homozygote for parent 2: aa
	H	Heterozygote: Aa
	C	Either heterozygote or homo for parent 2: Aa, aa
	D	Either heterozygote or homo for parent 1: Aa, AA
	–	Missing data
RI self	A	Homozygote for parent 1: AA
	B	Homozygote for parent 2: aa
	–	Missing data

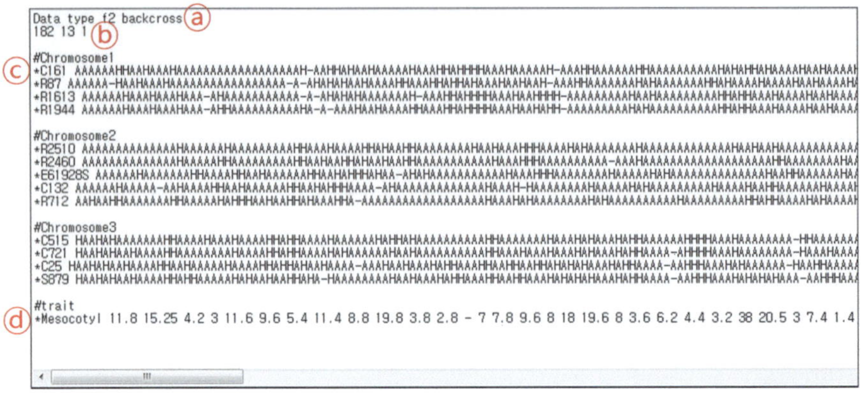

Fig. 3.5 Example of a *.raw file. ⓐ "f2 backcross" for BC1 and "f2 intercross" for the F2 population. ⓑ Numbers of samples, markers, and traits, respectively. ⓒ Genotype data for each marker. Marker name should start with a "★" and can have a maximum eight characters. ⓓ Phenotype data

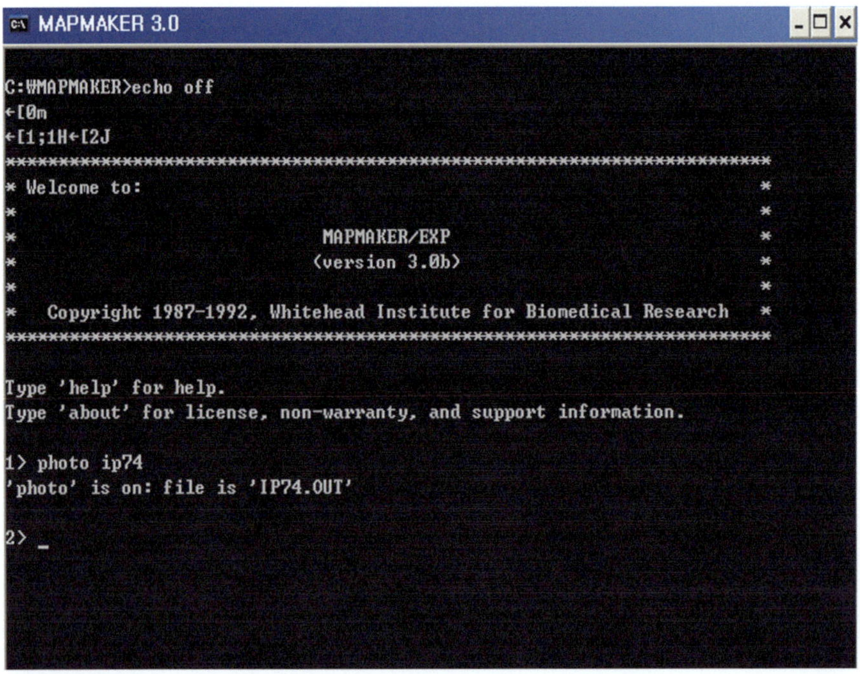

Fig. 3.6 Screenshot of output upon execution of a "photo" command

3.1.4.4 Construction of a Linkage Map

To construct a linkage map, run MAPMAKER and type in 'photo' followed by the file name (Fig. 3.6). This 'photo' command generates an ip74. OUT file in the 'Mapmaker' folder (Fig. 3.7).

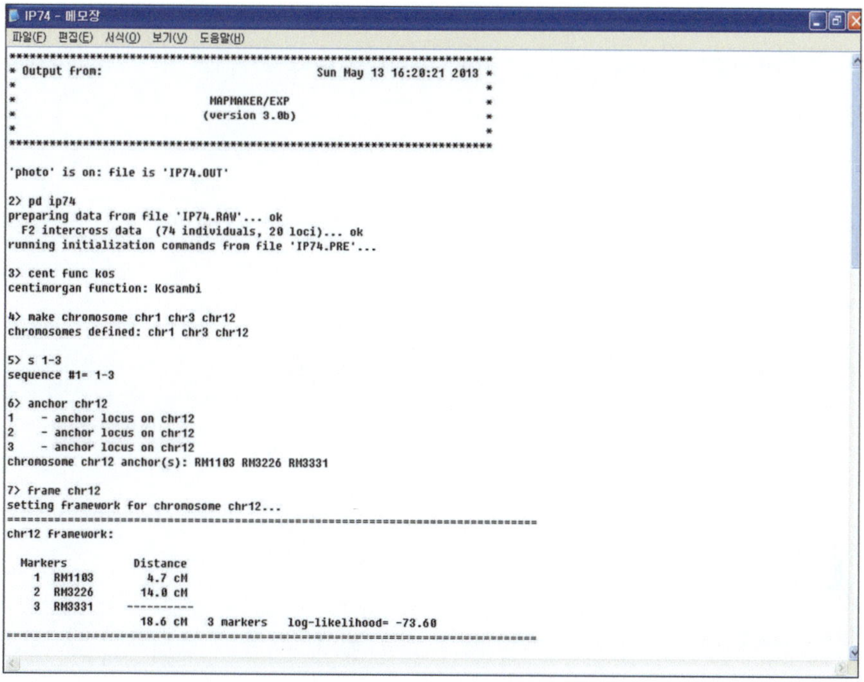

Fig. 3.7 Example of an *.out file

Open and check the ip74. OUT file for errors. Then, execute a 'prepare data' command and check that the *.MAP file has been made in the Mapmaker folder (Fig. 3.8).

3.1.4.5 Composite Interval Mapping Using QTL Cartographer

Select and execute 'Import' in the menu and select the data file for analysis. Select "Mapmaker/QTL format" to use the *.MAP and *.raw files (Fig. 3.9).

If the "Source Data Import—Step 2 of 2" window pops up, click "Map File" menu, select the map file, and select raw file in "Cross Data" menu (Figs. 3.10 and 3.11). Information on the linkage map and the mapping population will be displayed. Figure 3.12 shows the screen after selecting the map and raw files.

Choose "Basic information" from the "Source data manipulations" menu, and check the conditions for "Cross type" and "Map function" (Fig. 3.13). See Table 3.2 for cross type codes. In this case, the cross type and map function are set as B1 and Kosambi map function, respectively.

Select the option CIM from the "Analysis" menu (Fig. 3.14).

A new page will open. Select the "Control" menu and set up the conditions for analysis (Fig. 3.15). You can choose one of three models (Model 1: All marker control; Model 2: Unlinked marker control; Model 6: Standard model) in the "CIM

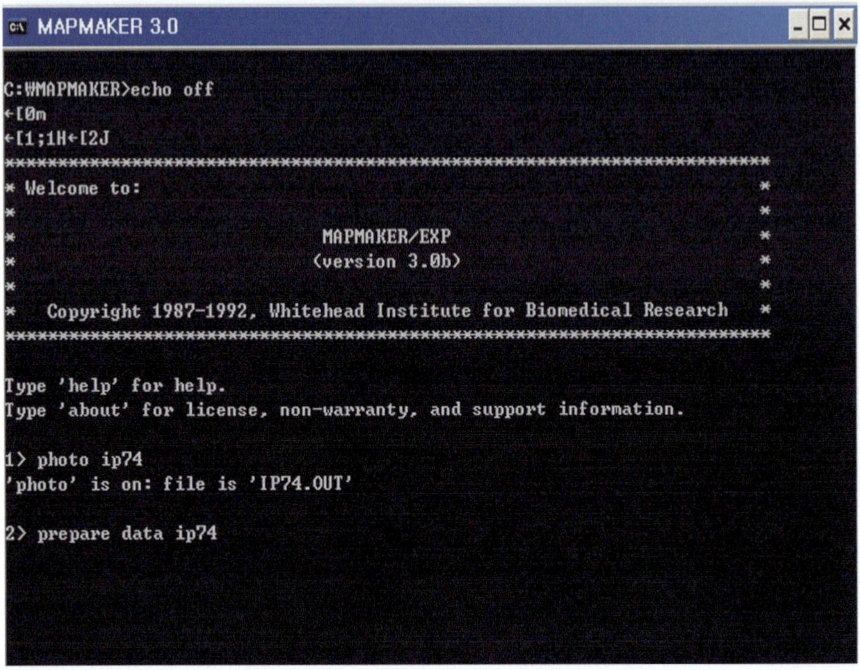

Fig. 3.8 Execution of "prepare data" in Mapmaker

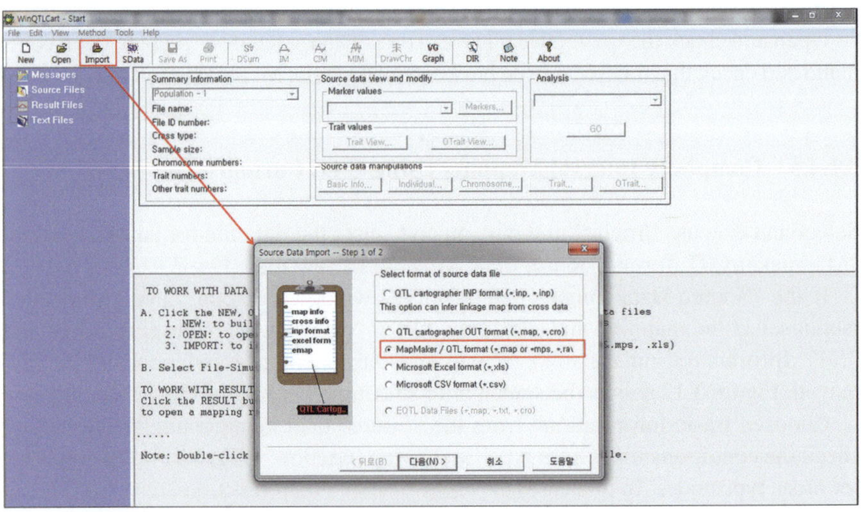

Fig. 3.9 Screenshot: selecting a *.data file for analysis

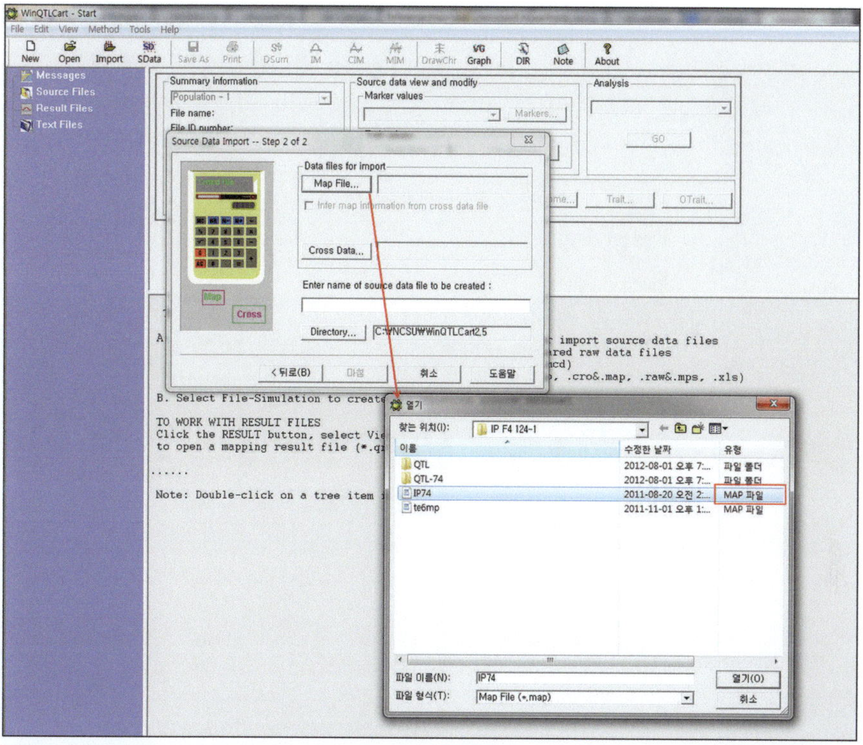

Fig. 3.10 Screenshot: selecting a *.map file

Model" menu and one of three methods (1: Forward regression method; 2: Backward regression method; 3: Forward & Backward method) in the "Regression methods" menu. You can type in a *P*-value of 0.05 or 0.01 in the "Probability for into" and "Probability for out" menu. Specific chromosomes or traits can be selected in the "Chromosome Selection" and "Trait Selection" menus, respectively. Here, a significance level of 0.05 with 1,000 permutations was selected (Fig. 3.16).

For the "Threshold Value Setting", designate the permutation number and significance level and click OK. Click "Start" and the program is initiated (Fig. 3.16).

A new window displaying the results on each chromosome pops up after completion of the analysis (Fig. 3.17). These results can be graphically displayed in a trait- and chromosome-based manner using a LOD (Log of ODDS) value. In the "Effect" menu in the graph window, you can obtain the additive value and R^2 (coefficient of determination). Figure 3.17 shows the results of QTL analysis for mesocotyl length with LOD peak per chromosome. The blue solid line (A) and red dotted line (B) are for two independent experiments. Two QTLs with high LOD values on the long arm of chromosomes 1 and 3 were detected for mesocotyl length (Fig. 3.18).

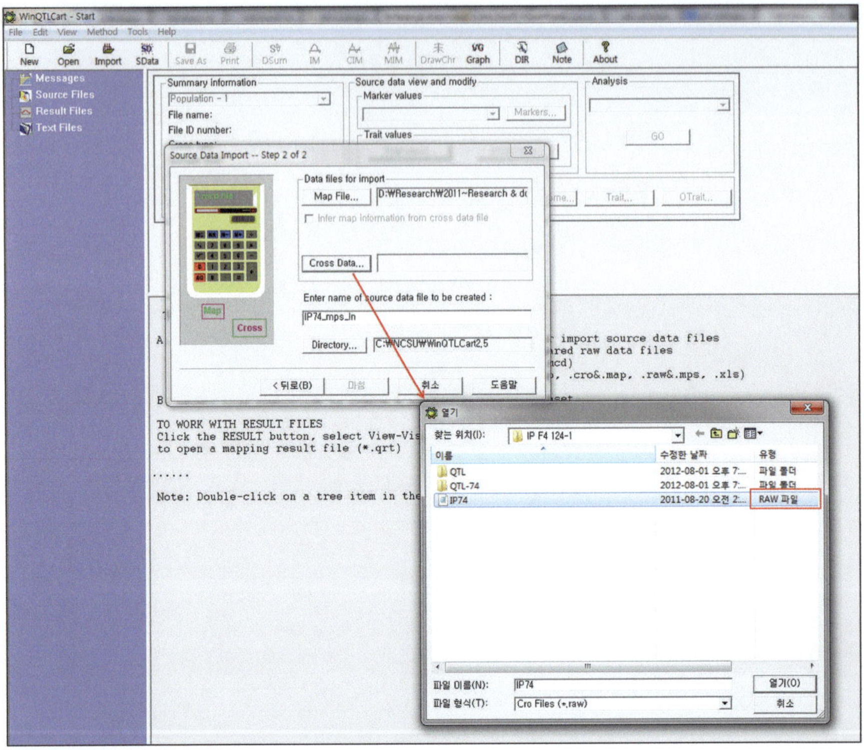

Fig. 3.11 Screenshot: selecting a *.raw file

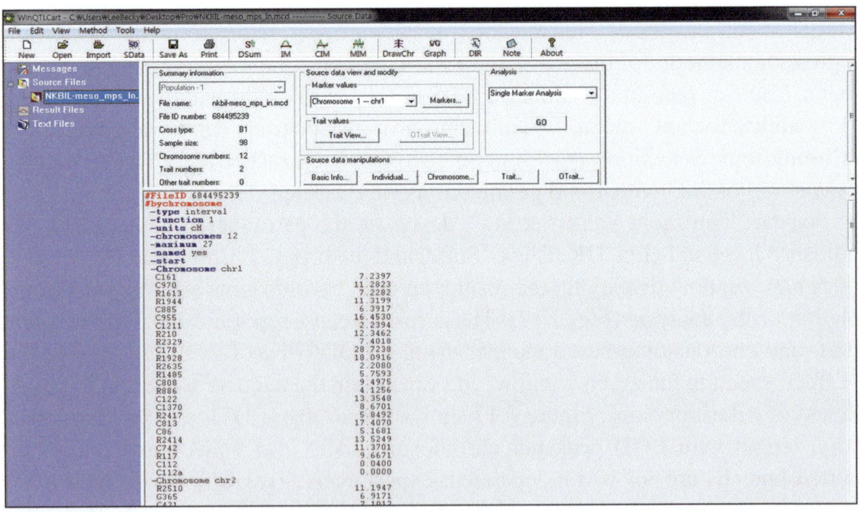

Fig. 3.12 Screenshot: selecting a *.map file and a *.raw file

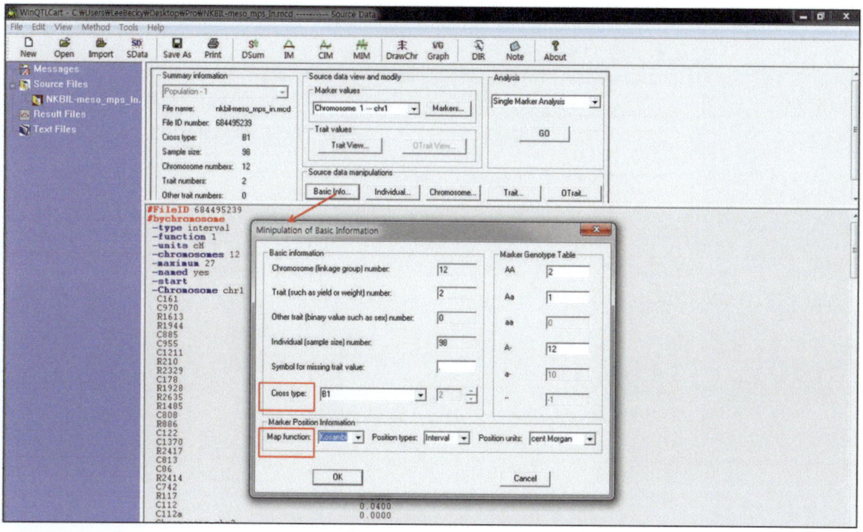

Fig. 3.13 Screenshot: execution of the "Source data manipulations" menu

Table 3.2 Codes for various cross types

Cross type	Code	Example
Backcross to P_i	B_i	B1
Backcross *j* times to P_i	B_{ij}	B13
Selfed generation *i* intercross	SF_i	SF3
Randomly mated generation *I* intercross	RF_i	RF2
Doubled haploid	RI_0	RI0
Recombinant inbred via selfing	RI_1	RI1
Recombinant inbred via sibling mating	RI_2	RI2
Testcross of SF_i to P_j	$T(B_j)SF_i$	T(B1)SF3
Testcross of SF_i to *j* generations	$T(SF_{i+j})SF_i$	T(SF4)SF3
Testcross of RF_i to P_j	$T(B_j)RF_i$	T(B1)RF3
Design III	$T(D3)SF_i$	T(D3)SF5

The results can be stored in an Excel file (Fig. 3.18). In the results, select an interval or marker with a LOD value exceeding defined threshold values. For example, four QTLs on chromosomes 1, 3, 7, and 12 were detected in experiment A with LOD values larger than the threshold of 3.1 at $P=0.05$. In the second experiment, three QTLs on chromosomes 1, 3, and 9 were detected (Fig. 3.19). These results are summarized in Table 3.3.

The linkage map is generated by clicking "DrawChr" in the page menu, and the saved graph file will be available in the other program (Fig. 3.20). The results are

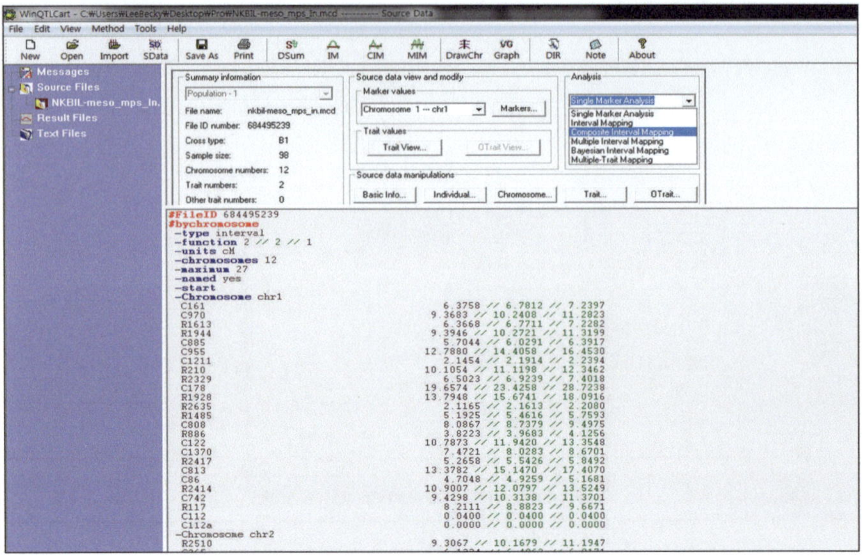

Fig. 3.14 Screenshot: selecting the analysis menu

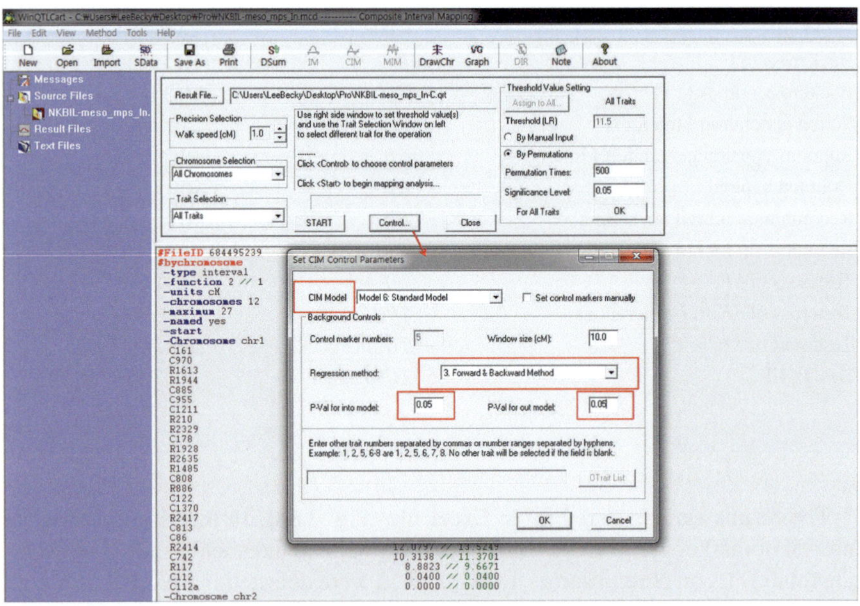

Fig. 3.15 Screenshot: the Control menu for CIM analysis

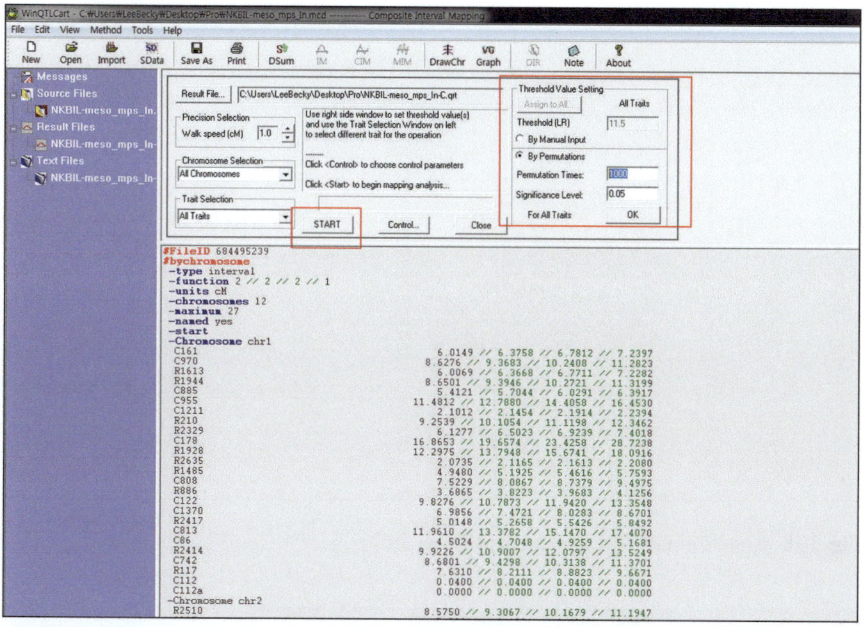

Fig. 3.16 Setting the “Threshold Value Setting” menu for CIM analysis

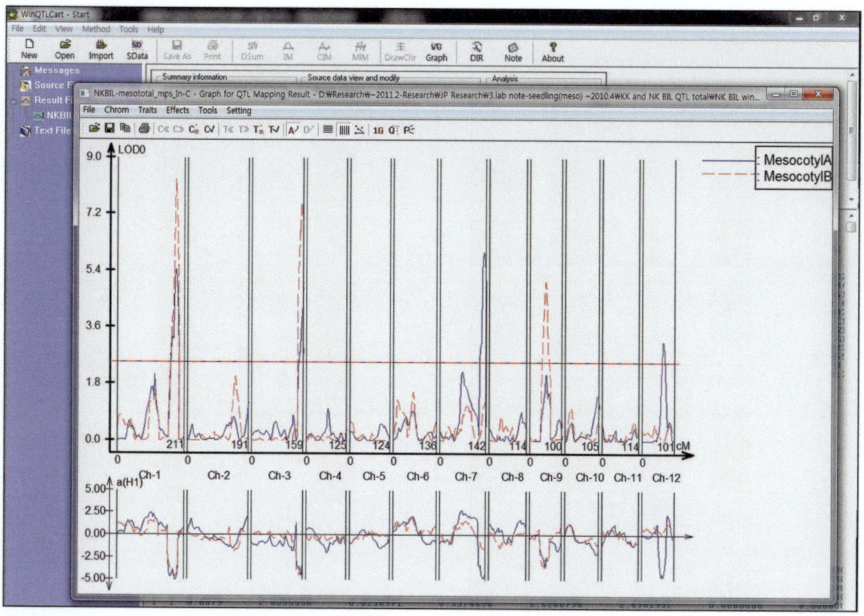

Fig. 3.17 Results of the CIM analysis

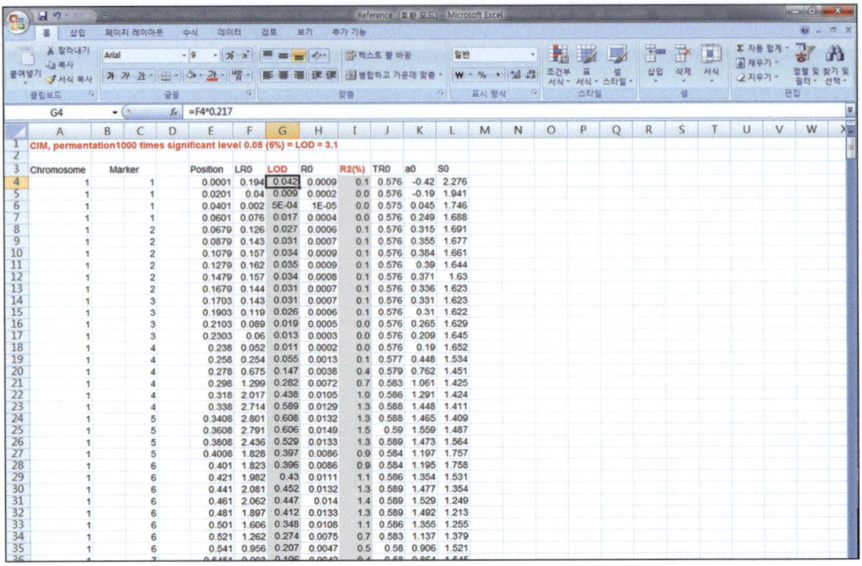

Fig. 3.18 Results of the CIM analysis in an Excel file

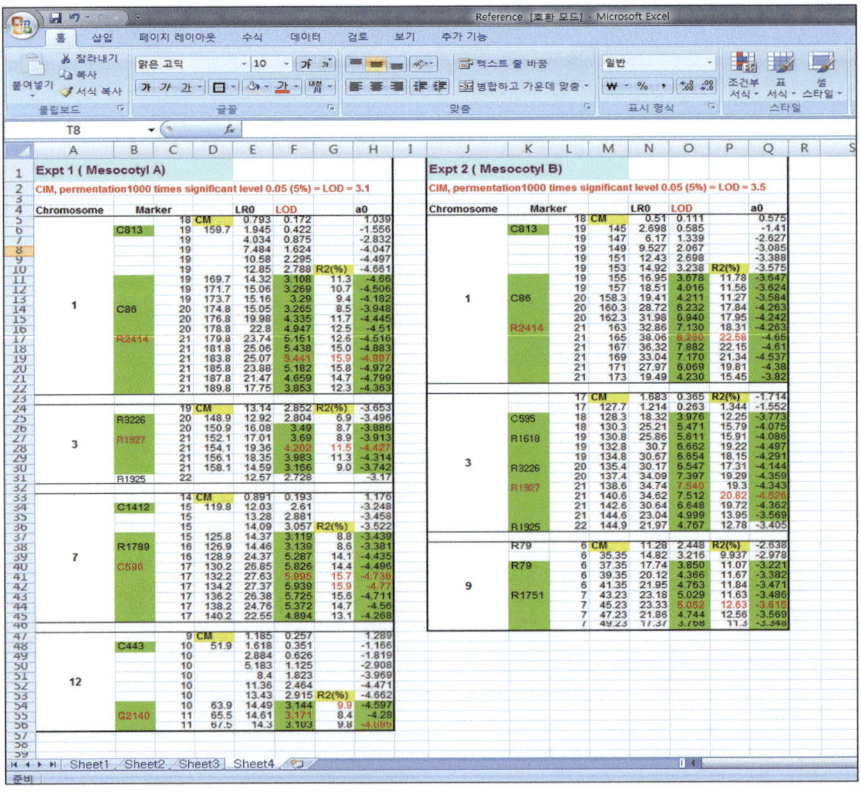

Fig. 3.19 Summary of quantitative trait loci for mesocotyl length

Table 3.3 Locations and effects of quantitative trait loci (QTLs) for mesocotyl length of backcross inbred lines (BILs) in two experiments

Locus[a]	Chr.	Marker interval[b]	Exp.	Mean Value NN		KK	LOD score[c]	R^2(%)[d]	Additive effect[e]
qMel-1	1	R2414	Expt 1	6.0	✓	11.8	5.4	15.9	2.9
		R2414	Expt 2				8.3	22.6	
qMel-3	3	R1927	Expt 1	6.0	✓	10.7	4.2	11.5	2.4
		R1927	Expt 2				7.5	20.8	
qMel-7	7	C596	Expt 1	6.1		10.4	6.0	15.9	2.2
qMel-9	9	R1751	Expt 2	6.4		11.2	5.1	12.6	2.4
qMel-12	12	G2140	Expt 1	6.8		5.4	3.2	9.9	−0.7

[a]QTLs were designated "*qMel-#*", where # is chromosome number
[b]The nearest RFLP marker to the QTL is underlined
[c]Putative QTLs with significant LOD scores tested at $P<0.05$
[d]Proportion of the phenotypic variance explained by the nearest marker of the QTL
[e]Estimated effect of replacing Nipponbare alleles by Kasalath alleles

stored in the *.qrt file format and are accessible in the Result file menu by clicking "Open a file" (Fig. 3.21).

3.1.5 *Experimental Results*

The analysis answered our first two questions:

- How many QTLs are responsible for the difference in mesocotyl length between these varieties?
- On what chromosome(s) are these QTLs located?

  We address the other questions below.

- What is the gene action (dominant, recessive, additive, overdominant or underdominant) of each QTL?

To answer this question, we must estimate the phenotypic means for each QTL genotype. In this example, we are only concerned about the *qMel-1* on chromosome 1. We know that this QTL is linked to the markers but not the precise location of *qMel-1* on the chromosome. We will therefore use the phenotypic means for the linked marker genotypes to estimate the phenotypic means for each QTL. We will refer to the QTL on chromosome 1 near C86 and C742 as *qMel-1* and the one on chromosome 3 near R3226 as *qMel-3*. We can estimate *qMel-1* based on the

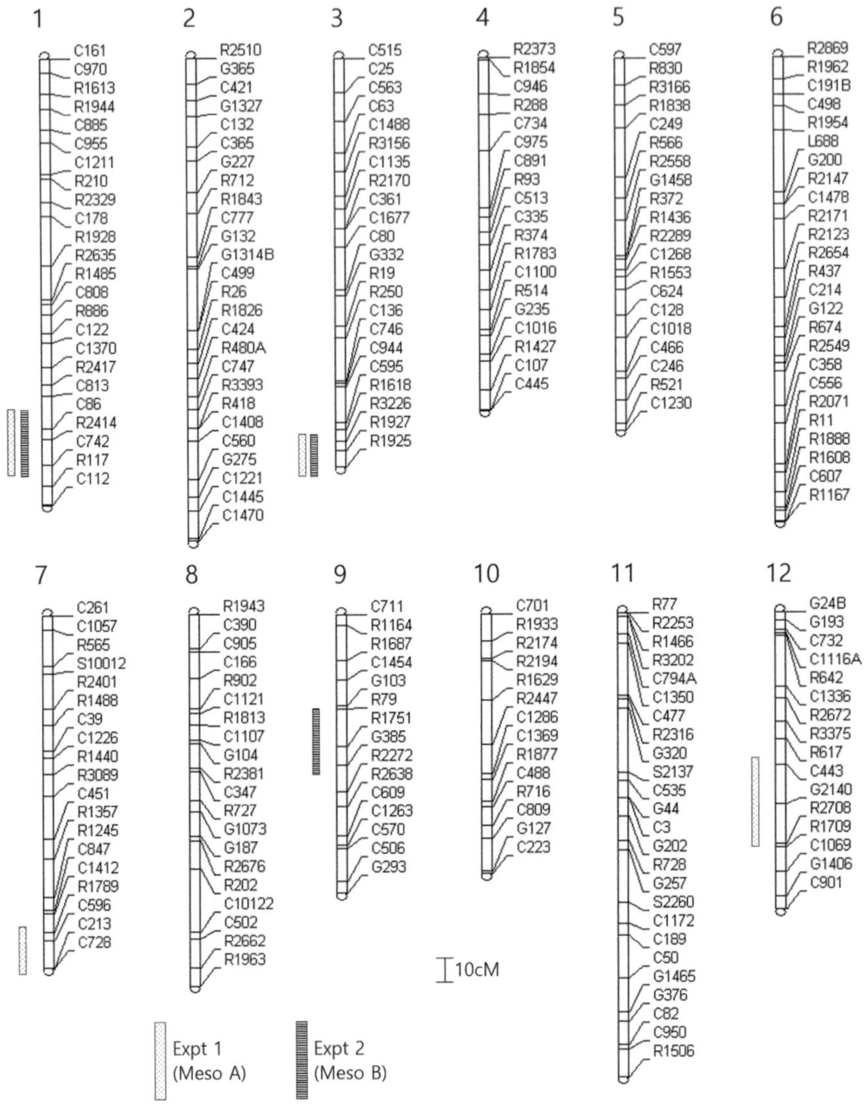

Fig. 3.20 Locations of quantitative trait loci on the linkage map for mesocotyl length (Lee et al. 2012)

genotypes of the flanking markers. The calculations for gene action (*d*/*a*) and additivity (*a*) are given below.

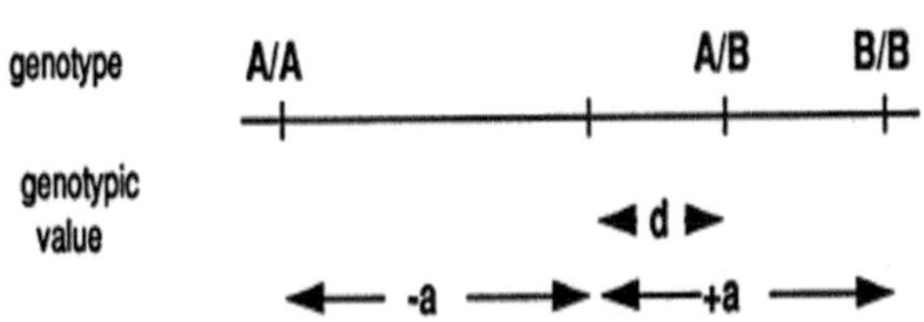

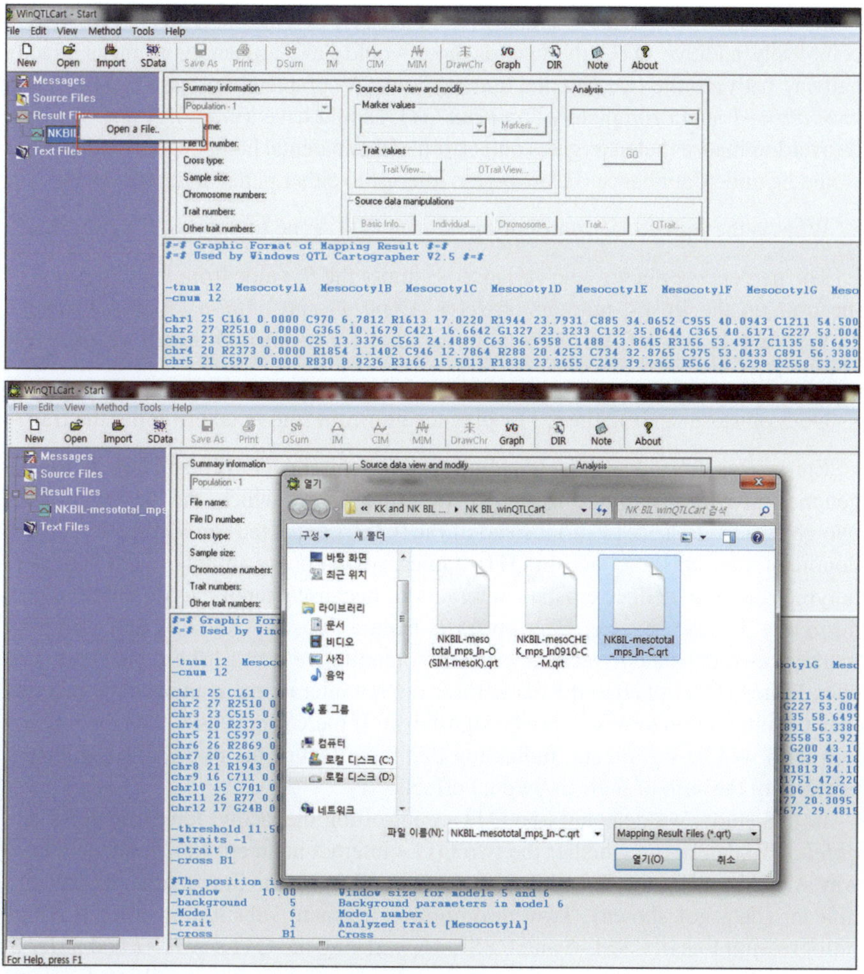

Fig. 3.21 Screenshot: re-executing a stored result file

A/A = mean phenotypic value of individuals homozygous for the A allele at a given marker

A/B = mean phenotypic value of individuals heterozygous for the A and B alleles at a given marker

B/B = mean phenotypic value of individuals homozygous for the B allele at a given marker

a = additive effect (of a single allele) = (B/B − A/A)/2

d = dominance deviation = A/B − [(A/A + B/B)/2]

d/a = degree of dominance ($-1 < d/a < 1$)

The additive effect estimate for *qMel-1* (Experiment 1) is $a = 2.9$. This means that the substitution of a Kasalath allele for *qMel-1* increases mesocotyl length by 2.9 mm.

In this example, we cannot estimate *d* or *d*/*a* because of a lack of heterozygotes. For a completely additive QTL, the heterozygote would have a phenotypic mean exactly halfway between the two parental homozygotes. A completely recessive QTL would have $d/a=-1$ and a completely dominant QTL should have a $d/a=1$. A $d/a>1$ would be overdominance (heterozygotes outperform either parental homozygote) and $d/a<-1$ would be underdominance (heterozygote inferior to either parental homozygote).

- What is the individual contribution of each QTL gene to the trait?

The percentage phenotypic variance is simply the R^2 value from the regression of the trait on the linked markers. *qMel-1* (Expt. 1) and *qMel-3* (Expt. 2) explain 15.9 % and 11.5 % of the phenotypic variance for mesocotyl length, respectively (Table 3.3).

- Does epistasis among the QTLs play a significant role in determining the trait?

Epistasis is defined as the interaction among genes in controlling a trait. Classical genetics provides familiar examples of epistasis in which the segregation of two genes lead to a 15:1 ratio instead of the 9:3:3:1 expected for segregation of two dominant, non-epistatic loci. In QTL studies, we cannot examine the ratios of phenotypic classes to deduce epistatic interactions because phenotypes do not segregate discretely. Instead, we can infer epistasis from a two-way ANOVA using pairs of markers as independent variables and the quantitative trait as the dependent variable. If the QTLs linked to the marker loci are not interacting in an epistatic manner, then the interaction term will not be significant. If there is epistasis, then the interaction term will be significant, indicating that the effect of the two QTLs together is not simply the sum of their individual effects.

In this study, we detected two QTLs controlling mesocotyl length: *qMel-1* and *qMel-3*. To determine whether the two QTLs interact in an epistatic manner, a two-way ANOVA using the linked markers was conducted; the interaction was not significant (data not shown). Two chromosome segment substitution lines (CSSLs; introgression lines), CSSL-6 and CSSL-15, carrying the QTLs *qMel-1* and *qMel-3*, respectively, were crossed to develop an F2:3 population. The resulting F_1 plants were selfed to obtain F_2 plants. Ninety-five F_2 plants were generated and used to confirm the interaction of the two QTLs.

In the F_2 population, two-way ANOVA revealed a non-significant digenic interaction between two markers, RM3602 and RM8277, linked to *qMel-1* and *qMel-3*, respectively ($P=0.31$) (data not shown). These results indicate that the two QTLs act additively in distinct or complementary pathways in controlling mesocotyl elongation.

- Are these QTLs expressed across different environments?

To answer this question, the same experiment must be conducted in different environments. F_3 seeds could be collected from each of the F_2 plants, and, if sufficient F_3 seeds are available, used in replicated field experiments in multiple locations. In this case, the experiment was conducted in two replicates. Two QTLs mapped on chromosomes 1 and 3, respectively, were detected in both experiments.

Three additional QTLs, *qMel-7*, *qMel-9*, and *qMel-12*, were each identified in only one experiment. Based on the results, *qMel-1* and *qMel-3* are the main QTLs.

- Do these QTLs determine differences in mesocotyl length among other rice varieties?

We do not have enough information to answer this question. We know that two cultivars differ for alleles at two QTLs (*qMel-1* and *qMel-3*). To determine whether these QTLs are important with respect to differences in mesocotyl length among other rice varieties, a similar experiment must be conducted using different rice varieties. If QTLs for the same trait map to the same or similar map positions in different crosses, the results would provide evidence that the same QTLs are responsible. Several studies in rice have shown that QTLs for mesocotyl length are colocalized with the two QTLs in this study (Cai and Morishima 2002), implying that these two QTLs are involved in controlling mesocotyl length indifferent varieties.

- For what does each of these genes code, and how do the gene products produce variation in the character?

To answer this question, one would need to clone the QTLs and determine the gene products. As a first step for QTL cloning, the construction of a high-resolution map of the QTL is a prerequisite. High-resolution mapping is also needed for more precise determination of the physical distance between the marker and QTL for MAS and for distinguishing between pleiotropy and linkage. A widely adopted strategy to more accurately estimate the position and effect of a coarsely mapped QTL is to create a new population by crossing nearly isogenic lines (NILs) or introgression lines differing only for the allelic constitution in the short chromosome segment harboring the QTL. NILs or introgression lines (segmental substitution lines) are produced by marker-assisted backcross introgression (Zamir 2001). Substitution mapping has been applied in diverse plant species to facilitate the fine mapping of QTLs (Wissuwa et al. 2002).

In this example, for fine mapping of *qMel-1* and *qMel-3*, two CSSLs (CSSL-6 and CSSL-15) were selected from the CSSLs developed from a cross between Nipponbare and Kasalath at the Rice Genome Resource Center, Japan (http://www.rgrc.dna.affrc.go.jp/ineNKCSSL54.html).CSSL-6 and CSSL-15, carrying the QTLs *qMel-1* and *qMel-3*, respectively, were crossed to develop an F2:3 population. The resulting three F_1 plants were selfed to obtain F_2 plants. Ninety-five F_2 plants were generated and used to confirm the target QTLs. Thirty-two F_2 plants with recombination breakpoints within the target QTL regions were selected and selfed to obtain F_3 seeds for substitution mapping. The mesocotyl length of each F_3 plant from each line was measured, and DNA of the seedlings was extracted for genotyping with simple sequence repeat (SSR) markers for substitution mapping of the target QTL.

CSSL-6 and CSSL-15 served as respective positive controls for the region as a whole. The mesocotyl lengths of lines 11 and 29 were not significantly different from that of CSSL-15. In addition, the three genotypes of lines 11 and 29 did not show significant differences, suggesting that neither contained a Kasalath allele affecting mesocotyl elongation in the introgressed segments. A comparison of

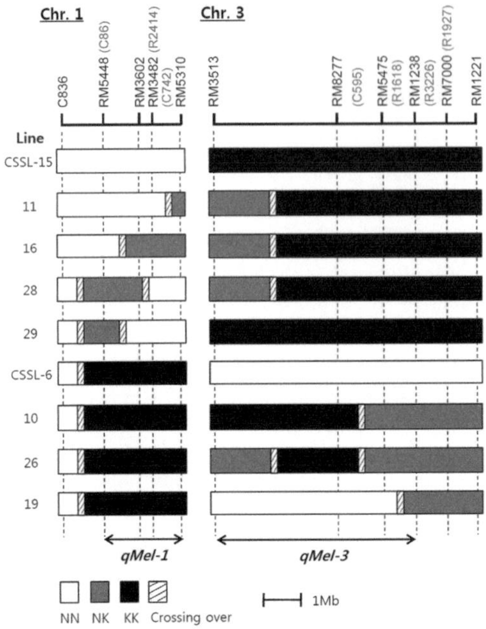

SSR Marker [1]	Mean of mesocotyl length (mm) [2]			R^2 (%)	n [3]
	NN	NK	KK		
RM3513~RM1221	-	-	6.1 (24)	-	24
RM5310	4.9 a (8)	6.0 a (10)	5.2 a (5)	-	23
RM3602	5.3 a (6)	7.7 a (10)	13.3 b (12)	46	28
RM3602	5.0 a (3)	8.8 a (13)	13.0 b (10)	42	26
RM5448	5.8 a (9)	7.0 a (12)	7.2 a (9)	-	30
RM5448~RM5310	-	-	1.2 (35)		35
RM5475	5.2 a (6)	7.0a (12)	12.7 b (7)	30	25
RM5475	11.0 a (4)	13.6 a (12)	21.5 b (6)	45	22
RM3513	12.8 a (6)	17.6 a (7)	15.1 a (9)	-	
RM1238	1.8 a (8)	2.5 a (15)	3.1 a (9)	-	32

Fig. 3.22 Substitution mapping of *qMel-1* and *qMel-3*. *White portions* of the graph indicate homozygous Nipponbare chromosome segments, *black regions* homozygous Kasalath chromosomes, *gray areas* heterozygous regions, and *slashed areas* where crossing-over occurred. The table to the *right* of the graphical genotypes indicates mean mesocotyl length for each of the three genotypes of F_3 lines and two chromosome segment substitution lines. One line was genotyped with two markers, RM5475 and RM3513. The broken vertical lines define the interval containing the *qMel-1* and *qMel-3* loci

[1]Markers within the heterozygous regions were tested, and the ones with the highest R^2 scores are shown

[2]Numbers followed by different letters in each row are significantly different at $P=0.05$ based on the Duncan's multiple range test. *NN* Nipponbare homozygotes, *NK* Nipponbare/Kasalath heterozygotes, *KK* Kasalath homozygotes

[3]*n* number of evaluated individuals in each line

*Numbers in parentheses indicate the numbers of F_3 plants in each genotype

mesocotyl length among the F_3 progeny showed significant differences among the three genotypes for the populations from lines 16 and 28. Based on the size of the chromosome 1 introgression in lines 16 and 28, *qMel-1* was inferred to be located in the interval RM5448–RM5310, a region of approximately 3,799 kb (Fig. 3.22) (http://archive.gramene.org/markers, Reference to Gramene Annotated Nipponbare Sequence 2009). RM5448 and RM5310 represented the outside borders of the introgression. Thus, we localized the *qMel-1* locus to a region <3,799 kb in size (Fig. 3.22). The same procedure can be applied to narrow down the location of *qMel-3*.

3.1.6 *Conclusions*

Recent progress in molecular biology and genomics has made possible the dissection of loci responsible for quantitative traits. QTL mapping is a well-established procedure in quantitative genetics (Lynch and Walsh 1997). It evolved from point analysis using a single marker, to interval analysis that tests sets of linked markers simultaneously, and finally to all-marker-based whole genome approaches. QTL analysis has been conducted using primary populations such as F_2, doubled haploid lines, and recombinant inbred lines (RILs) from bi-parental crosses and can now be extended to any population, including those derived from multiple parents. The emergence of high-resolution molecular-marker technology generated from whole-genome sequencing of many crops facilitates large-scale QTL analyses (IRGSP 2005).

Whole-genome sequences and technical progress in genomics have made possible the cloning of many important QTLs in crop species (Li et al. 2010; Miura et al. 2011; Song et al. 2007). The majority of cloned QTLs were isolated via positional cloning. More than 30 QTLs controlling heading date, yield-related traits, disease resistance, and abiotic stresses have been cloned using NIL-derived populations in rice. The constant improvement of molecular platforms, new types of genetic materials, an improved phenotyping method, and the availability of tools for testing candidate genes will offer more opportunities to isolate QTLs. The cloned QTLs will have the potential to greatly improve crop production.

3.2 Meta-analysis of QTLs

3.2.1 *Meta-analysis in Plants*

Meta-analysis is a statistical method used to analyze data from independent studies and identify associations in the results (Glass 1976; Rosenberg et al. 2004). This method has been effectively applied to social, medical, and behavioral sciences (Hedges and Olkin 1985). QTL mapping is now used to identify which genomic regions control complex traits by analyzing integrated datasets of large biological studies (Gyenis et al. 2007). In addition, meta-analysis has been useful in marker-assisted breeding for crop improvement using the QTL consensus map position (Bernardo and Charcosset 2006). Recently, MQTL analysis has generated some interesting results associated with agronomical traits, i.e., flowering time and ear-rot resistance in maize (Chardon et al. 2004; Xiang et al. 2010), earliness in wheat (Hanocq et al. 2007), blast resistance and root traits in rice (Ballini et al. 2008; Courtois et al. 2009; Khowaja et al. 2009), abiotic stress responses in barley (Li et al. 2013), and seed protein concentration QTL in soybean (Qi et al. 2011). Of note, Said et al. (2013) reported that possible QTL clusters and hotspots had been identified for various traits such as fiber quality, yield, yield-related morphological traits, drought tolerance, and disease resistance in tetraploid cotton via meta-analysis and suggested that these results might be useful in plant breeding and in future studies. In addition, Swamy et al. (2011) constructed an integrated consensus map by meta-analysis that consisted of 15 QTL maps and 541 markers from different studies and reliably applied MQTLs estimated for MAS to a panel of random drought-tolerant lines.

3.2.2 *Analysis of MQTLs Combined with Gene Expression Profiling Datasets*

The identification of QTL genes controlling useful traits is important for improving crops. Genome-wide expression profiling permits various genetics data to be combined; merged approaches such as QTL/microarray and genetical genomics have succeeded in detecting candidate genes in linked QTL regions (Jansen and Nap 2001; Wayne and McIntyre 2002).

The QTL/microarray approach uses combined information derived from traditional QTL mapping and microarray profiling to reduce the number of candidate genes in linked QTL regions (Arbilly et al. 2006; Drake et al. 2006; Matthews et al. 2005). Verdugo et al. (2010) summarized the common procedures used in different studies for QTL/microarray analysis: (1) identification of candidate QTL genes within a confidence interval detected by QTL mapping, (2) verification of genes that are differentially expressed between parental strains, (3) acquisition of integrated data of candidate genes identified by steps 1 and 2, (4) knowledge- or hypothesis-driven filtering of acquired data, (5) testing differential gene and protein expression using qRT-PCR, northern, or western blot analysis, and (6) experimental validation of QTL genes. Similarly, Marone et al. (2013) reported that the QTL genes associated with MQTLs for resistance to powdery mildew in wheat were predicted by comparative analysis of transcript levels within conserved regions between rice and wheat. Furthermore, Matsuda et al. (2012) developed a method to detect trait-associated pathways in animals using integrated information on linked regions of QTLs.

The genetical genomics approach has been used to study the genetic basis of gene expression and to detect the genetic contribution of QTLs to phenotypic differences that are induced by functional changes in proteins (Jansen and Nap 2001; Li and Burmeister 2005). Furthermore, several earlier studies reported that the level of gene expression was inherited (Brem et al. 2002; Cheung et al. 2003). This method has been successfully applied to yeast (Brem et al. 2002; Yvert et al. 2003), fly (Wayne and McIntyre 2002), plant, and human genetics (Schadt et al. 2003).

3.2.3 *Tools and QTL Databases for Meta-analysis*

The accumulation of genotypic data regarding molecular markers and phenotypic characteristics is essential for QTL mapping. Genetic maps produced using a standard linkage-mapping approach generally lead to QTL locations with a confidence interval (CI) of around 10 cM owing to the limited number of individuals and generations in each experiment (Kearsey and Farquhar 1998). A confidence interval of 10 cM equates to 300 kbp and 6,000 kbp in *Arabidopsis* and wheat, respectively.

Individual QTL data are now available from public databases, i.e., GRAMENE (http://archive.gramene.org/qtl/) for 10 plant species (*Avena sativa*, *Hordeum vulgare*, *Oryza sativa*, *Pennisetum glaucum*, *Setaria italica*, *Sorghum bicolor*, *Triticum aestivum*, *Triticum turgidum*, *Zea mays* subsp. *mays*, and *Zizania palustris*), MaizeGDB for maize (http://www.maizegdb.org/qtl.php), Q-TARO for rice (http://qtaro.abr.affrc.go.jp/), and PGDB for 18 plant species (*Brachypodium distachyon*, *Brassica oleracea*, *Cajanus cajan*, *Carica papaya*, *Eucalyptus globulus*, *Fragaria × ananassa*, *Jatropha curcas*, *Lactuca sativa*, *Malus × domestica*, *Manihot esculenta*, *Medicago truncatula*, *Nicotiana tabacum*, *Populus nigra*, *Populus trichocarpa*, *Prunus persica*, *Solanum tuberosum*, *Theobroma cacao*, and *Vitis vinifera*) (http://pgdbj.jp/en/dna-marker-linkage-map/plant-qtl-list.html). Public databases have provided statistical information by means of comparative analyses of trait ontology and different QTL results. In particular, the GRAMENE database is a comparative information resource for integrated data sets (genetic and physical maps, sequences, markers, germplasm resources, genes, proteins, pathways, and phenotype) for plants across other databases, such as MaizeGDB (Jaiswal et al. 2006; Liang et al. 2008; Ni et al. 2009; Ware et al. 2002). However, the comparative information in this database is often based on simple descriptive statistics (Veyrieras et al. 2007).

To survey the linked QTL region associated with useful traits, QTL meta-analysis software provides more information by statistically analyzing integrated QTL data. Arcade et al. (2004) evaluated QTL locations using BioMercator (Sosnowski et al. 2012), which has been used to study the MQTLs as QTL hotspots in different plant species, i.e., in tetraploid cotton (Said et al. 2013), *B. napus* (Wang et al. 2013), and soybean (Qi et al. 2011). Furthermore, the MQTL program constructs the consensus QTL map using the iterative projection procedure with likelihood estimation of QTL clusters (Goffinet and Gerber 2000).

3.2.4 Case Study: Meta-analysis of QTLs Associated with Abiotic Stresses and Genome-Wide Expression Profiling in Rice

Rice is one of the most important cereal crops in the world, and many researchers have studied QTLs for useful rice traits using molecular markers. Thus, rice is a good model plant species for MQTL analysis. Here, to introduce meta-analysis of QTLs we demonstrate a methodology to detect QTL hotspots associated with abiotic stresses in rice using BioMercator (Sosnowski et al. 2012). This program is available free online (https://urgi.versailles.inra.fr/Tools/BioMercator-V4).

To detect the MQTLs related to abiotic stresses, we collected 11 QTL map sets (QTL UCD M-202/IR50 RI SSR QTL 2003, Rice-KRGRP Milyang23/Gihobyeo AFLP-SSLP-RFLP QTL 1998, Rice-Cornell IR64/Azucena DH QTL, CTIR CT9993/IR6226 QTL 2000, MU CT9993/IR6226 QTL 2004, IRRI IR64/Azu DH QTL 2003, JRGP Nip/Kas F2 QTL 2000, IGCN ZYQ18/JX17 DH QTL 1998, IRRI IR74/Jalmagna RI AFLP/RFLP QTL 2001, UCD IR40931/PI543851 QTL 1996, and CNYU Naj11/Bal DH QTL 1999) from GRAMENE (http://archive.gramene.

Table 3.4 Data set of rice quantitative trait loci (QTLs) associated with abiotic stresses that were used in the present study

No. of initial QTLs	Population no.[a]	Parent 1	Parent 2	Population size	Analysis method[b]	Population type	Reference
15	1	M-202	IR50	191	CIM	RIL	Andaya and Mackill (2003a)
9	1	M-202	IR50	191	CIM	RIL	Andaya and Mackill (2003b)
19	2	IR64	Azucena	135	ANOVA	DH	Hemamalini et al. (2000)
2	3	Milyang 23	Gihobyeo	164	IM	RIL	Lee et al. (2006)
19	4	IR62266-42-6-2	CT9993-5-10-1-M	154	ANOVA	DH	Nguyen et al. (2002)
2	5	ZhaiYeQing 8	JingXi 17	127	CIM	DH	Teng et al. (2002)
1	6	Nipponbare	Kasalath	186	CIM	F2	Wissuwa et al. (2002)
41	7	IR62266-42-6-2	CT9993-5-10-1-M	154	ANOVA	DH	Zhang et al. (2001)

[a]The numbers represent the population characters used for the QTL experiment
[b]Composite interval mapping (CIM) and analysis of variance (ANOVA) indicate the statistical method used for QTL mapping

org/qtl/). QTL information contained in different QTL map sets was retrieved using a QTL search of GRAMENE (http://archive.gramene.org/qtl/). The BioMercator program requires the R^2 (variance percentage), confidence interval (CI), and logarithm of odds (LOD) score of each QTL in the input file. If the CI is not available, it can be calculated by the following formula (Darvasi and Soller 1997):

$$\mathrm{CI} = 163 / \left(\mathrm{N} \times R^2\right) \tag{3.1}$$

$$\mathrm{CI} = 530 / \left(\mathrm{N} \times R^2\right) \tag{3.2}$$

Equations 3.1 and 3.2 were used for RILs and backcross and F_2 populations, respectively.

To identify the R^2 and LOD scores of QTLs contained in the 11 map sets, we retrieved data from past studies. Here, we analyzed the MQTLs of rice using the QTL information obtained from different studies associated with abiotic stress; a total of 105 QTL have been described in eight published reports (Table 3.4). These rice QTLs were observed in association with aluminum tolerance (Nguyen et al. 2002), cold stress (Andaya and Mackill 2003a, b), drought resistance (Teng et al. 2002; Zhang et al. 2001), low moisture stress (Hemamalini et al. 2000), phosphorus stress (Wissuwa et al. 2002), and salt tolerance (Lee et al. 2006) using a standard linkage mapping approach.

To run BioMercator, the MAP input file (*.map) requires a header of 13 fields and marker positions on each chromosome, and the QTL input file (*.qtl) contains the QTL information in 12 columns, i.e., QTL IDs, trait name, trait ontology ID, place of experiments, year of experiments, chromosome name, linkage group name, LOD score, R^2, QTL most likely position (cM), and start and end positions (cM) of CI.

For example:

<MAP input files (*.map)>

```
mapName=V.C. Andaya and D.J. Mackill
Organism Genus=unknown
Organism Species=unknown
crossType=RIL
popSize=191
mappingCrossType=unknown
mappingFunction=unknown
mapUnit=cM
mapExpansion=0
mapQuality=0
locusLocation=1
chr=1
lg=1
1     RM323    0
2     RM86     2.6
3     RM84     2.9
4     RM220    4.8
5     RM151    12.3
6     RM259    26.8
7     RM243    28.2
8     RM312    40
9     RM292    50.1
10    RM294A   53.8
11    RM23     59.4
12    RM185    64
13    RM113    69.6
14    RM306    85
15    RM237    98.2
16    RM246    106.3
17    RM128    121.7
18    RM297    123.7
19    RM319    129.6
20    RM302    132.6
21    RM104    159.3
22    RM14     164.9
```

<QTL input files (*.qtl)>

mapName=V.C. Andaya and D.J. Mackill

qCTS 8-2b	Cold	ID:0	P1	Y1	8	8	4.41	12.1	0.8	0.1	1.5
qCTS11-2	Cold	ID:0	P1	Y1	11	11	4.77	13	78.55	73	84.1
qCTS12b	Cold	ID:0	P1	Y1	12	12	18.51	41.7	43.4	36.7	50.1
qCTS4-2	Cold	ID:0	P1	Y1	4	4	4.79	11.5	97.25	91.6	102.9
qCTS6-2	Cold	ID:0	P1	Y1	6	6	3.54	8.7	49.65	49.2	50.1
qCTS8-2a	Cold	ID:0	P1	Y1	8	8	5.29	12.7	0.8	0.1	1.5
qCTS11-1	Cold	ID:0	P1	Y1	11	11	3.81	9.4	0.3	0	0.6
qCTS12a	Cold	ID:0	P1	Y1	12	12	20.34	40.6	43.4	36.7	50.1
qCTS4-3	Cold	ID:0	P1	Y1	4	4	5.3	14.2	75.1	67.4	82.8
qCTS1	Cold	ID:0	P1	Y1	1	1	3.53	9.3	126.65	123.7	129.6
qCTS3	Cold	ID:0	P1	Y1	3	3	5.28	13.8	166	161.5	170.5
qCTS4-1	Cold	ID:0	P1	Y1	4	4	8.36	20.8	2.6	0	5.2
qCTS6-1	Cold	ID:0	P1	Y1	6	6	6.01	15.3	26.7	24.8	28.6
qCTS8-1	Cold	ID:0	P1	Y1	8	8	5.68	14.5	12.35	1.5	23.2
qCTS10	Cold	ID:0	P1	Y1	10	10	3.98	10.4	5	8.5	1.5

The input file is loaded as follows (Fig. 3.23):

1. Execute BioMercator, then click on "File" and "Genetic data loading".
2. Choose a project and click on "Next".
3. Click "Browse" and select the map and QTL input files for each dataset, then click "Open".
4. Click "Next" and "Finish".

To construct a consensus map for different QTL results, we compiled the different markers into one QTL map using the BioMercator package ConsMap. Previously

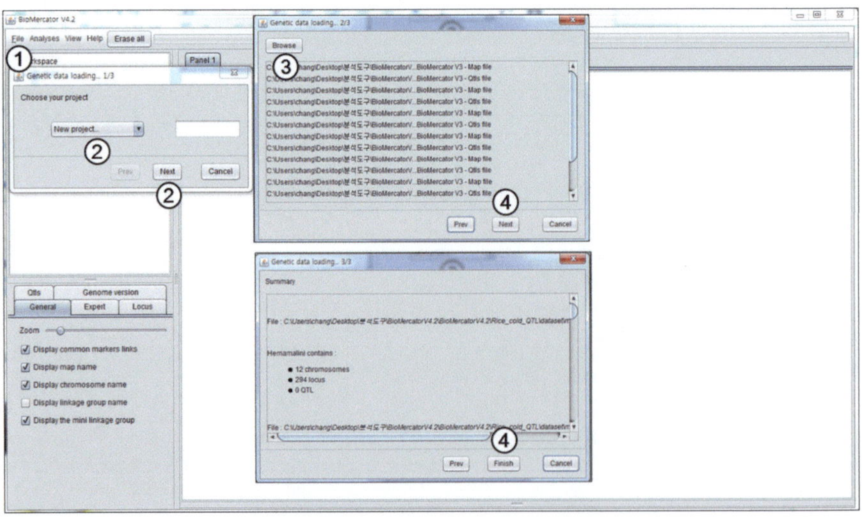

Fig. 3.23 Screenshot: loading an input file

studied QTL regions were then visualized using the BioMercator package QTLProj (Fig. 3.24).

The consensus map is constructed as follows (Fig. 3.25):

1. Click the buttons “Analyses”, “Map compilations”, and “Map compilation (ConsMap)”.
2. Choose a project, and then select the input files to compile different maps.
3. Create the result file name in the Result map box and click on “Next” then “Finish”.
4. Launch QTLProj, then click “Analyses”, “Map compilations”, and “QTL projection (QTLProj)”.
5. Choose a project, and click on “Next”.
6. Choose a different dataset including QTL information and the result file compiled from ConsMap as a reference map.
7. Configure the Ratio (minimal value of the ratio of the flanking marker interval distances) and *p* value (homogeneity test of the flanking marker interval distances between the original and reference map).
8. Click on “Next” and “Finish”.

A total of 105 QTLs were identified on the consensus map. To detect which MQTLs were associated with abiotic stresses, we identified 73 MQTLs using

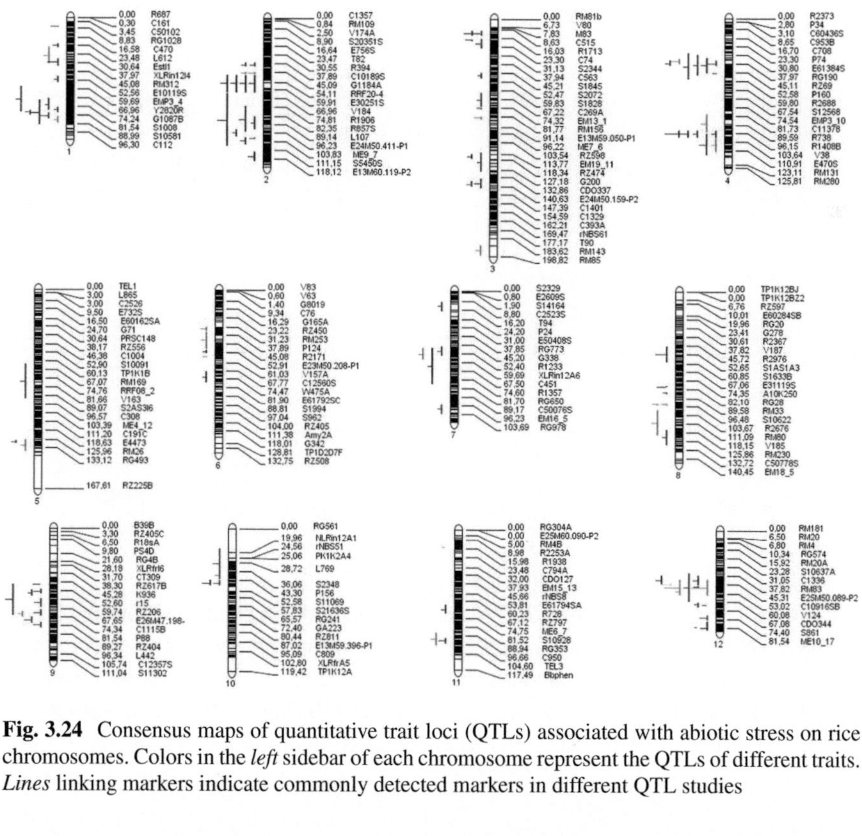

Fig. 3.24 Consensus maps of quantitative trait loci (QTLs) associated with abiotic stress on rice chromosomes. Colors in the *left* sidebar of each chromosome represent the QTLs of different traits. *Lines* linking markers indicate commonly detected markers in different QTL studies

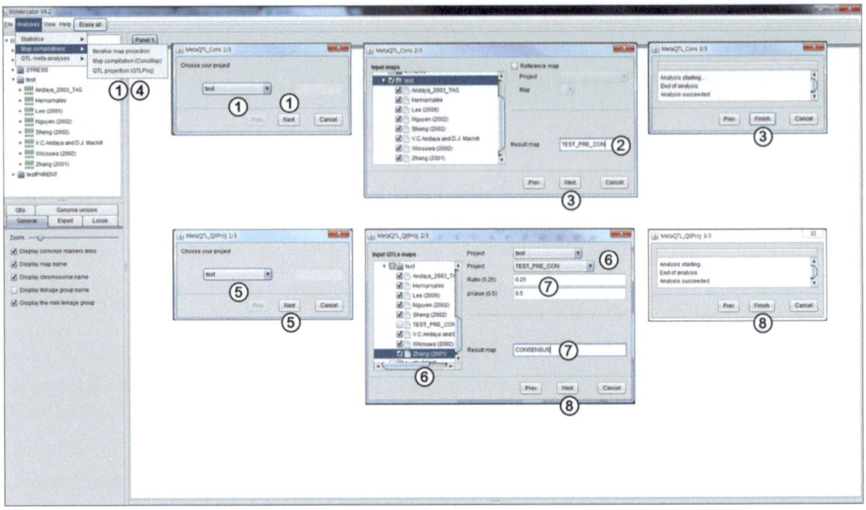

Fig. 3.25 Screenshot: constructing a consensus map

Table 3.5 Numbers of detected meta-quantitative trait loci (MQTLs) associated with abiotic stress on the consensus map

Chr.	No. of initial QTLs on consensus map	No. of MQTL associated with abiotic stress						
		Aluminum	Cold	Drought	Low moisture	Osmotic	Phosphorus	Salt
1	12	3 (0)[a]	2 (0)	1 (3)	2 (0)	3 (0)	0	1 (3)
2	15	1 (7)	2 (5)	2 (6)	6 (0)	4 (6)	0	0
3	13	1 (5)	2 (6)	3 (0)	4 (0)	2 (5)	0	1 (3)
4	14	1 (3)	3 (3)	4 (0)	3 (0)	3 (4)	0	0
5	3	0	1 (0)	2 (0)	0	0	0	0
6	6	1 (0)	3 (0)	1 (0)	0	1 (0)	0	0
7	5	2 (0)	1 (0)	0	1 (0)	1 (0)	0	0
8	7	1 (0)	3 (0)	0	1 (0)	2 (0)	0	0
9	9	2 (0)	1 (0)	0	1 (0)	5 (0)	0	0
10	4	2 (0)	1 (0)	1 (0)	0	0	0	0
11	4	0	2 (0)	1 (0)	0	1 (0)	0	0
12	13	2 (4)	3 (0)	2 (5)	1 (0)	4 (0)	1 (5)	0
Total	105	16 (19)	24 (14)	17 (14)	19 (0)	26 (15)	1 (5)	2 (6)

[a]Numbers are the initial QTL and MQTL (in brackets)

the N-QTLs model (models 1, 2, 3, 4, and n) method (Goffinet and Gerber 2000) (Table 3.5). We detected overlapping MQTLs on chromosomes 1, 2, 4, and 12, and the MQTLs were mainly located on chromosomes 2 and 3. However, we found no MQTLs on chromosomes 5, 6, 7, 8, 9, 10, and 11 because of the low number of initial QTLs. The most MQTLs (19) were detected for aluminum stress; seven were located on chromosome 2. Six cold-stress MQTLs were detected on chromosome 3. The MQTLs associated with drought (6 MQTLs) and osmotic stress (6) were mainly located on chromosome 2. We did not detect the MQTLs associated with low-moisture stress. We identified more MQTLs than initial QTLs, i.e., on chromosomes 2, 3, 4, and 12 for aluminum stress, on chromosomes 2 and 3 for cold stress, on chromosomes 2 and 12 for drought stress, on chromosomes 2 and 3 for osmotic stress, and on chromosomes 1 and 3 for salt stress. These results suggest additional MQTL regions can be detected during meta-analyses of integrated data sets from different QTL studies. In this study, the meta-analysis also showed that MQTL detection was highly dependent on the number of initial QTLs.

To determine the most likely number of "real" QTLs as input QTLs, this statistical method can detect the N-QTLs in independent experiments. Goffinet and Gerbers (2000) considered the four QTL models described and did not extend the N-QTL model algorithm to intermediary models (5, 6, 7, 8, …). The QTL clustering (Meta-analysis 1/2) is performed as follows (Fig. 3.26):

1. Click on "Analyses", "QTL meta-analyses", and "Meta-analysis 1/2 (Veyrieras)".
2. Choose a project, map (consensus mapfile constructed by CONproj), chromosome number, and linkage group.
3. Choose an algorithm configuration (kMax, ci mode, ci miss, emrs, emeps).
4. Choose traits that you want to analyze.
5. Click on "Next" then "Finish".

QTL clustering generates three result files: *_res.txt (clustering result summary), *_model.txt (model choice criteria value), and *_res.txt (optimal number of QTL location). In particular, the *_model.txt file provides the best values for the N-model method using the Meta-analysis 1/2 (Veyrieras) procedure. We identified the different best values for abiotic stresses on each chromosome using the AIC statistical

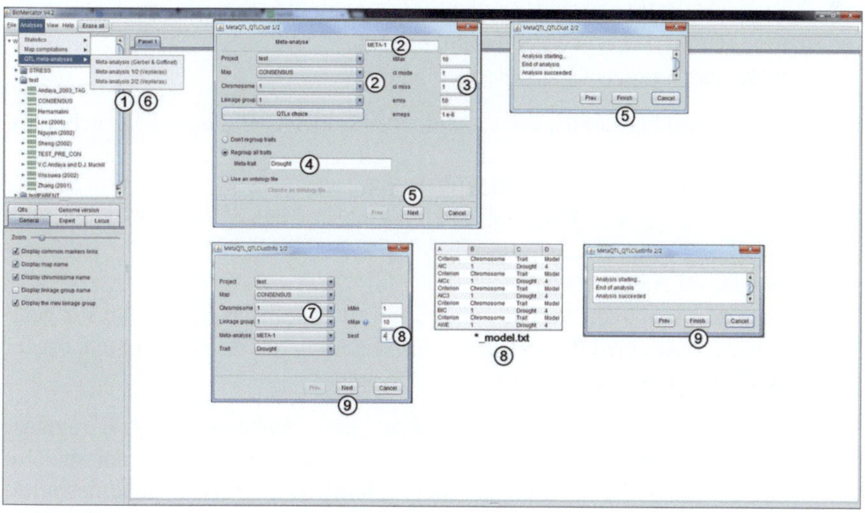

Fig. 3.26 Screenshot: quantitative trait loci clustering analysis

test. For example, the three best values of QTLs associated with drought stress on chromosome 3 as found with different statistical tests (AIC, AICc, AIC3, BIC and AWE) were used for the Meta-analysis 2/2 (Veyrieras) procedure.

The QTL clustering (Meta-analysis 2/2) was performed as follows:

1. Click on "Analyses", "QTL meta-analyses", and "Meta-analysis 2/2 (Veyrieras)".
2. Choose a project, map (consensus mapfile constructed by CONproj), chromosome number, linkage group, and trait.
3. Choose an algorithm configuration, such as the best value detected in the Meta-analysis 1/2 (Veyrieras) procedure.
4. Click on "Next" then "Finish".

The output file (*_best value_table.txt) provides the summary table (position, weight, distance, CI, and UCI of MQTL) and initial QTLs among MQTLs.

To detect hotspots, we identified the overlapping MQTLs associated with different abiotic stresses within common QTL regions on each chromosome (Table 3.6). This result showed that 19 MQTL clusters (except 10 unique MQTL clusters) overlapped within common QTL regions. The R^2 value is the most important parameter for efficient estimation of MAS versus conventional phenotypic selection (Bernardo and Charcosset 2006). Here, the detected MQTLs exhibited R^2 values ranging from 8.30 to 27.62 %. The MQTL D10 on chromosome 12 had the highest R^2 value, with flanking markers at 43.3 cM and 46.73 cM and a CI of 2.81 cM. This information can be used for effective MAS of M-202×IR50, Nipponbare×Kasalath and IR62266-42-6-2×CT9993-5-10-1-M. The detailed information of MQTL described in Table 3.6.

To study the MQTL genes associated with abiotic stress in rice, we retrieved the genome sequences of 95 QTLs within the initial QTL region from the GRAMENE database (http://archive.gramene.org/qtl/). These sequences were then used to search for rice genes on each chromosome associated with QTLs by BLASTN with e^{-20}. The local blastall package is available from National Center for Biotechnology Information (NCBI; http://www.ncbi.nlm.nih.gov/).

The blast search proceeds as described below:

1. Install the Local blastall package downloaded from NCBI (http://www.ncbi.nlm.nih.gov/Ftp/).
2. Construct the database for the BLAST search.
3. Execute the command: Formatdb –i (input file for database) –p F (nucleotide sequence) or T (amino acids).
4. Perform the BLASTN search.
5. Execute the command: blastall –p (choose a program) –i (query file for gene search) –d (constructed BLAST database) –e (choose the e-value) –m (choose an output file form from 1 to 8) –o (output file name).

In the results, we detected 20 candidate genes within 40 MQTL regions. However, only 17 MQTL genes were retrieved from microarray datasets by GENEVESTIGATOR because of the unavailability of probes for gene detection (Hruz et al. 2008) (Fig. 3.27).

Table 3.6 Characteristics of detected meta-quantitative trait loci (MQTLs) associated with abiotic stress

Features[a]	MQTL[b]	Chr.	Number of initial QTL[c]	CI	Position (cM)[d]	Left (cM)[e]	Coordinate (cM)[e]	Right (cM)[e]	Coordinate (cM)[e]	R^{2f}	Traits[g]	No. of populations[h]
O	D1	1	2	3.82	24.39	C101, S1651, C668	21.52	E31375S	26.74	13.75	OACAOS, CT	1, 7
	S1	1	2	3.77	24.4	G359	21.59	E31375S	26.74	13.75	CT, OACAOS	1, 7
U	D2	1	2	4.24	45.13	Y3683R	42.98	Y1895L	47.27	18.53	ST, PRTAOS	3, 7
U	S2	1	1	5.83	52.23	R108	49.15	RM185	55.31	9.30	PRTAOS	7
U	S3	1	7	0.4	80.36	G1133	79.96	RZ569A	80.66	14.84	RLLM, CT, RLRAS, PRTAOS, SRLAS, TRDWDS, PHLM	1, 2, 4, 7
U	D3	1	6	0.51	81.34	RZ569A	80.66	RZ801	81.63	15.77	CT, SRLAS, PRTAOS, RLRAS, RLLM, PHLM	1, 2, 4, 7
O	A1	2	2	2.47	39.47	Y2724R, V259	38.1	RM71	40.73	11.80	TRDWDS, PRTAOS	7
	C1	2	2	4.87	39.47	E25M60.168-P1	36.88	P86	41.94	11.80	TRDWDS, PRTAOS	7
	D4	2	1	3.5	39.47	G1184C-2	37.6	M130	41.85	12.60	PRTAOS	7
	O1	2	1	3.5	39.47	G1184C-2	37.6	M130	41.85	11.00	TRDWDS	7
O	A2	2	2	1.84	46.49	R1989	45.54	RM301	47.64	13.65	CT	1
	D5	2	2	1.84	46.49	R1989	45.54	RM301	47.64	13.65	CT	1
	O2	2	2	1.84	46.49	R1989	45.54	RM301	47.64	13.65	CT	1

(continued)

Table 3.6 (continued)

Features[a]	MQTL[b]	Chr.	Number of initial QTL[c]	CI	Position (cM)[d]	Left (cM)[e]	Coordinate (cM)[e]	Right (cM)[e]	Coordinate (cM)[e]	R^{2}[f]	Traits[g]	No. of populations[h]
O	C2	2	4	8.49	55.32	EM14_4	50.48	E30251S	59.91	15.98	RVLM, TRNLM, RLLM	2
	A3	2	4	7.53	55.33	NLRin1213	51.59	RM341	59.25	15.98	RVLM, TRNLM, RLLM	2
	D6	2	4	7.53	55.33	EM14_4	50.48	RM341	59.25	15.98	TRNLM, RVLM, RLLM	2
	O3	2	4	7.52	55.36	NLRin12I3	51.59	RM341	59.25	15.98	RVLM, TRNLM, RLLM	2
O	D7	2	3	10.2	87.4	RZ103	81.46	R3393	92.7	16.67	OACAOS, PRTAOS, TRNLM	2, 7
	C3	2	3	11.06	87.42	RZ103	81.46	R2216	93.24	16.67	OACAOS, PRTAOS, TRNLM	2, 7
	A4	2	3	10.31	87.43	RZ103	81.46	R3393	92.7	16.67	OACAOS, PRTAOS, TRNLM	2, 7
O	A5	2	1	6.46	96.61	R2216	93.24	C520	100.19	9.00	RPFDS	7
	O4	2	3	6.52	96.61	R2216	93.24	C520	100.19	17.05	TRNLM, RPFDS	2, 7
O	C4	2	3	4.47	101.13	RM318	98.85	ME9_7	103.83	11.47	BRTAOS, RLRAS, RPFDS	4, 7
	O5	2	1	1.7	101.27	R2643	100.28	C11895S	102.71	13.40	RLRAS	4

	D8	2	2	1.64	101.41	E13M59.504-P2	100.51	C10187S	102.3	12.70	BRTAOS, RLRAS	4, 7
	A6	2	1	6.29	103.29	R3128	99.79	RM240	106.58	12.00	BRTAOS	7
O	O6	2	0	4.27	116.52	G275	114.17	RM138	118.82	17.70	TRNLM	2
	A7	2	1	4.27	116.53	G275	114.17	G317A	118.76	17.70	TRNLM	2
	C5	2	1	5.41	116.53	RG151	113.39	RZ913	119.32	17.70	TRNLM	2
	D9	2	1	4.27	116.53	G275	114.17	RM138	118.82	17.70	TRNLM	2
O	C6	3	2	3.12	12.64	C51477S	10.63	V134	14.63	21.10	RDWLM, RVLM	2
	O7	3	2	3.12	12.64	C51477S	10.63	C51476S, G8004	14.33	21.10	RVLM, RDWLM	2
O	A8	3	4	1.26	17.19	R1713	16.03	C814B	17.93	18.33	RVLM, TRNLM, RDWLM, RPFDS	2, 7
	S4	3	4	1.26	17.19	R1713	16.03	C814B	17.93	18.33	RDWLM, RPFDS, RVLM, TRNLM	2, 7
O	C7	3	2	1.39	18.09	EM11_9	17.31	L837	19.23	15.55	TRNLM, RPFDS	2, 7
	O8	3	2	1.39	18.09	EM11_9	17.31	L837	19.23	15.55	RPFDS, TRNLM	2, 7
O	A9	3	1	5.2	47.9	S1845, S2680	45.21	C316A	50.65	9.90	OACAOS	7
	C8	3	1	5.2	47.9	S1845, S2680	45.21	C316A	50.65	9.90	OACAOS	7

(continued)

Table 3.6 (continued)

Features[a]	MQTL[b]	Chr.	Number of initial QTL[c]	CI	Position (cM)[d]	Left (cM)[e]	Coordinate (cM)[e]	Right (cM)[e]	Coordinate (cM)[e]	R^{2f}	Traits[g]	No. of populations[h]
U	S5	3	2	4.79	50.96	R2293	48.2	C68	53.84	10.40	RPIDS, OACAOS	7
O	C9	3	2	1.49	74.97	E26M47.162-P2	74.11	R2170	75.83	11.40	RPIDS, RPFDS	7
	O9	3	2	1.49	74.97	E26M47.162-P2	74.11	R2170	75.83	11.40	RPFDS, RPIDS	7
	A10	3	3	1.34	76.5	R2170	75.83	S11493	77.76	13.10	RPIDS, RPFDS, CT	1, 7
U	O10	3	1	3.1	83.11	RZ284	81.13	TP1D2D7B	85.06	16.50	CT	1
O	C10	3	2	4.42	116.39	EM19_11	113.77	RRK08_2	120.82	9.58	BRTAOS, ST	3, 7
	A11	3	3	3.93	121.07	RZ474	118.34	C136	123.29	11.75	BRTAOS, ST, PHLM	2, 3, 7
	S6	3	4	0.57	135.44	V120	135.09	S10087	136.05	13.18	RLRAS, BRTAOS, CT, PHLM	1, 2, 4, 7
U	C11	3	2	1.18	138.86	E3011S	137.4	PS4B	139.77	14.45	RLRAS, PHLM	2, 4
U	O11	3	4	0.91	147.44	G249	145.83	R1618, R2632	148.23	12.97	RLRAS, CT, ST, PHLM	1, 2, 3, 4
U	A12	3	1	5.27	194.32	RM200	189.82	RM85	198.82	13.80	CT	1
O	A13	4	5	4.69	37.78	EM16_3	35.07	ME2_5	40.47	14.48	TRDWDS, CT, DSLM, TRNLM	1, 2, 7

	O12	4	6	4.22	37.97	EM16_3	35.07	ME10_13	40.1	15.42	TRDWDS, CT, RLRAS, DSLM, TRNLM	1, 2, 4, 7
	C12	4	5	4.56	38.61	EM16_3	35.07	RG908	41.15	14.34	RLRAS, TRDWDS, DSLM, TRNLM	2, 4, 7
	A14	4	1	3.22	46.12	ME7_7	44.32	EMP2_2	47.88	8.30	RPIDS	7
	C13	4	1	3	46.12	ME7_7	44.32	rNBS56a	47.67	8.30	RPIDS	7
	O13	4	1	3.13	46.12	ME7_7	44.32	EMP2_2	47.88	8.30	RPIDS	7
O	O14	4	2	6.01	96.17	C79	92.25	RG329	99.38	12.60	CIYT, RPIDS	1, 7
	C14	4	5	0.59	97.52	rNBS19	96.66	V125	99.14	22.26	RPFDS, PRDWAOS, RPIDS, PRTAOS, BRTAOS	7
	A15	4	7	0.53	97.81	V65	97.34	V125	99.14	19.57	RPFDS, PRDWAOS, RPIDS, CIYT, CIWT, PRTAOS, BRTAOS	1, 7
U	O15	4	2	6.43	106.79	E24M50. M002-P2	103.05	E470S	110.91	15.70	RPFDS, CIWT	1, 7
O	P1	12	2	6.07	42.82	E60377A	39.71	PK1K214	46.04	13.85	PRTAOS, PRDWAOS	7

(continued)

Table 3.6 (continued)

Features[a]	MQTL[b]	Chr.	Number of initial QTL[c]	CI	Position (cM)[d]	Left (cM)[e]	Coordinate (cM)[e]	Right (cM)[e]	Coordinate (cM)[e]	R^{2f}	Traits[g]	No. of populations[h]
	A16	12	3	2.86	44.97	C53024S	43.3	RG9	46.73	18.60	PHOUP, PRTAOS, PRDWAOS	6, 7
	D10	12	5	2.81	44.98	C53024S	43.3	RG9	46.73	27.62	PRTAOS, PRDWAOS, PHOUP, CINT, CIWT	1, 6, 7
	A17	12	5	2.17	48.11	RG361A	46.95	C11831S	49.42	23.44	CINT, CIWT, DT, CT, RPIDS	1, 5, 7
	P2	12	5	2.14	48.16	L1087B	47.03	C11831S	49.42	23.44	DT, CINT, CIWT, RPIDS, CT	1, 5, 7
	D11	12	1	6.01	51.08	E26M47.282-P2	47.7	S894	54.24	11.60	CT	1
O	A18	12	2	0.4	63.26	ME6_6	62.8	E20994S	63.86	12.15	PRTAOS, BRTAOS	7
	D12	12	2	0.4	63.26	MRGH	63.06	E20994S	63.86	12.15	PRTAOS, BRTAOS	7
	P3	12	2	0.4	63.26	mRGH	63.06	E20994S	63.86	12.15	BRTAOS, PRTAOS	7
O	D13	12	1	4.6	73.71	RG403	71.4	B264	76.12	19.70	RLRAS	4
	P4	12	1	4.6	73.71	RG403	71.4	B264	76.12	19.70	RLRAS	4

O	D14	12	2	0.01	81.45	R2501B	81.37	ME10_17	81.54	12.95	SRLAS, DSLM	2, 4
	P5	12	2	0.01	81.45	R2501B	81.37	ME10_17	81.54	12.95	SRLAS, DSLM	2, 4
	A19	12	1	3.91	82.09	C51116S	79.56	EM15_15	84.49	16.10	DSLM	2

[a]Features of MQTL clusters: *O* overlapping MQTL in common QTL region, *U* unique MQTL in common QTL region

[b]Symbols represent the MQTLs associated with different abiotic stresses: *A* aluminum stress, *C* cold stress, *D* drought stress, *O* osmotic stress, *P* phosphorus stress, *S* salt stress

[c]The numbers of initial QTLs contained in the MQTL

[d]Position of MQTL on the consensus map

[e]Flanking markers of detected MQTLs

[f]Average R^2 values of the initial QTL contained in the MQTL

[g]Traits associated with initial QTL: *CT* cold tolerance, *BRTAOS* basal root thickness, *DSLM* drought score in low moisture stress, *LRLM* leaf rolling in low moisture stress, *TRNLM* total root number in low moisture stress, *TRNLM* number of tillers in low moisture stress, *OACAOS* osmotic adjustment capacity, *PHLM* plant height in low moisture stress, *PRDWAOS* penetrated root dry weight, *PRLAOS* penetrated root length, *PRTAOS* penetrated root thickness, *PHOUP* phosphorus uptake, *RLRAS* root length ratio, *SRLAS* stress root length in aluminum stress, *CL* control root length, *CINT* cold-induced necrosis tolerance, *CIWT* cold-induced wilting tolerance, *CIYT* cold-induced yellowing tolerance, *DT* drought tolerance, *ST* salt tolerance, *RDWLM* root dry weight in low moisture stress, *RLLM* root length in low moisture stress, *RPFDS* root pulling force, *RPIDS* root penetration index, *RVLM* root volume in low moisture stress, *TRDWDS* total root dry weight

[h]Populations of detected MQTLs: 1, M-202 × IR50; 2, IR64 × Azucena; 3, Milyang 23 × Gihobyeo; 4, IR62266-42-6-2 × CT9993-5-10-1-M; 5, ZhaiYeQing 8 × JingXi 17; 6, Nipponbare × Kasalath; 7, IR62266-42-6-2 × CT9993-5-10-1-M

Chr	MQTL	Positon	CI	Genes	Annotations
1	D1, S1	23.83	3.88	LOC_Os01g09340	hypothetical protein
1	D1, S1	32.16	14.5	**LOC_Os01g10504**	OsMADS3
				LOC_Os01g13360	phosphatidylinositol 3- and 4-kinase family protein
1	D2, S2	52.21	5.8	**LOC_Os01g26130**	expressed protein
1	D3, O2, S3	86.24	5.92	**LOC_Os01g58270**	OsSub7 - Putative Subtilisin homologue
1	O5	101.28	1.69	**LOC_Os01g03490**	heavy-metal-associated domain
2	A1, C1, O4	39.47	3.49	**LOC_Os02g28830**	tetratricopeptide repeat domain
2	A2, D5, O2	46.47	2.59	**LOC_Os02g15594**	protein phosphatase 2C
2	A4, C3, D7	85.29	22.5	**LOC_Os02g48290**	thioredoxin reductase 2
2	A4, C3, D7	88.04	11.68	**LOC_Os02g46510**	adaptor complexes medium subunit family
2	A5, C4, O9	96.75	6.51	LOC_Os02g49190	SWI/SNF
2	C4, D8, O11	101.28	1.69	LOC_Os02g49190	SWI/SNF
				LOC_Os02g48290	thioredoxin reductase 2
2	A6, C4, D8, P3	103.32	6.31	LOC_Os02g48290	thioredoxin reductase 2
3	A9, C8, S5	47.9	5.21	**LOC_Os03g15370**	ubiquitin fusion protein
				LOC_Os03g12590	coatomer subunit gamma-1
3	A10, C9, O15	75.07	1.49	**LOC_Os03g25940**	cystathionine gamma-synthase
3	A11, C10, S6	116.05	4.59	**LOC_Os03g47800**	RNA recognition motif containing protein
3	A11, C10	120.44	15.99	**LOC_Os03g27310**	histone H3
				LOC_Os03g41510	aldo/keto reductase family protein
3	C11, S6	138.89	4.9	LOC_Os03g47800	RNA recognition motif containing protein
4	A15, C14	106.58	18.38	**LOC_Os04g58640**	cleavage and polyadenylation specificity factor subunit 5
12	A16, D10, O11	45.43	3.29	**LOC_Os12g23180**	3-beta hydroxysteroid dehydrogenase/isomerase family
				LOC_Os12g19040	expressed protein
12	A17, D10, O1, P2	48.45	13.41	**LOC_Os12g15470**	OsSCP65

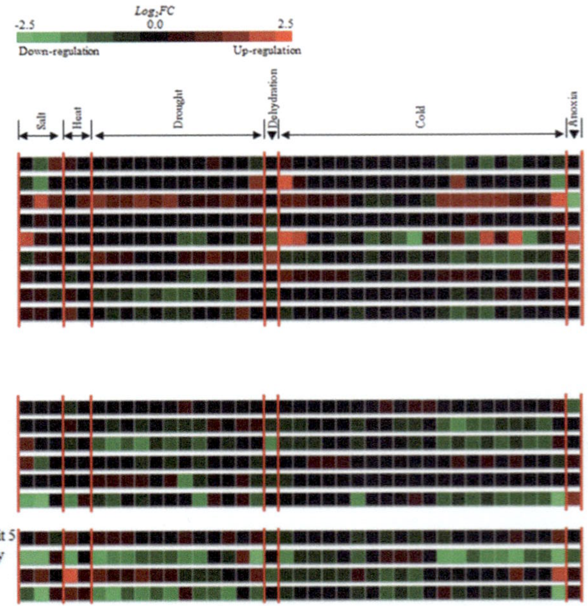

Fig. 3.27 Expression patterns of candidate meta-quantitative trait locus (MQTL) genes associated with abiotic stress. Colors of expression patterns represent the transcriptional differences between wild-type and stress-treated plants: *red* up-regulation, *green* down-regulation. MQTL symbols represent the MQTLs associated with different abiotic stresses: *A* aluminum stress, *C* cold stress, *D* drought stress, *O* osmotic stress, *P* phosphorus stress, *S* salt stress. Bold fonts indicate the MQTL genes associated with abiotic stress that were detected from microarray datasets

GENEVESTIGATOR provides useful web-based tools (condition, gene, and similarity search tools) for meta-analysis of the transcriptome. The GENEVESTIGATOR analysis proceeds as described below:

1. Login to the GENEVESTIGATOR web-based tool (https://www.genevestigator.com/gv/).
2. Launch GENEVESTIGATOR.
3. Click on "New" sample selection, and select the microarray datasets to use for transcriptome analysis.
4. Click on "New" gene selection.
5. Choose a suitable search tool for transcriptome analysis.

For example, in this study, we used the bi-clustering analysis within the similarity search tool, and analysis of the gene expression pattern within the MQTL region was performed in the perturbations category, in which the different microarray datasets were divided into four categories (samples, anatomy, development, and perturbations) by GENEVESTIGATOR.

Transcriptional changes were detected in 17 MQTL genes in response to different abiotic stresses (salt, heat, drought, dehydration, cold, and anoxia stress). Of note, the gene (LOC_Os01g26130) associated with the MQTL of drought- and salt-related traits was induced under salt- and drought-stress conditions in rice. Furthermore, the gene (LOC_Os12g23180) located within the MQTL region associated with aluminum, drought, and oxidative stress exhibited low expression levels during six abiotic stress tests. The gene (LOC_Os03g41510; aldo-ketoreductase) within the MQTL region associated with aluminum and cold stress was down-regulated during salt- and cold-stress treatments. A role of aldo-ketoreductase genes has been suggested under various abiotic stresses, including cold, high temperature, drought, and heavy metal stress (Gavidia et al. 2002; Hegedus et al. 2004; Lee and Chen 1993; Oberschall et al. 2000).

Here, we identified 19 clusters of commonly detected MQTLs as hotspots among different abiotic stresses (aluminum, cold, drought, low moisture, osmotic, phosphorus, and salt) by meta-analysis. Furthermore, some MQTLs were newly identified as linked regions associated with other abiotic stresses as compared with the initial QTLs on each chromosome. In addition, the candidate genes controlling abiotic stress-related traits within several MQTLs were detected using an MQTL analysis compiled from gene-expression profiling datasets. However, this method may be limited by the number of integrated QTL studies. With an increase in the amount of information available from QTL studies, meta-analysis may improve the accuracy and efficiency of MQTLs for detecting "real" QTLs.

References

Andaya VC, Mackill DJ (2003a) Mapping of QTLs associated with cold tolerance during the vegetative stage in rice. J Exp Bot 54:2579–2585

Andaya VC, Mackill DJ (2003b) QTLs conferring cold tolerance at the booting stage of rice using recombinant inbred lines from a japonica x indica cross. Theor Appl Genet 106:1084–1090

Arbilly M, Pisante A, Devor M et al (2006) An integrative approach for the identification of quantitative trait loci. Anim Genet 37:7–9
Arcade A, Labourdette A, Falque M et al (2004) BioMercator: integrating genetic maps and QTL towards discovery of candidate genes. Bioinformatics 20:2324–2326
Ballini E, Morel JB, Droc G et al (2008) A genome-wide meta-analysis of rice blast resistance genes and quantitative trait loci provides new insights into partial and complete resistance. Mol Plant Microbe Interact 21:859–868
Bernardo R, Charcosset A (2006) Usefulness of gene information in marker-assisted recurrent selection: a simulation appraisal. Crop Sci 46:614–621
Brem RB, Yvert G, Clinton R et al (2002) Genetic dissection of transcriptional regulation in budding yeast. Science 296:752–755
Cai HW, Morishima H (2002) QTL clusters reflect character associations in wild and cultivated rice. Theor Appl Genet 104:1217–1228
Chardon F, Virlon B, Moreau L et al (2004) Genetic architecture of flowering time in maize as inferred from quantitative trait loci meta-analysis and synteny conservation with the rice genome. Genetics 168:2169–2185
Cheung VG, Conlin LK, Weber TM et al (2003) Natural variation in human gene expression assessed in lymphoblastoid cells. Nat Genet 33:422–425
Courtois B, Ahmadi N, Khowaja F et al (2009) Rice root genetic architecture: meta-analysis from a QTL database improves resolution to a few candidate genes. Rice 2:115–128
Darvasi A, Soller M (1997) A simple method to calculate resolving power and confidence interval of QTL map location. Behav Genet 27:125–132
Drake TA, Schadt EE, Lusis AJ (2006) Integrating genetic and gene expression data: application to cardiovascular and metabolic traits in mice. Mamm Genome 17:466–479
Gavidia I, Perez-Bermudez P, Seitz HU (2002) Cloning and expression of two novel aldo-keto reductases from Digitalis purpurea leaves. Eur J Biochem/FEBS 269:2842–2850
Glass G (1976) Primary, secondary, and meta-analysis of research. Educ Res 5:3–8
Goffinet B, Gerber S (2000) Quantitative trait loci: a meta-analysis. Genetics 155:463–473
Gyenis L, Yun SJ, Smith KP et al (2007) Genetic architecture of quantitative trait loci associated with morphological and agronomic trait differences in a wild by cultivated barley cross. Genome 50:714–723
Hanocq E, Laperche A, Jaminon O et al (2007) Most significant genome regions involved in the control of earliness traits in bread wheat, as revealed by QTL meta-analysis. Theor Appl Genet 114:569–584
Hedges L, Olkin I (1985) Statistical methods for meta-analysis. Academic, Orlando
Hegedus A, Erdei S, Janda T et al (2004) Transgenic tobacco plants overproducing alfalfa aldose/aldehyde reductase show higher tolerance to low temperature and cadmium stress. Plant Sci 166:1329–1333
Hemamalini GS, Shashidha HE, Hittalmani S (2000) Molecular marker assisted tagging of morphological and physiological traits under two contrasting moisture regimes at peak vegetative stage in rice (*Oryza sativa* L.). Euphytica 112:69–78
Hruz T, Laule O, Szabo G et al (2008) Genevestigator v3: a reference expression database for the meta-analysis of transcriptomes. Adv Bioinform 2008:420747
International Rice Genome Sequencing Project (2005) The map-based sequence of the rice genome. Nature 436:793–800
Jaiswal P, Ni J, Yap I et al (2006) Gramene: a bird's eye view of cereal genomes. Nucleic Acids Res 34:D717–D723
Jansen RC, Nap JP (2001) Genetical genomics: the added value from segregation. Trends Genet 17:388–391
Kao CH, Zeng ZB, Teasdale RD (1999) Multiple interval mapping for quantitative trait loci. Genetics 152:1203–1216
Kearsey MJ, Farquhar AG (1998) QTL analysis in plants; where are we now? Heredity 80:137–142
Khowaja FS, Norton GJ, Courtois B et al (2009) Improved resolution in the position of drought-related QTLs in a single mapping population of rice by meta-analysis. BMC Genomics 10:276

Lander ES, Botstein D (1989) Mapping mendelian factors underlying quantitative traits using RFLP linkage maps. Genetics 121:185–199
Lee SP, Chen TH (1993) Molecular cloning of abscisic acid-responsive mRNAs expressed during the induction of freezing tolerance in bromegrass (Bromus inermis Leyss) suspension culture. Plant Physiol 101:1089–1096
Lee SY, Ahn JH, Cha YS et al (2006) Mapping of quantitative trait loci for salt tolerance at the seedling stage in rice. Mol Cells 21:192–196
Lee HS, Sasaki K, Higashitani A et al (2012) Mapping and characterization of quantitative trait loci for mesocotyl elongation in rice (*Oryza sativa* L.). Rice 5:13
Li J, Burmeister M (2005) Genetical genomics: combining genetics with gene expression analysis. Hum Mol Genet 14(Spec No. 2):R163–R169
Li Q, Li L, Yang X et al (2010) Relationship, evolutionary fate and function of two maize co-orthologs of rice GW2 associated with kernel size and weight. BMC Plant Biol 10:143
Li WT, Liu CJ, Liu YX et al (2013) Meta-analysis of QTL associated with tolerance to abiotic stresses in barley. Euphytica 189:31–49
Liang C, Jaiswal P, Hebbard C et al (2008) Gramene: a growing plant comparative genomics resource. Nucleic Acids Res 36:D947–D953
Lynch M, Walsh B (1997) Genetics and analysis of quantitative traits. Sinauer Assoc, Sunderland
Marone D, Russo MA, Laido G et al (2013) Genetic basis of qualitative and quantitative resistance to powdery mildew in wheat: from consensus regions to candidate genes. BMC Genomics 14:562
Matsuda H, Taniguchi Y, Iwaisaki H (2012) QTL/microarray approach using pathway information. Algorithm Mol Biol 7:1
Matthews DB, Bhave SV, Belknap JK et al (2005) Complex genetics of interactions of alcohol and CNS function and behavior. Alcohol Clin Exp Res 29:1706–1719
Miura K, Ashikari M, Matsuoka M (2011) The role of QTLs in the breeding of high-yielding rice. Trends Plant Sci 16:319–326
Nguyen VT, Nguyen BD, Sarkarung S et al (2002) Mapping of genes controlling aluminum tolerance in rice: comparison of different genetic backgrounds. Mol Genet Genomics 267:772–780
Ni J, Pujar A, Youens-Clark K et al (2009) Gramene QTL database: development, content and applications. Database (Oxford) 2009:bap005
Oberschall A, Deak M, Torok K et al (2000) A novel aldose/aldehyde reductase protects transgenic plants against lipid peroxidation under chemical and drought stresses. Plant J Cell Mol Biol 24:437–446
Qi ZM, Sun YN, Wu QO et al (2011) A meta-analysis of seed protein concentration QTL in soybean. Can J Plant Sci 91:221–230
Rosenberg MS, Garrett KA, Su Z et al (2004) Meta-analysis in plant pathology: synthesizing research results. Phytopathology 94:1013–1017
Said JI, Lin Z, Zhang X et al (2013) A comprehensive meta QTL analysis for fiber quality, yield, yield related and morphological traits, drought tolerance, and disease resistance in tetraploid cotton. BMC Genomics 14:776
Schadt EE, Monks SA, Drake TA et al (2003) Genetics of gene expression surveyed in maize, mouse and man. Nature 422:297–302
Song XJ et al (2007) A QTL for rice grain width and weight encodes a previously unknown RING-type E3 ubiquitin ligase. Nat Genet 39:623–630
Sosnowski O, Charcosset A, Joets J (2012) BioMercator V3: an upgrade of genetic map compilation and quantitative trait loci meta-analysis algorithms. Bioinformatics 28:2082–2083
Swamy BPM, Vikram P, Dixit S et al (2011) Meta-analysis of grain yield QTL identified during agricultural drought in grasses showed consensus. BMC Genomics 12
Tanksley SD (1993) Mapping polygenes. Annu Rev Genet 27:205–233
Teng S, Qian Q, Zeng DL et al (2002) Analysis of gene loci and epistasis for drought tolerance in seedling stage of rice (*Oryza sativa* L.). Acta Genet Sin 29:235–240

Verdugo RA, Farber CR, Warden CH et al (2010) Serious limitations of the QTL/Microarray approach for QTL gene discovery. BMC Biol 8:96

Veyrieras JB, Goffinet B, Charcosset A (2007) MetaQTL: a package of new computational methods for the meta-analysis of QTL mapping experiments. BMC Bioinform 8:49

Wang X, Wang H, Long Y et al (2013) Identification of QTLs associated with oil content in a high-oil Brassica napus cultivar and construction of a high-density consensus map for QTLs comparison in *B. napus*. PLoS One 8, e80569

Ware DH, Jaiswal P, Ni J et al (2002) Gramene, a tool for grass genomics. Plant Physiol 130:1606–1613

Wayne ML, McIntyre LM (2002) Combining mapping and arraying: an approach to candidate gene identification. Proc Natl Acad Sci U S A 99:14903–14906

Wissuwa M, Wegner J, Ae N et al (2002) Substitution mapping of Pub1: a major QLT increasing phosphorus uptake of rice from a phosphorus-deficient soil. Theor Appl Genet 105:890–897

Xiang K, Zhang ZM, Reid LM et al (2010) A meta-analysis of Qtl associated with ear rot resistance in maize. Maydica 55:281–290

Yvert G, Brem RB, Whittle J et al (2003) Trans-acting regulatory variation in Saccharomyces cerevisiae and the role of transcription factors. Nat Genet 35:57–64

Zamir D (2001) Improving plant breeding with exotic genetic libraries. Nat Rev Genet 2:983–989

Zeng ZB (1994) Precision mapping of quantitative trait loci. Genetics 136:1457–1468

Zhang J, Zheng HG, Aarti A et al (2001) Locating genomic regions associated with components of drought resistance in rice: comparative mapping within and across species. Theor Appl Genet 103:19–29

Chapter 4
Marker-Assisted Breeding

Jae Bok Yoon, Soon-Wook Kwon, Tae-Ho Ham, Sunggil Kim, Michael Thomson, Sherry Lou Hechanova, Kshirod K. Jena, and Younghoon Park

Abstract Marker-assisted breeding (MAB) is based on the theories and technology of molecular biology to improve the efficiency of conventional breeding. Molecular markers can be applied to marker-assisted selection (MAS), which capitalizes on genetic linkage between target gene(s) or quantitative trait loci (QTLs) and phenotypes. Constructing a molecular genetic map and detecting the target trait-linked markers should be performed prior to MAS; for quantitative traits, the fundamental procedure is to detect DNA markers linked to the quantitative trait via QTL mapping. Molecular markers can be also used for backcross breeding, which is widely used to transfer a single agronomically-important trait from donor plants into elite recipient plants in a relatively short time. Cost-effective high-throughput assays for broad usage of marker-assisted backcrossing (MABC) systems must be developed,

---

Author contributed equally with all other contributors.

J.B. Yoon
R&D Unit, Pepper & Breeding Institute, Suwon, Republic of Korea
e-mail: yoonjb2@snu.ac.kr

S.-W. Kwon
Department of Plant Biosciences, Pusan National University, Miryang, Republic of Korea
e-mail: swkwon@pusan.ac.kr

T.-H. Ham
Department of Agricultural Sciences, Korea National Open University, Seoul, Republic of Korea
e-mail: kg780516@knou.ac.kr

S. Kim
Department of Plant Biotechnology, Biotechnology Research Institute, Chonnam National University, Gwangju, Republic of Korea
e-mail: dronion@jnu.ac.kr

M. Thomson • S.L. Hechanova • K.K. Jena
Plant Breeding, Genetics and Biotechnology Division, International Rice Research Institute, Metro Manila, Philippines
e-mail: m.thomson@irri.org; k.jena@irri.org

Y. Park (✉)
Department of Horticultural Biosciences, Pusan National University, Miryang, Republic of Korea
e-mail: ypark@pusan.ac.kr

H.-J. Koh et al. (eds.), *Current Technologies in Plant Molecular Breeding*,
DOI 10.1007/978-94-017-9996-6_4

and strategies for optimizing such systems have been proposed. In addition, MAB uses molecular markers to analyze the phylogenetic relationships of germplasm and in cultivar identification for property rights protection. In this chapter, the general application of molecular markers in plant breeding is explained, with a focus on MAS, MABC, phylogenetic analysis, cultivar identification, introgression of genes from exotic germplasm, and techniques for high-throughput marker genotyping. Several case studies show practical applications in different crop species.

4.1 Application of Molecular Markers to Plant Breeding

4.1.1 Concept of Molecular Marker-Assisted Breeding (MAB)

Conventional breeding is the process of intentional selection and hybridization among plants to develop new cultivars with improved agricultural traits (e.g., high yield, disease resistance, high sugar content, and chilling tolerance). In contrast, marker-assisted breeding (MAB) takes advantage of the theories and technology of molecular biology, especially molecular markers, to guide breeding. In general plant breeding programs, phenotype screening and selection for superior plants within a large number of progeny derived from crosses are cyclically repeated; developing new cultivars in this way can take between 7 and 20 years, depending on plant species. The use of molecular markers improves the efficiency for each step of breeding in terms of labor and time. Major applications of molecular markers in breeding are:

- Assessing genetic diversity
- DNA fingerprinting and genotype screening
- Assessing genetic distance among populations, cultivars, and breeding materials
- Detecting qualitative trait loci (monogenic) and quantitative trait loci (QTLs)
- Marker-assisted selection (MAS)
- Identifying nucleotide sequences of useful candidate genes

Molecular markers used in breeding can identify specific phenotypes or genotypes and are referred to as "tags". Molecular markers are stably inherited from generation to generation. Generally, markers are classified as phenotypic (morphological variation scorable on an individual plant), biochemical (variation in size or net charge of a specific protein, or variation in chemical composition of a metabolite), or molecular (variation in DNA sequence). In current MAB, DNA markers are most widely used and have the following advantages:

- No environmental effects
- No limits to the number of markers that can be developed
- Easier to analyze and more objective than other marker types
- Lower costs for large-scale sample analysis than other marker types.

A typical application of molecular markers is marker-assisted selection (MAS), which is based on the concept of genetic linkage. When two independent loci or alleles are located in proximity on a chromosome, they have a tendency to be inherited together. By observing the frequency of segregation in a progeny population (F2) derived from a cross, the genetic distance between these loci can be calculated in centimorgans (cM). In MAS, markers that are tightly linked to the genes for a specific trait (<0.1 cM) can be used to select target genes via a marker genotyping assay.

4.1.2 Advantages of Marker-Assisted Breeding

4.1.2.1 Faster Breeding Process

As molecular markers that are linked to useful traits are developed, they can facilitate the detection of genetically superior individual plants. DNA samples can be purified from cotyledons or true leaves at the early-seedling stage, so marker analysis is conducted much earlier than in conventional breeding of plants that must phenotypically express the traits before they can be assessed, which occurs in the adult stage for most traits of interest. Therefore, molecular markers can reduce the required time for selecting plants with a specific trait by several months or more. Most breeding programs require multiple consecutive generations and, especially for trees with long life cycles, breeding can be shortened by several years.

4.1.2.2 Less Expensive Breeding for Certain Traits

Field sites and greenhouse space are costly to maintain, and labor is needed for phenotype assays. Furthermore, for some types of plant diseases, a phenotype assay is not possible. Thus, expenses can be greatly reduced by detecting the presence of an allele linked to a trait using molecular markers rather than using a phenotypic screening.

4.1.2.3 Not Affected by the Environment

With MAS, targeted traits can be selected regardless of cultivation conditions (place and climate), as molecular markers are not affected by environmental factors. Additionally, newly developed cultivars can be traced and appraised via unique genetic fingerprints (cultivar-specific molecular markers). Using molecular markers also enables in-depth studies of how target traits and particular genes interact. If a molecular marker is developed from the gene directly controlling a trait, more efficient selection can occur. For example, the marker can be used for gene cloning and studying the gene's activity. In tomato, a major QTL affecting fruit weight has been cloned and its role in controlling the number of cells in carpels early in fruit maturation has been determined.

4.1.3 Disadvantages of Marker-Assisted Breeding Versus Conventional Breeding

4.1.3.1 High Initial Investment

The use of molecular markers requires specific experimental instruments, such as a PCR machine, gel electrophoresis apparatus, and gel imager for gel-based markers or high-throughput genotyping systems for single nucleotide polymorphism (SNP) markers. Therefore, setting up a MAB laboratory may be expensive, but savings in the subsequent breeding program can offset the costs in the long term. Furthermore, the costs per sample for marker genotyping are declining.

4.1.3.2 Requires Special Techniques

Along with the experimental instruments needed for molecular marker work, knowledge of laboratory techniques and data analyses are required. Although acquiring this knowledge is not difficult, MAB is not possible without additional education and training beyond that needed for conventional breeding.

4.1.4 Properties of Molecular Markers Required for MAB

The types of molecular markers are diverse, and which type should be used in a breeding program is associated with several factors, including:

- The goal of the breeding program
- The genetic diversity of the germplasm
- The properties of the population to be evaluated
- The level of accuracy required for marker assays
- The existence of previous research findings.

Known molecular markers have different properties depending on their types, and fundamental characteristics of the markers that are desirable for MAB are as follows:

- Polymorphism
- Reproducibility
- Even distribution in the genome (not clustered in a specific genomic region)
- Affordable price
- Convenient to assay
- Codominance (heterozygosity must be discernible from homozygosity).

Another consideration in deciding marker type is the instrument platform to be used to run markers. Platforms depend on the marker type and include agarose gels, polyacrylamide gels, sequencing equipment, SNP genotyping systems, and real-time PCR.

4.2 Marker-Assisted Selection

DNA marker technology has led to groundbreaking changes in plant genetics and breeding. There are various ways to apply DNA markers in plant breeding, most notably MAS for developing new varieties. MAS is a selection strategy for individual plants that contain a molecular marker that is closely linked to a target gene(s) or QTL; thus, the selection criterion is not based on the phenotype but on the genotype of the target gene linked marker(s).

4.2.1 Marker Development for MAS

The first step in applying MAS for plant breeding is to develop and detect the DNA markers closely linked to the target traits. Thus, constructing a molecular genetic map and detecting linked markers should be performed prior to MAS.

In rice, a morphological genetic map for 12 linkage groups with 185 phenotype markers has been constructed over the last nine decades. In 1988, a genetic map was constructed with 135 restriction fragment length polymorphism (RFLP) markers by a Cornell University group (McCouch et al. 1998). In 2001, a genetic map of 3,267 DNA markers was reported in Japan (http://rgp.dna.affrc.go.jp/publicdata/geneticmap2000), and recently-developed next-generation sequencing (NGS) technology has expedited the construction of high density fine maps.

For quantitative traits to be applied in MAS, the basic procedure is to conduct QTL mapping to detect DNA markers linked to the trait. The results of QTL mapping are affected by the type and size of the mapping population, the error in collecting and evaluating phenotype or genotype data, and environmental effects. Therefore, the candidate DNA markers obtained through QTL mapping must be tested before applying them to MAS (Collard and Mackill 2008) (Fig. 4.1).

In rice breeding, there have been many reports of MAS since the development of DNA markers (Table 4.1).

Recently developed NGS technology can be applied in various molecular biology fields. In particular, high-density molecular genetic maps and association analyses with more candidate genes can be conducted to improve breeding efficiency with more accurate DNA markers.

Huang et al. (2009) reported that the multiplex re-sequencing method allowed them to construct a high resolution genetic map with average intervals of ~40 Kbp using 150 recombinant inbred lines (RILs) derived from across between two *Oryza sativa* subspecies, *indica* and *japonica* (Fig. 4.2).

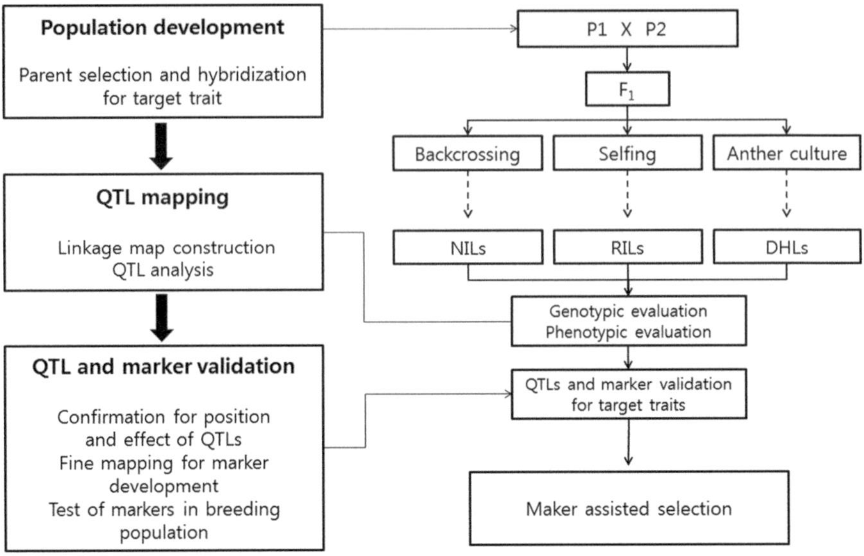

Fig. 4.1 Summary of marker-assisted selection

Takagi et al. (2013) reported that they were able to detect flanking single nucleotide polymorphism (SNP) markers linked with quantitative traits by applying whole genome re-sequencing to bulk DNA samples composed of pooled DNA from individual plants showing opposite traits. The SNP index (defined as the ratio between the number of reads of a mutant SNP and the total number of reads corresponding to the SNP) obtained by comparing the re-sequenced and reference genomes was used to detect flanking SNP markers (Fig. 4.3).

4.2.2 *Application of MAS*

4.2.2.1 Accumulating Useful Genes

In rice, 8,648 QTLs for agronomic traits had been reported in the Gramene database (http://archive.gramene.org/qtl/) as of January 2015. Most QTLs were for yield and plant morphological type—for example, 1,011 for plant height, 618 for heading date, and 353 for spikelet number. QTLs for biotic stress are mainly used in MAS to select offspring derived by crossing parents with different resistance genes. By screening DNA markers for all the resistance genes from the parents, new varieties with multiple resistance genes can be developed.

In general, the accumulation of resistance genes via MAS is performed by backcross breeding and gene pyramiding. Usually, an elite commercial variety with insufficient resistance genes serves as the recurrent parent and other parents with specific resistance gene(s) are used as donor parents. PR106, an *indica* rice cultivar,

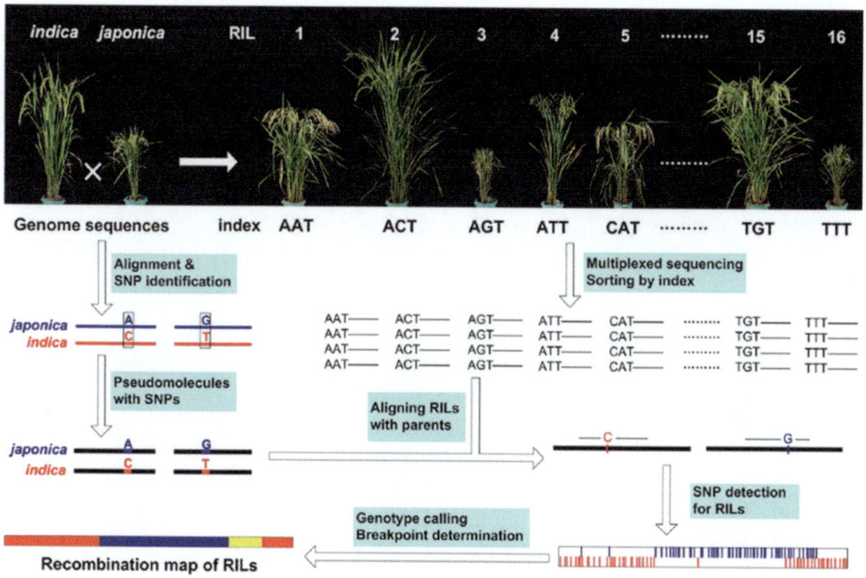

Fig. 4.2 Genotype analysis based on next-generation sequencing (Huang et al. 2009)

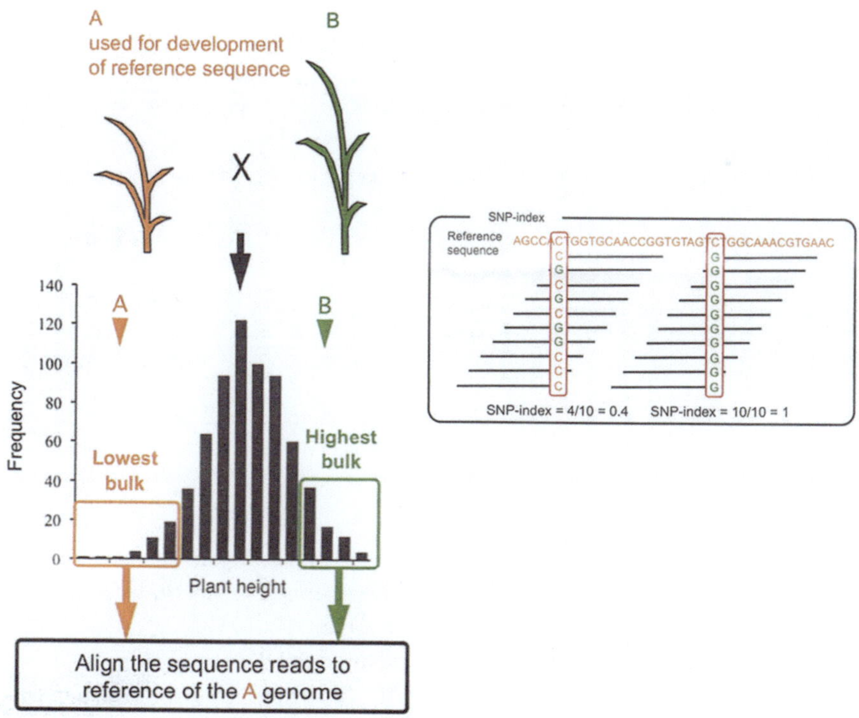

Fig. 4.3 Quantitative trait locus sequencing (QTL-seq) in rice (Takagi et al. 2013)

Table 4.1 Application of marker-assisted selection to rice breeding

Gene/QTLs	Traits or germplasm	Marker used	Application	Reference
Pi5(t)	Blast disease, predom.	PCR/DNA gel blot	Gene surveys in parental material	Yi et al. (2004)
Pi-z	Blast disease	SSR	Gene surveys in parental material	Fjellstrom et al. (2006)
Pi-ta	Blast disease	Gene-specific marker	Gene surveys in parental material	Wang et al. (2007)
Pi1	Blast	SSR and ISSR	MABC	Liu et al. (2003)
xa5	Bacterial blight	STS	MABC	Toenniessen et al. (2003)
xa5, xa13, Xa21	Bacterial blight	STS	MABC	Kottapalli et al. (2010)
xa13, Xa21	Bacterial blight + quality	STS, SSR, and AFLP	MABC	Joseph et al. (2004)
Sub QTL, *Xa21*, *Bph* and blast	Submergence tolerance, disease resistance, quality	SSR and STS	MABC	Toojinda et al. (2005)
Sub1 QTL	Submergence tolerance	SSR	MABC	Neeraja, et al. (2007)
QTLs for *Hd1*, *Hd4*, *Hd5*, or Hd6)	Heading date	RFLP, STS, SSR, CAPS, dCAPs	MABC	Takeuchi et al. (2006)
Waxy	Quality	RFLP and AFLP	MABC	Zhou et al. (2003)
Xa4, xa5, Xa10	Bacterial blight	RFLP and RAPD	Pyramiding	Yoshimura et al. (1995)
Pi1, Piz-5, Pi2, Pita	Blast disease	RFLP, STS	Pyramiding	Hittalmani et al. (2000)
xa5, xa13, *Xa21*	Bacterial blight	STS and CAPS	Pyramiding	Sanchez et al. (2000)
xa5, xa13, Xa21	Bacterial blight	SSR and STS	Pyramiding	Davierwala et al. (2001)
xa5, xa13, Xa21, Wx	Bacterial blight and waxy genes	SSR, STS, and CAPS	Pyramiding	Ramalingam et al. (2002)
Xa21 and *Bt*	Insect resistance and bacterial blight	STS	Pyramiding	Jiang et al. (2004)
Xa7 and *Xa21*	Bacterial blight	STS	Pyramiding	Zhang et al. (2006a)
Xa4, Xa7, and *Xa21*	Bacterial blight	STS	Pyramiding	Perez et al. (2008)
tms2, tgms, tms5	Thermosensitive genic male sterility (TGMS) genes	SSR	Pyramiding	Nas et al. (2005)
Pi-z and Xa21	Blast and bacterial blight	STS	Pyramiding/transgene selection	Narayanan et al. (2002)
xa5, xa13, Xa21	Bacterial blight	STS	Pyramiding/transgene selection	Swamy et al. (2006)

was developed through backcross breeding by accumulating the resistance genes *xa5*, *xa13*, and *Xa21* against bacterial blight disease. In the Philippines, *Xa4*, *xa5*, and *Xa21* resistance genes were introduced into the elite *indica* variety IR64 (Singh et al. 2001). For the efficient gene accumulation, near-isogenic lines (NILs) that share same genomic background can be used (such as the IRBB lines containing bacterial blight resistance genes, Ogawa et al. 1991).

Ogawa et al. (1991) accumulated several resistance genes (*Xa1*, *Xa2*, *Xa3*, *xa5*, *Xa7*, *xa8*, *Xa10*, and *Xa11*) in different genomic backgrounds such as IR24 (*indica*), Milyang 23 (*indica*/*japonica*), and Toyonishike (*japonica*). In Korea, similar progress was reported by Shin et al. (1994, 2000) in accumulating *Xa1*, *Xa2*, *Xa3*, *Xa22*, and *Xa7* into Suwon 345. Recently, Kim et al. (2011) developed NILs that share a common genomic background with the addition of one resistance gene. There were three resistance genes: *xa5* for resistance against the blight races K1, K2, and K3 and partial residence against K3a; *Xa4* for resistance against K3a; and *Xa21* for susceptibility to K1 but resistance against K2, K3, and K3a. The resistance-linked DNA markers and pathogen inoculation were used to select monogenic lines for *Xa4* in BC_4F_3, *xa5* in BC_3F_{10}, and *Xa21* in BC_5F_7 (Fig. 4.4).

Suh et al. (2013) accumulated the bacterial leaf-blight resistance genes *Xa4*, *xa5*, and *Xa21* in the susceptible variety Mangeum. The linked markers were MP1 and MP2 for *Xa4*, 10603 and T10Dw (digested with *Rsa*I) for *xa5*, and U1 and I1 for Xa21. Mangeum was crossed to IRBB57, containing *Xa4*, *xa5*, and *Xa21*, to produce 288 BC_1F_1 plants. Twenty-eight plants with the linked markers were selected. The same markers were applied to BC_2F_1 and BC_3F_1 to obtain 42 BC_3F_1 plants, which were self-pollinated to produce 1,260 BC_3F_2. The marker genotypes and disease resistance after pathogen inoculation were evaluated in subsequent generations to develop three lines containing all three genes (*Xa4*, *xa5*, and *Xa21*) in BC_3F_5 (Fig. 4.5).

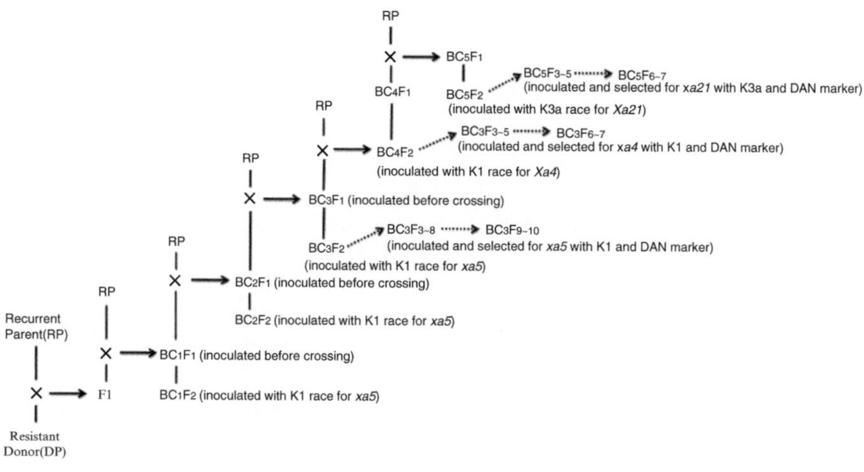

Fig. 4.4 Procedure for developing bacterial leaf blight resistant near-isogenic lines

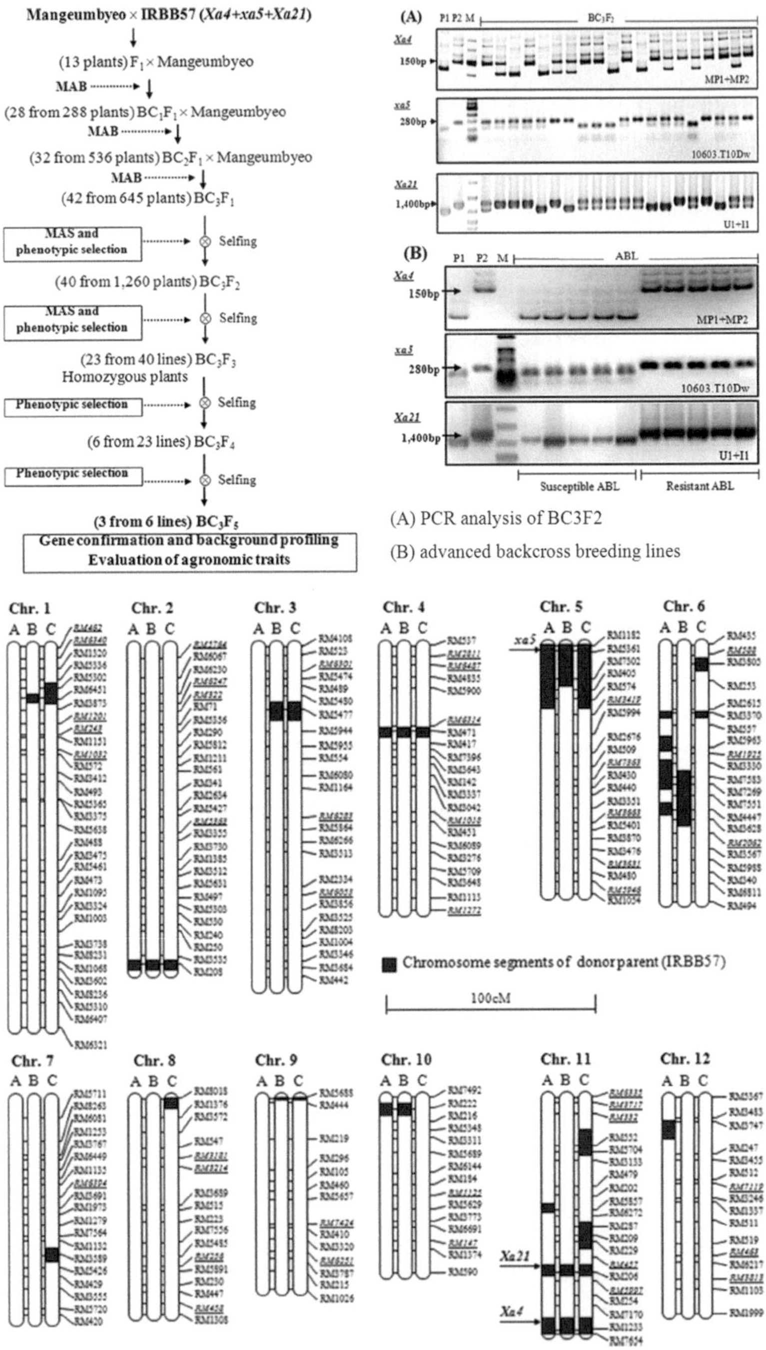

Fig. 4.5 Accumulating *Xa4*, *xa5*, and *Xa21* by marker-assisted selection

4.2.2.2 Selection for Target Traits in Early Generations

In pedigree breeding, MAS can be used for selection in early generations, especially F2 or F3. Using MAS, many individuals or lines with unwanted genotypes can be discarded early in the process, saving labor, breeding space, and the costs of trait evaluation (Collard and Mackill 2008).

For early-generation selection of rice eating quality in *japonica* rice, Lestari et al. (2009) constructed a multiple regression model between markers and palatability, which was measured by the Toyo taste meter and sensory test. The R^2 of the multiple regression analysis was 0.990 (Fig. 4.6), indicating significant correlation between the genotypes and palatability. Evaluating eating quality requires many grains, time, and expense, so MAS is particularly useful for this type of trait. The results of Lestari et al. (2009) showed that DNA markers can be used in early-generation selection for rice eating quality.

	palatability by Toyo taste meter (P)			palatability by sensory tast (ST)		
PCR primer	parameter estimate	*t* value	R^2	parameter estimate	*t* value	R^2
G4	−16.97 ± 1.19	−14.22**	0.087	−1.20 ± 0.06	−21.77**	0.212
M11	−1.94 ± 0.60	−3.25**	0.096	−0.14 ± 0.03	−5.03**	0.010
E30	26.55 ± 0.83	32.12**	0.104	0.86 ± 0.04	19.62**	0.059
M2CG	−2.40 ± 0.56	−4.33**	0.060	−0.38 ± 0.03	−15.21**	0.041
GPA	−21.14 ± 1.11	−19.12**	0.129	−0.82 ± 0.05	−17.28**	0.021
S3cl	−1.62 ± 0.62	−2.60*	0.017	−0.38 ± 0.03	−13.41**	0.005
P5	19.01 ± 1.32	14.44**	0.307	1.09 ± 0.04	26.81**	0.288
B1	6.42 ± 0.77	8.30**	0.047	0.41 ± 0.03	12.90**	0.015
CBG	13.45 ± 1.11	12.12**	0.087	0.68 ± 0.07	10.27**	0.228
J6	3.87 ± 0.74	5.21**	0.083			
WK9	2.62 ± 0.59	4.42**	0.003			
A7	−12.33 ± 1.27	−9.72**	0.031			
AMs	−8.72 ± 1.56	−5.58**	0.021			
G81				0.27 ± 0.03	10.41**	0.021
F6				0.32 ± 0.03	10.36**	0.033
SSIla				−0.27 ± 0.03	−8.06**	0.001
G28				0.33 ± 0.04	8.81**	0.005
AcPh				−0.48 ± 0.04	−11.44**	0.060
intercept	76.66 + 2.71	28.29**		−0.54 ± 0.11	−4.74**	
total			0.990			0.990
eq	*Y* = 76.66 − 16.97(G4) − 1.94(M11) + 26.55(E30) − 2.40(M2CG) − 21.14(GPA) − 1.62(S3cl) + 19.01(P5) + 6.42(B1) + 13.45(CBG) + 3.87(J6) + 2.62(WK9) − 12.33(A7) − 8.72(Ams)			*Y* = − 0.54 − 1.20(G4) − 0.14(M11) + 0.86(E30) − 0.38(M2CG) − 0.82(GPA) − 0.38(S3cl) + 1.09(P5) + 0.41(B1) + 0.68(CBG) + 0.27(G81) + 0.32(F6) − 0.27(SSIIa) + 0.33(G28) − 0.48(AcPh)		

[a] ** and *, significant at 1% and 5% level, respectively.

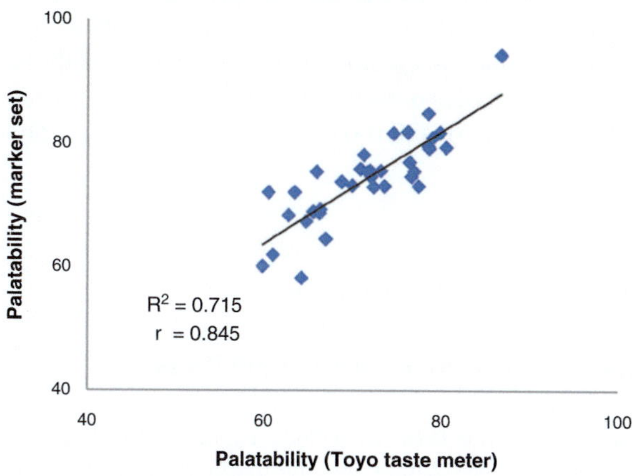

Fig. 4.6 Multiple regression between marker sets and eating quality

Kwon et al. (2011) reported that they successfully applied QTL markers for rice eating quality in different genomic backgrounds to other breeding populations. Three QTLs for eating quality were detected from QTL mapping with 190 RILs derived from the cross between Suwon365 (*japonica*) and the high-quality variety Chucheong (*japonica*) (Table 4.2). The QTL (*qGCR6*) originated from Suwon 365 and *qGCR7* and *qGCR8* originated from Chucheong. MAS was used to develop QTL-NILs (near isogenic lines) for these three QTLs to confirm their effects (Table 4.3). In backcross lines derived from the parent of the low-quality variety Palgong, the BC_1F_3 lines with higher quality could be selected with the flanking markers for the QTLs detected in the Suwon365 × Chucheong cross. In addition, the same markers could be used to select progeny from the backcross of the high-quality variety Hopyeong. The selected progeny showed improved eating quality (Table 4.4).

4.2.2.3 Developing Submergence-Resistant Rice by MAB

A major QTL (*Sub1*) has been identified on chromosome 9 in the submergence-tolerant cultivar FR13A. By backcross breeding, this QTL was successfully introduced into the variety Swarna (Vasista/Mahsuri) (Neeraja et al. 2007), which is widely cultivated in India and Bangladesh (Fig. 4.7).

In the breeding program, Swarna was crossed with IR40931-33-1-3-2, which has the submergence resistance gene *Sub1*, to produce F_1 seed. The F_1 plants were backcrossed to Swarna to produce 697 BC_1F_1 plants. The RM464A marker linked to *Sub1* at a 0.7-cM interval was used to select 376 plants. To discard recombinant plants, the RM219 marker (linked to *Sub1* at a 1.5 cM interval) was used to select 20 BC_1F_1 plants. These plants were backcrossed again to produce 320 BC_2F_1 plants. Using the same strategy for the BC_1F_1 plants, the markers RM464A and RM316 (1.5-cM interval with RM464A) were used to select two BC_2F_1 plants and produce BC_3F_1. One of BC_3F_1 plants, including *Sub1*, was selected and self-pollinated to produce 432 BC_3F_2 plants. After selection for the genomic background of the recurrent parent (Swarna) with 32 unlinked SSR markers (background selection), the progeny containing *Sub1* in the genomic background of Swarna could finally be selected. Septiningsih et al. (2009) introduced *Sub1* using the backcross strategy into the varieties Samba Mahsuri (India), CR1009 (India), IR64 (IRRI), and Thadokkham 1 (TDK1; Laos).

4.2.3 *Marker-Assisted Selection Case Studies*

4.2.3.1 Molecular Markers for Disease Resistance in Chili Pepper

Many molecular markers associated with disease resistance, especially to viral diseases, have been developed in chili pepper. Allele-specific markers of the *L* locus, which is the Tomato Mosaic Virus (TMV) resistance gene, were developed and used

Table 4.2 Quantitative trait locus analysis for eating quality of 190 recombinant inbred lines

Trait	QTLs	Chr.	Flanking marker	Year	Interval mapping			Composite interval mapping			Positive parent[a]
					LOD	RSq(%)	Add.	LOD	RSq(%)	Add.	
Glossiness of cooked rice (GCR)	*qGCR6*	6	RM589-RM253	2006	3.80	9.8	1.25	7.90	17.6	1.69	S
				2007	–	–	–	3.80	6.5	0.96	
	qGCR7	7	RM8261-RM3555	2006	6.04	14.1	1.50	–	–	–	C
				2007	4.09			–	–	–	
	qGCR8	8	RM5556-RM547	2006	11.44	31.2	2.20	14.44	32.9	2.28	C
				2007	9.87	22.4	1.74	10.92	19.2	1.68	

[a] *S* Suwon365, *C* Chucheongbyeo

Table 4.3 Confirming quantitative trait locus effects with QT-NILs (quantitative trait-near isogenic lines)

	Genotype on QTLs			2007		2008	
Lines[a]	*qGCR6*	*qGCR7*	*qGCR8*	GCR	OE	GCR	OE
SCQ04	CC	SS	CC	80.5	0.2	79.8	0.2
SCQ07	SS	CC	SS	75.8	−0.4	73.3	−0.2
SCQ14	CC	SS	SS	66.2	−1.4	68.3	−1.2
Suwon365				67.1	−1.2	64.6	−1.0
Chucheongbyeo				75.8	0.0	74.5	0.0

[a]BC_3F_4 in 2007 and BC_3F_5 in 2008 from a Suwon365*4 /Chucheongbyeo

Table 4.4 Developing quantitative trait locus introgression BC_1F_3 lines for eating quality with marker-assisted backcrossing

	QTLs[b]					GCR	
Lines[a]	*qGCR6*	*qGCR7*	*qGCR8*	Gen.	No. of line[c]	Mean	Range
Palgongbyeo*2/ SR23577-F_{12}-17	PP	CC	CC	BC_1F_3	4(3)	78.8	69.7–84.8
Palgongbyeo*2/ SR23577-F_{12}-100	PP	PP	CC	"	12(6)	71.8	63.7–80.5
Palgongbyeo*2/ SR23577-F_{12} 108	SS	CC	PP	"	10(2)	68.9	65.0–78.2
Palgongbyeo*2/ SR23577-F_{12} 118	PP	CC	CC	"	17(9)	72.9	64.1–79.4
Palgongbyeo*2/ SR23577-F_{12} 147	SS	CC	CC	"	8(2)	68.4	62.9–74.6
Palgongbyeo*2/ SR23577-F_{12} 154	SS	CC	CC	"	5(3)	73.2	68.2–76.8
Palgongbyeo*2/ SR23577-F_{12} 196	PP	PP	CC	"	18(8)	72.7	65.8–81.2
Hopyeongbyeo*2/ SR23577-F_{12}-17	HH	CC	CC	"	16(12)	79.8	69.7–88.7
Chucheongbyeo	CC	CC	CC			72.4	
Suwon365	SS	SS	SS			66.5	
Palgongbyeo	PP	PP	PP			64.5	
Hopyeongbyeo	HH	HH	HH			73.6	

[a]SR23577 is the number of a cross between Suwon365 and Chucheongbyeo
[b]*CC* Chucheongbyeo, *HH* Hopyeongbyeo, *PP* Palgongbyeo, *SS* Suwon365
[c]The numbers in the parentheses indicate the number of lines that had higher GCR values than that of Chucheongbyeo

in Korea (Lee et al. 2012a; Yang et al. 2012). The *L* locus has five different alleles: L^+, L^1, L^2, L^3 and L^4. L^+ is the susceptible allele, and the others are resistance alleles to different pathotypes of TMV. Markers linked to the resistance genes to potyvirus, such as *pvr1* (=*pvr2*, eIo4E), *pvr6* (eIF[iso]4E), and *Pvr4*, are known (Yeam et al. 2005; Ruffel et al. 2006; Kim et al. 2011). Kang et al. (2010) reported the *Cmr1-linked* marker resistant to Cucumber Mosaic Virus (CMV), and Kim et al. (2008)

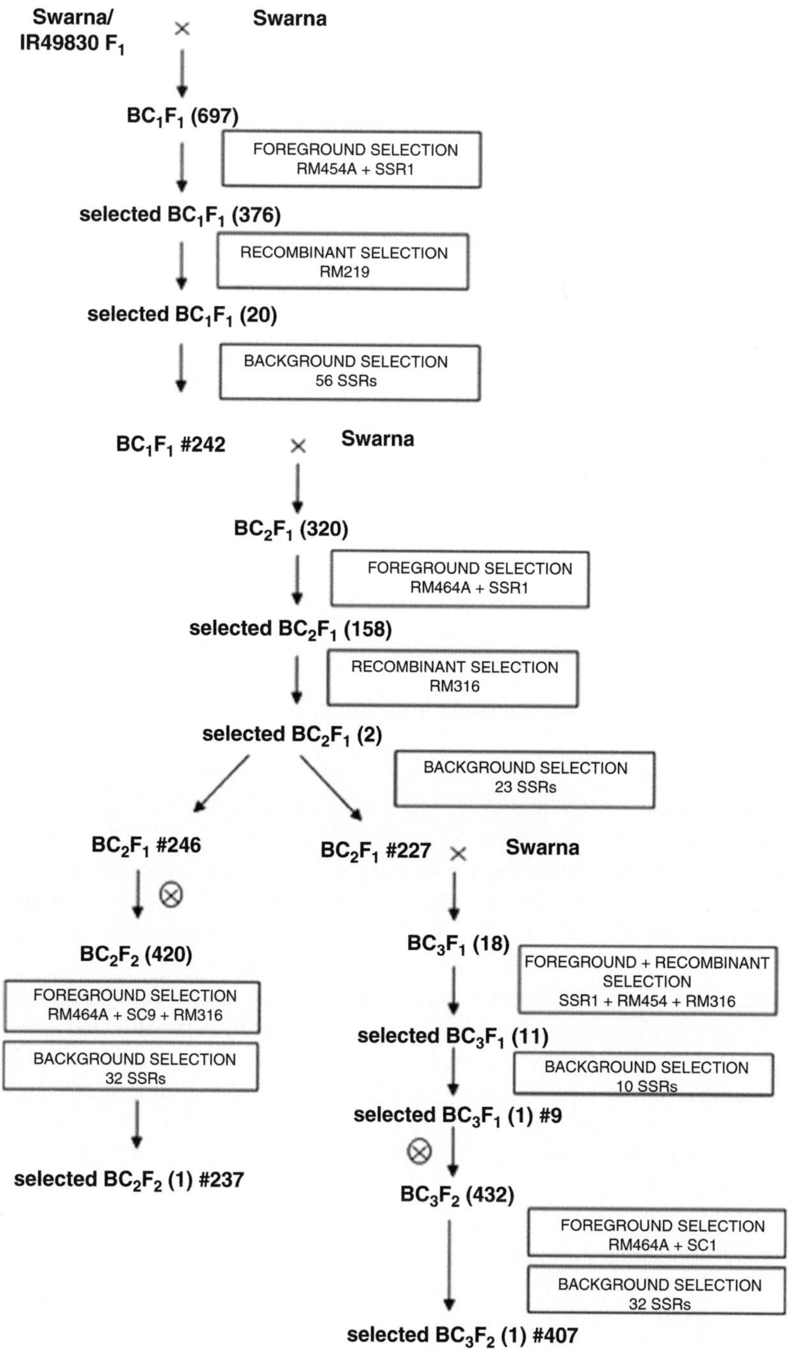

Fig. 4.7 Development of the submergence-tolerant Swarna-*Sub1* with details of markers used for foreground, recombinant, and background selection (Neeraja et al. 2007)

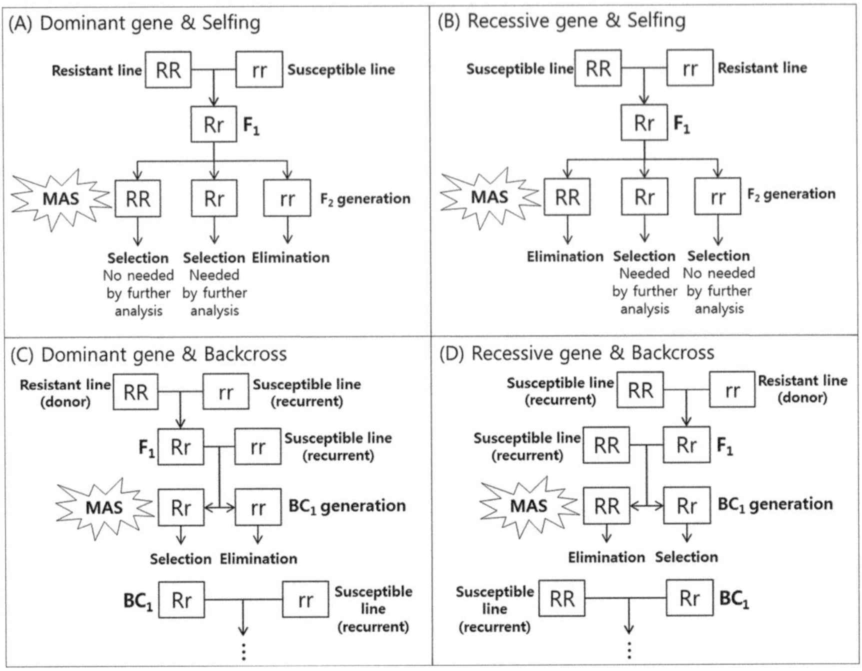

Fig. 4.8 Strategies for using molecular markers according to gene mode and breeding method

identified the *Tsw1*-linked marker resistant to Tomato Spot Wilt Virus (TSWV). In addition to the molecular markers for resistance to viral diseases, there are several gene-based markers derived from the cloning of *BS2* and *BS3* genes responsible for resistance to bacterial spot disease (Tai et al. 1999; Römer et al. 2007). Molecular markers linked to the major loci resistant to *Phytophthora* root rot (Lee et al. 2012b) and anthracnose (Lee et al. 2011) have also been developed.

Plant selection using molecular markers linked to the disease resistance depends on the inheritance modes and breeding methods. For a self-pollinated line, the homozygous resistant plants (dominant RR or recessive rr) are simply selected in early generation (Fig. 4.8a, b), after which the marker is no longer needed. However, with backcross breeding, heterozygous plants (Rr) must be selected in every generation and used as donor parents for consecutive backcrossing (Fig. 4.8c, d).

Practical Application of the *Cmr1*-linked Marker

DNA Extraction

1. Fresh leaf discs were ground with two beads in a Tissue-Lyser (Qiagen, Venlo, Netherlands).
2. DNA was extracted from the ground tissue by the CTAB method (Kang et al. 2010).

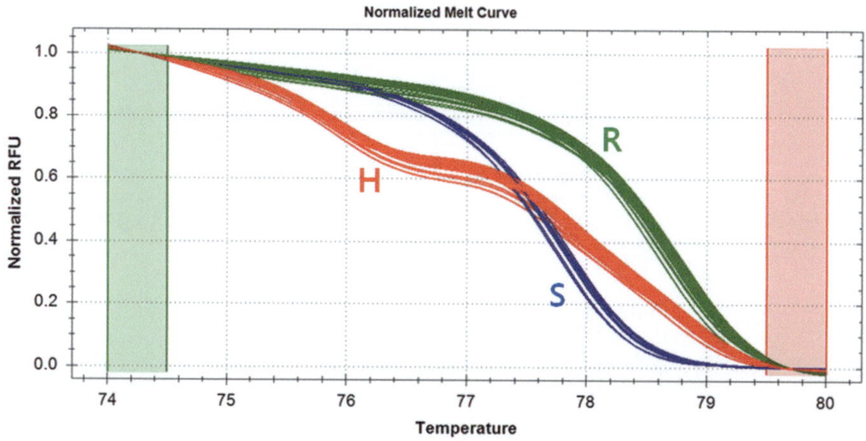

Fig. 4.9 A graph for normalized high resolution melting (HRM) curve. R, RR (homozygous resistant); H, Rr (heterozygous resistant); S, rr (susceptible)

3. The final DNA concentration was adjusted to 20 ng·μL^{-1} with a NanoDrop®ND-1000 (Nanodrop, Wilmington, DE, USA).

HRM Analysis

1. High-resolution melting (HRM) analysis was conducted with a CFX96 Touch™ Real-Time PCR Detection System (Bio-Rad, Hercules, CA, USA).
2. The forward and reverse primers were 5′-TGGGCTATCCCGTAAGCCAGCTCAT-3′ and 5′-GCCGAATCCCTTCATCCTATTTCTCCT-3′.
3. PCR solutions were made by mixing 2 μL of 10 ng·μL^{-1} genomic DNA, 2 μL of 10× PCR buffer, 1 μL of 10 mM dNTP mixture, 0.1 μL of rTaq PLUS polymerase (ELPIS Bio, Daejon, Korea), 1-μL LCGreen® Plus+Melting Dye (BioFire Diagnostics, Salt Lake City, UT, USA), 0.5 μL of each 10-pmol·μL^{-1} specific primer, and 12.9-μL autoclaved distilled water.
4. PCR profiles were: pre-denaturation at 95 °C for 5 min and 40 cycles of denaturation at 95 °C for 30 s, annealing at 58 °C for 30 s, and extension at 72 °C for 30 s.
5. HRM melting curves were drawn by increasing the temperature from 65 to 90 °C at 0.2 °C intervals. Then genotypes were analyzed with the Precision Melt Analysis program (Bio-Rad) (Fig. 4.9).

4.2.3.2 Molecular Marker for Pungency in Chili Pepper

Pun1 is a key gene in the capsaicinoids synthesis pathway, and only the homozygous recessive *pun1/pun1* genotype is non-pungent. Stewart Jr. et al. (2005) developed markers from the *Pun1* gene and also found some difference in gene structure between *Capsicum* species (Stewart Jr. et al. 2007; Stellari et al. 2010). In addition,

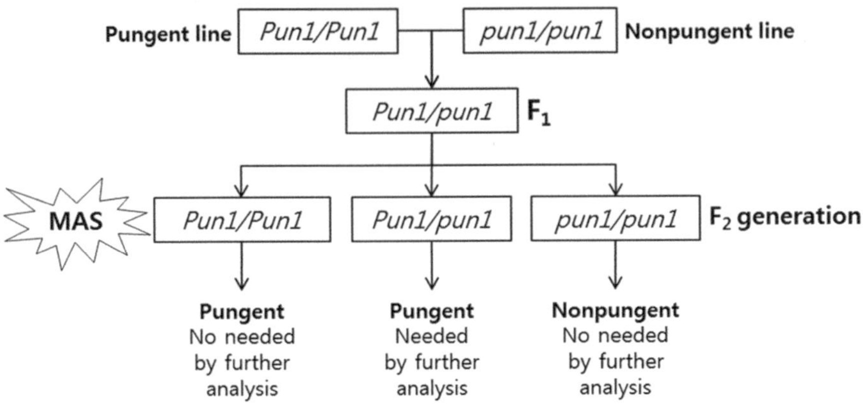

Fig. 4.10 Breeding strategy using the *Pun1* gene marker

Wyatt et al. (2012) reported molecular markers to distinguish *pun1*^{1}, *pun1*^{2}, and *pun1*^{3}. These markers are very useful for selecting genotypes from the progeny of crosses between pungent and non-pungent chili pepper (Fig. 4.10).

Practical Use of *Pun1*^{1} Gene-Based SCAR Marker

DNA was extracted as described in Sect. 4.2.3.1.

SCAR Analysis

1. Two forward (5′-TCCTCATGCATCTCTTGCAG-3′ and 5′-GCTCCACGGAAAAGACTCAT-3′) and one reverse (5′-CAAATGGCAGTTTCCCTTCTCTCATT-3′) primers were used.
2. PCR solutions were made by mixing 10 μL of 20 ng·μL^{-1} genomic DNA, 2.5 μL of 10×PCR buffer, 1 μL of 2.5 mM dNTP mixture, 0.25 μL of 5 U/μL Tag polymerase, 0.5 μL of each 10 μM primer, and 11.5-μL autoclaved distilled water.
3. PCR profiles were: pre-denaturation at 94.5 °C for 4 min; 35 cycles of denaturation at 94 °C for 30 s, annealing at 60 °C for 60 s, and extension at 72 °C for 120 s; and final extension at 72 °C for 10 min.
4. The PCR product was loaded onto a 1 % agarose gel with EtBr (Fig. 4.11).

4.2.3.3 Identification of Cytoplasm Types Using Molecular Markers in Radish

Male sterility, the inability of a plant to produce viable pollen, can arise through genetic and environmental factors. Male sterility caused by genetic factors is further divided into genic male sterility (GMS) and cytoplasmic male sterility (CMS), depending on whether the causal genes are nuclear (GMS) or mitochondrial (CMS).

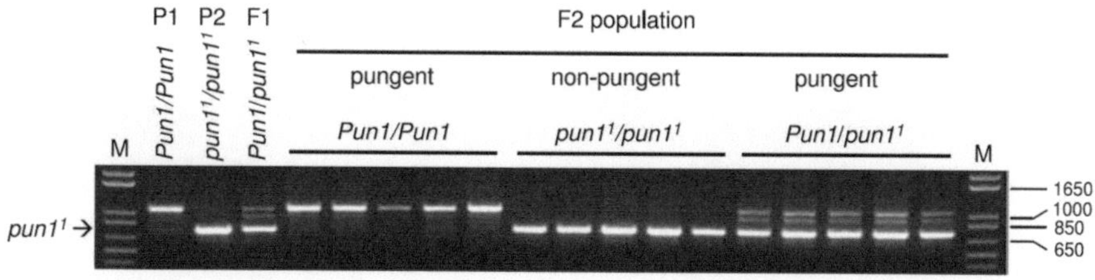

Fig. 4.11 Agarose gel image for genotyping of *pun1*^{1}-SCAR molecular marker. *Pun1/Pun1*, homozygous pungent; *Pun1/pun1*, heterozygous pungent; *pun1/pun1*, homozygous non-pungent

CMS has been widely used to produce F_1 hybrid seeds in many crops, and it has also been an ideal model for studying communication between the nuclear and mitochondrial genomes. In radish, F_1 hybrid seeds have been produced using self-incompatibility (SI) for several decades, but SI does not completely prevent self-pollination. As a result, F_1 hybrid cultivars were of low quality because of self-pollinated contaminants. For this reason, production of F_1 hybrid seed of radish cultivars usually harvested in the spring season is currently performed using CMS instead of SI.

Ogura (1968) first reported CMS in radish, and a new type of CMS (Dongbu CMS: DCGMS) was reported by Lee et al. (2008). Although these two types of CMS show different phenotypes of pollen shedding, at least 2 years of plant growth were required to distinguish them. Therefore, molecular markers to distinguish the cytoplasm types were developed using polymorphic sequences of mitochondrial (Kim et al. 2007) and chloroplast (Kim et al. 2009b) genomes. Radish germplasm from diverse countries were analyzed using these markers, and all accessions were clearly classified into four types of cytoplasm, including Ogura CMS and DCGMS (Fig. 4.12).

4.2.3.4 Development of Molecular Markers to Identify CMS Types and to Genotype Restorer-of-fertility Genes

Onion is the second most important vegetable crop in the world after tomato. The size of the onion seed market in Korea in the last 5 years has ranked third, after hot pepper and radish. Despite its economic importance, a molecular breeding system for onion lags far behind those of hot pepper and tomato. Several characteristics of onion, such as a biennial growth habit, severe inbreeding depression, and huge genome (16,400 Mb/1C), hinder the establishment of a molecular breeding system.

Open-pollinated onion varieties were predominantly cultivated before the 1960s, but the proportion of F_1 hybrid cultivars has been steadily increasing; more than 90 % of cultivars in Korea are now F_1 hybrids. F_1 hybrid seeds of onion are produced by CMS, because no SI has been reported and mechanical emasculation is almost impossible because of the complex flowering pattern. Two types of CMS have been reported: CMS-S and CMS-T. A single nuclear restorer-of-fertility gene restores

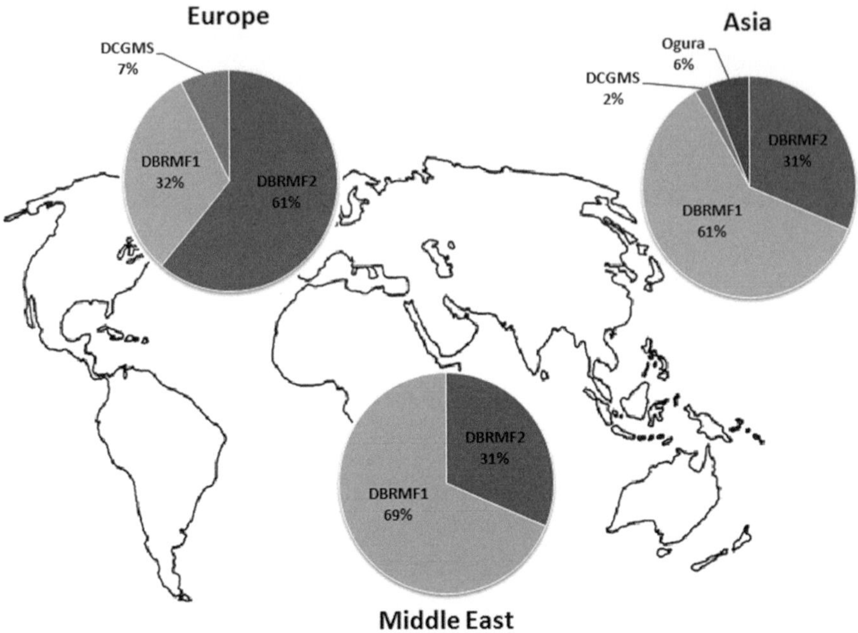

Fig. 4.12 Geographic distribution of four types of radish cytoplasm analyzed using molecular markers based on polymorphic chloroplast sequences (Kim et al. 2009b)

male fertility in CMS-S, but more than three independent loci control fertility restoration in CMS-T. Because CMS-S and CMS-T cannot be distinguished visually and progeny tests require at least 5 years, molecular markers that can distinguish the CMS types quickly are essential tools in onion F_1 hybrid breeding. Although a couple of molecular markers exist for polymorphic chloroplast (Havey 1995) and mitochondrial (Sato 1998) sequences, they could not identify the CMS-T cytoplasm. Kim et al. (2009a) developed a simple PCR marker that clearly distinguished three onion cytoplasm types (Fig. 4.13), and Kim (2014) recently developed a molecular marker (jnurf13) for genotyping of a restorer-of-fertility gene (*Ms*). This marker was in perfect linkage disequilibrium with the *Ms* locus and could be successfully used to select maintainer and restorer lines.

4.2.3.5 Molecular Markers for Genotyping *DFR-A* Alleles Responsible for Bulb Color Differences

Bulb color is a major trait in onion and has been used as a criterion classifying onion cultivars. Red, yellow, and white are common, but chartreuse and gold are also known. Most cultivars grown in Korea are yellow varieties, but the proportion of red varieties is increasing in Korea and worldwide as health benefits of red onions have been reported. Anthocyanin is a pigment for red coloration in onion. This compound

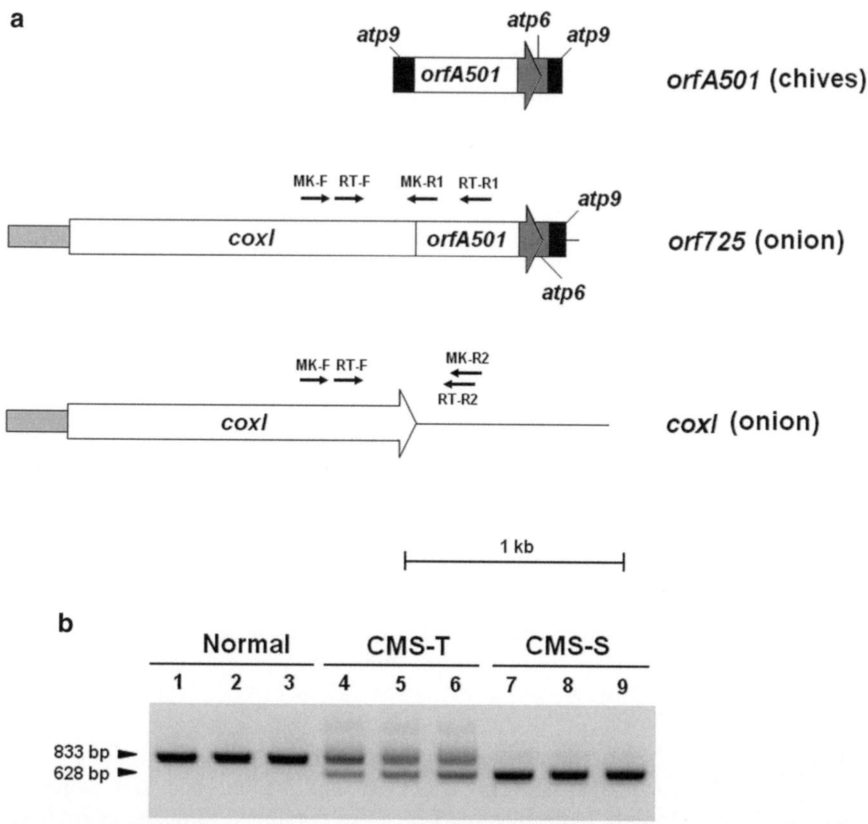

Fig. 4.13 Organization of a chimeric gene, *orf725*, and normal *coxI* in the onion mitochondrial genome (**a**) and the molecular marker developed based on differential copy numbers of the *orf725* and *coxI* genes for identifying cytoplasm types (**b**) (Kim et al. 2009a)

is known to function as a UV protectant in plants, but it has many health-promoting effects such as anti-oxidant and anti-cancer properties.

Despite the health benefits of red onions, some red cultivars have undesirable features, such as low storability. To complement this inferiority of red cultivars, desirable alleles have been introduced from elite yellow cultivars by backcross breeding. However, breeders have had trouble selecting an individual that is homozygous for red pigmentation gene from segregating populations because they could not distinguish homozygous and heterozygous red onions by visual examination.

Kim et al. (2005) revealed that the color difference between red and yellow onions was caused by mutations in the *DFR-A* gene encoding dihydroflavonol 4-reductase, which is involved in anthocyanin biosynthesis. Based on these critical mutations, simple PCR markers were developed for efficient selection of homozygous red onions from segregating populations. Later, four functional alleles and nine independent mutant alleles containing diverse SNPs and indel mutations were

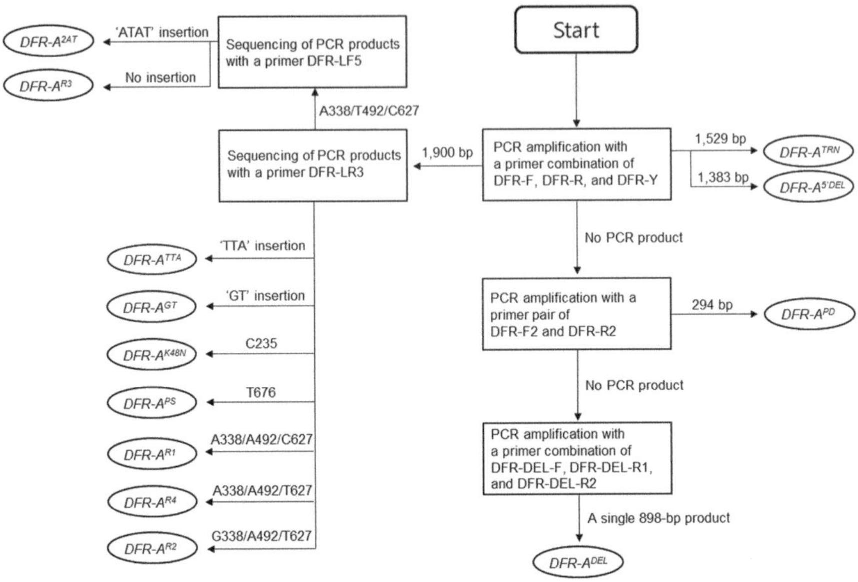

Fig. 4.14 Process showing the sequential steps of PCR amplification and product sequencing for genotyping 13 *DFR-A* alleles in onions (Song et al. 2014)

identified from diverse onion germplasm. To identify specific alleles, Song et al. (2014) devised a process consisting of serial PCR amplifications and direct sequencing (Fig. 4.14). Multiple molecular markers were combined to distinguish all 13 *DFR-A* alleles.

4.3 Marker-Assisted Backcrossing

Since Harlan and Pope first devised the backcross breeding method in 1922, this methodology has been widely used to transfer a single agronomically-important trait from donor plants into elite recipients in a relatively short time. The population sizes handled by breeders can also be minimized compared with strategies such as pedigree and bulk methods. In addition to efficiently transferring a single trait to elite breeding lines or cultivars, backcross breeding has been used to develop cultivars with resistance to multiple diseases or with a combination of several useful traits by 'gene pyramiding'. Alternatively, 'multiline' cultivars have also been developed in which specific portions of seeds of several isogenic lines containing different alleles for disease resistance are mixed together to maintain durable disease resistance. In particular, backcross breeding was recently used to introduce useful traits, such as disease resistance, from closely-related species by inter-specific hybridization.

Although plant breeding originated approximately 10,000 years ago with the domestication of wild species by our ancestors, the history of commercial breeding is not more than 100 years long. However, this relatively short time frame has seen the development of a multitude of significantly improved cultivars and great achievements such as the 'Green Revolution'. In South Korea, the size of vegetable seed markets has grown exponentially since the F_1 hybrid breeding system was established in the 1960s. Thanks to extensive breeding efforts since then, most fundamental morphological traits, such as product shape, have steadily improved and reached almost ideal levels. Therefore, most leading varieties today have resistance to intractable diseases or increased amounts of health-promoting compounds. Such leading varieties were mostly developed by backcross breeding, in which useful resistance genes were introduced from landraces or wild species into elite breeding lines or cultivars. Thus, most seed companies worldwide have invested substantial resources into developing efficient backcross breeding techniques.

Molecular markers and genetic transformation techniques have evolved from an initial testing stage to become essential tools in breeding programs. In fact, major global seed companies have made strategic investments in developing techniques for MAS and marker-assisted backcrossing (MABC). Although MABC plays a pivotal role in developing leading varieties in a short period, the installation of instruments and analysis of enormous numbers of samples is expensive. Thus, only a few seed companies in Korea have established basic MABC systems. Therefore, cost-effective HT assays for broader usage of MABC systems by small-scale seed companies are necessary. The following strategies for optimizing MABC systems have been proposed.

4.3.1 Combining a High-Throughput Marker System and a Single-Marker System for Efficient Selection in MABC

Herzog and Frisch (2011) proposed an efficient selection strategy for MABC by computer simulation. They used a combination of a single-marker (SM) system consisting of simple sequence repeat (SSR) markers and an HT assay system in which automated SNP detection was used. Key results of this simulation study are summarized below.

- Background selection using HT assays only in the BC_1 generation and using SM assays in BC_2 and BC_3 generations was more economical than using HT assays in all three generations.
- The optimal average interval between molecular markers used in background selection was 10 cM; costs could not be reduced with shorter average intervals.
- Molecular markers positioned in linkage maps with similar intervals were more efficient for background selection than randomly distributed markers. With only 50 % equally spaced markers, selection effects were similar to those with randomly distributed markers. Therefore, selecting equally spaced markers from relatively high-resolution linkage maps was recommended.

- A three-stage selection strategy including recombinant selection was more efficient than a two-stage selection strategy because the number of markers to be analyzed was significantly reduced. Therefore, recombinant selection using SM assays was recommended followed by selection of desirable plants by background selection using HT assays of the selected plants at the previous recombinant selection step.
- The optimal distance between the target gene and flanking markers used in recombinant selection was 20 cM when the population size was less than 60, and no severe linkage drag existed. However, in populations larger than 60, a distance of 10 cM produced better results. Thus, unless there is severe linkage drag around the target gene, the optimal distances between the target gene and flanking markers was 10–20 cM.
- Three-stage selection in the BC_1 generation rather than BC_3 was recommended. In addition, analyzing twice as many BC_1 plants as BC_2 and BC_3 plants was more economical if three-stage selection was applied in the BC_1 generation.

4.3.2 Comparison of MABC Results Produced by Computer Simulation and Practical Experiments

Many computer-simulation studies have been carried out to identify optimal conditions for the number of molecular markers, kinds of linkage maps, and selection strategies, but a comparison simulation and empirical studies was first performed by Prigge et al. (2008). They compared the results of computer simulation with those of experiments in which the *Sub1* gene controlling resistance to submergence in rice was introduced into two elite cultivars by MABC. The results showed that the proportions of the recurrent parent genome in all BC generations were similar in both simulation and actual selection experiments. In addition, the kinds of mapping functions and linkage maps used in simulation studies did not significantly affect the outcomes. This comparison proved a high correlation between simulated and actual results of MABC and that computer simulation was sufficiently reliable. In conclusion, computer simulations area valuable tool for establishing optimal selection strategies for MABC in advance.

4.3.3 Case Studies of Developing Elite Cultivars Using MABC

4.3.3.1 Development of Rice Cultivars Containing Resistance to Submergence

Development of rice cultivars' resistant to submergence is the most salient example of the efficiency of MABC. The Flood Resistance 13A (FR13A) accession was found 28 years ago, and several resistant cultivars have been developed by conventional backcross breeding. However, these cultivars were not welcomed by farmers and consumers because of their inferior grain quality and morphological traits. These pitfalls were overcome by MABC based on cloning the *Sub1* gene and constructing high-density linkage maps. The *Sub1* gene on chromosome 9 of the rice genome was a major QTL and conferred complete resistance to submergence for as long as 14 days. Three genes coding for ethylene-responsive transcription factors were identified at the *Sub1* locus. The *Sub1* locus was successfully introduced into the cultivar 'Swarna', a leading variety in India, by MABC within only two BC generations (Neeraja et al. 2007).

Later, several submergence-resistant cultivars grown in India, the Philippines, Laos, and Bangladesh were developed using MABC. These resistant cultivars were almost morphologically identical to the susceptible cultivars used as elite recipient parents. Furthermore, their yields were twice those of susceptible controls when submerged (Septiningsih et al. 2009).

4.3.3.2 Development of Introgression Lines (IL) Using MABC in Rye

Falke et al. (2009) successfully developed an IL library using MABC in rye. They used a landrace (Altevogt 14160) that was similar to wild species as a donor parent and the elite breeding line L2053-N as a recipient parent. Through MABC, they selected a variety of IL lines containing diverse donor genome fragments at the BC_2S_3 generation. After field performance tests of these IL lines, they could select five IL lines showing five-fold higher yields than the recipient parent.

4.3.3.3 Development of a Chickpea Breeding Line Showing Double Podding and Resistance to Ascochyta Blight

Taran et al. (2013) successfully introduced a quantitative trait, resistance to ascochyta blight, and a qualitative double-podding trait into elite breeding lines using MABC. Two major QTLs conferring resistance to ascochyta blight were selected using two foreground selection markers, and 16–22 SSR markers were used for background selection. Within only two BC generations, they successfully developed elite lines with ascochyta blight resistance and double podding.

4.4 Genotypic Analysis of Germplasm and Varieties

4.4.1 Importance of Germplasm in Breeding

Germplasm is defined as biological materials containing useful genes that can be used in current or future breeding programs. The gene pool of germplasm consists of wild species, closely-related species, landraces, breeding lines, current cultivars, and obsolete cultivars. In particular, landraces are the result of long-term selection by humans and the environment and are adapted to diverse environments while maintaining specific traits. Therefore, landraces are precious reservoirs of valuable traits, such as resistance to disease, environmental stress tolerance, and diverse health-promoting compounds, not found in current cultivars. Wild and closely related species have been difficult to use for breeding purposes because of a lack of techniques for inter-specific hybridization and inferior performance of inter-specific progeny. However, these limitations have been resolved by MAB methods such as MABC that allow the development of unprecedented cultivars containing useful traits introduced from wild and related species. Therefore, collecting, preserving, and evaluating diverse germplasm and making databases for systematic management is an important mission. Only when this mission is fulfilled can we protect germplasm from 'genetic erosion' caused by the decreasing number of wild species and landraces and expansion of the growing areas of elite cultivars such as F_1 hybrids.

4.4.2 Genetic Diversity and Phylogenetic Relationships of Germplasm

Although much germplasm is available, most accessions have not yet been evaluated for the diverse traits contained within. Germplasm can be evaluated at two levels. At the basic level, major morphological traits can be examined. Evaluation of resistance to diseases and environmental stresses and analysis of diverse secondary metabolites beneficial to human health can be carried out at the second level. In addition, identification of novel useful genes and mining diverse alleles of those genes is also important. Useful alleles of important genes that are found in landraces but absent in current cultivars can be used to develop leading cultivars in the future.

Genetic diversity in genes, traits, and genomes is important to estimate the phylogenetic relationships of diverse germplasm. In the past, such genetic diversity has been evaluated using morphology, crossability, chromosome karyotyping, and isozyme analysis, but these methods were not sophisticated or objective enough. Recently, molecular markers and numerous polymorphisms present across entire genomes have been used to analyze phylogenetic relationships of germplasm.

4.4.3 Molecular Markers and Polymorphic Sequences Used to Analyze Phylogenetic Relationships of Germplasm

Past analyses of genetic diversity among related species was based on morphological traits, but polymorphic sequences present in intergenic regions of the chloroplast genome and the nuclear internal transcribed spacer have become more widely used for such purposes with the development of more advanced sequencing technologies. Polymorphic mitochondrial sequences have not been widely used in phylogenetic analysis because the plant mitochondrial genome rearranges frequently and the level of polymorphism is much lower than that in the chloroplast genome. In addition, the unstable nature of plant mitochondrial genomes also hinders reliable analysis of genetic diversity. In contrast, some intergenic regions of the chloroplast genome show extremely high polymorphism in some species, and these hypervariable regions have been used in phylogenetic analysis of germplasm of those same species. For example, hypervariable intergenic sequences in the chloroplast genomes of radish and onions were used to analyze diverse accessions (Kim et al. 2009b; Kim 2013).

4.4.4 Phylogenetic Analysis of Intraspecific Variation Using Molecular Markers

Polymorphisms in the chloroplast and mitochondrial genomes have not been frequently used for intraspecific analyses because the level of variation was low, and most agronomically important traits are controlled by nuclear genes. Instead, molecular markers such as RFLP, RAPD, AFLP, SSR, and ISSR have been broadly used to analyze intraspecific variation in germplasm. With the advent of NGS technologies, genetic variation in germplasm can be analyzed byre-sequencing whole genomes of core germplasm collections of model plants, such as rice, for which complete whole genome sequences are available. In addition, genome haplotyping is also used to analyze diverse accessions. As the number of whole genome sequences increases, genetic diversity of germplasm will be mainly assessed using re-sequencing of whole genomes for more accurate analysis. Genome-wide haplotyping technologies will be fundamental for realizing the goal of Breeding by Design™ proposed by Peleman and van der Voort (2003).

4.4.5 Case Studies of Phylogenetic Analysis of Germplasm Using Molecular Markers

4.4.5.1 Genetic Diversity Analysis of Chickpea Using ISSR Markers

Diverse molecular markers such as RFLP, RAPD, AFLP, and SSR have been used to analyze the genetic diversity of chickpea. RFLP, RAPD, and AFLP markers showed very little polymorphism among cultivars, although higher polymorphism was observed among wild species. SSR markers, however, showed a relatively high level of polymorphism. ISSR markers also showed high polymorphism among wild species and cultivars that could be used to analyze the phylogenetic relationships of germplasm and gene flow from wild species into cultivated varieties (Choudhary et al. 2013).

4.4.5.2 Genetic Diversity Analysis of *Lagerstroemia* Using SSR Markers

Although *Lagerstroemia* is one of the most important floricultural plants, few studies of its genetic diversity have been carried out. He et al. (2012) phylogenetically analyzed 81 varieties, five closely related species, and 10 interspecific hybrids using SSR markers. A total of 275 alleles, with an average of nine per SSR marker, were identified. Phylogenetically, the 96 genotypes were clearly divided by SSR markers into three groups that matched well with the genetic backgrounds and origins of the genotypes.

4.4.5.3 Genetic Diversity Analysis of Maize Using High-Throughput SNP Genotyping

SNP genotyping based on the GoldenGate® technology has allowed the rapid analysis of a large number of SNP markers in a single assay. Such an assay of 1,536 SNP markers was used to evaluate the genetic diversity of 154 maize inbred lines. All but 20 of the SNP markers showed phylogenetically informative polymorphism. In addition, a high-resolution linkage map was constructed via the same SNP genotyping technique using RIL populations (Yan et al. 2010).

4.4.5.4 Genetic Diversity Analysis of Grape Using Re-sequencing and High-Throughput SNP Genotyping

Re-sequencing technology was used to identify numerous SNPs from 10 major cultivars of grape and seven wild species. Using an array of 9,000 SNP markers, genetic diversity and the level of linkage disequilibrium were analyzed. Phylogenetic relationships based on SNP array data agreed well with the geographical origins of cultivars, and the level of linkage disequilibrium was relatively low in cultivated varieties compared with wild species (Myles et al. 2010).

4.4.6 Development and Application of Molecular Markers to Identify Cultivars

4.4.6.1 Use of Molecular Markers to Identify Cultivars

Variety protection is a regulatory system that guarantees breeders of new cultivars exclusive commercial rights. Cultivar protection ensures investment recovery, encourages active research into new cultivars, and promotes the import/export of superior breeding materials and germplasm, all of which will improve plant breeding.

For protection, new varieties must first be registered. The requirements for registration are novelty, distinctness, uniformity, and stability. Particularly in the cases of novelty and distinctness, applicants for registration must be clearly discriminated from generally known varieties. The existing methods to evaluate these requirements mainly rely on observations of phenotypic traits via test cultivation. Recently, however, molecular markers have been of great help.

Using molecular markers to assess varieties is advantageous for several reasons, described below. However, this method requires expensive equipment and expert knowledge and the outcomes may differ from those of traditional assessment. As yet, molecular markers cannot completely supersede test cultivation and are being used in a limited way as an important reference for cultivar identification. The advantages are:

- Environmental effects can be ignored.
- Objective judgment is possible.
- Year does not affect the outcome.
- Time and labor are saved.
- An unlimited number of samples can be assayed.
- The growth stage of the plant does not affect the outcome.
- A database can be easily built.

4.4.7 *Case Studies for Using Molecular Markers for Cultivar Identification*

According to the Korea Seed & Variety Service, protocols for cultivar identification using molecular markers developed in South Korea have been established for 3,200 cultivars of 14 crop species and are being used to resolve disputes about cultivar similarity. Molecular markers are used to select control cultivars and evaluate distinctness, uniformity, and stability at the time of cultivar protection application and to confirm the maintenance of the characteristics of registered cultivars. Selection of control cultivars and certification of new varieties using molecular markers proceeds as follows:

- Establishment of high-accuracy database for molecular markers in diverse varieties
- Genotype analysis and setup of genetic similarity for applied/control/reference varieties using molecular markers
- Test cultivation using control varieties with high DNA similarity
- Decision of variety identity after overall evaluation of DNA test and test cultivation results
- Selection of control varieties for the new candidate variety for protection

4.4.7.1 Example of Control Variety Selection Using Molecular Markers

Generally, marker polymorphism between a group of existing varieties and a new candidate for protection is analyzed using allele-specific codominant markers, such as SSR and SNP markers. Based on marker polymorphism, dendrograms (phenetic trees) are constructed via genetic similarity analysis. As shown below (Fig. 4.15) for cucumber and pepper, the efficiency of cultivar identification can be improved by selecting control varieties from a group related to the new variety (red circle, Fig. 4.15).

4.4.7.2 Authenticity Determination of Varieties Using Molecular Markers

The authenticity of a variety can be determined using molecular markers that are unique to a specific variety. Figure 4.16 below shows an example of pepper plants sampled from a grower that were not ‘Sunchaksoon’ but ‘Ochu’, as determined using two cultivar-specific SSR markers (CAM-51 and HpmF015). Figure 4.17 shows another example; a DNA test for authenticity of a tomato cultivar ‘Green Power’ exhibiting root maldevelopment and fusarium disease indicated that tomato samples collected from the grower were not ‘Green Power’. Thus, cultivar-specific molecular markers can not only efficiently distinguish varieties that would be difficult to identify phenotypically but also provide intellectual protection of registered cultivars by preventing illegal duplication.

4.4.7.3 Purity Test for F1 Hybrid Using Molecular Markers

The quality of F_1 hybrid seeds declines significantly with self-pollination of the maternal parent and mixing of off-type seeds during seed gathering. Therefore, F_1 seeds must be tested for purity. By using molecular markers specific to parental

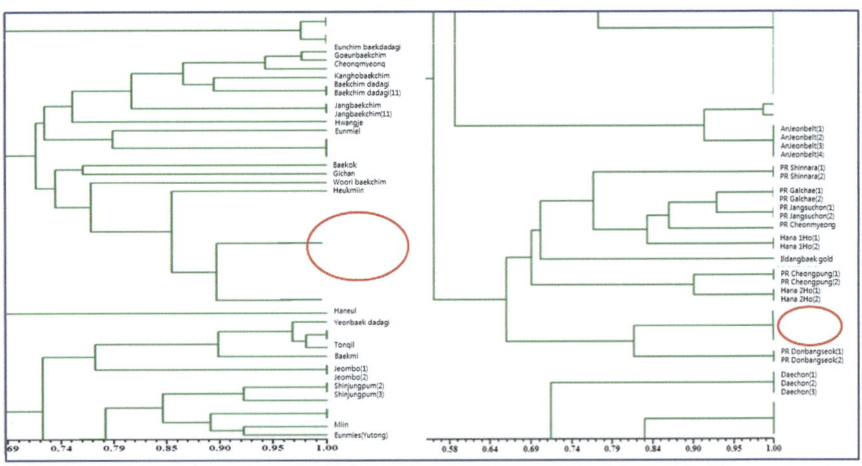

Fig. 4.15 Example of selecting a control variety for a new candidate variety (*red circles*) for protection (*Left*: cucumber (2011), *Right*: pepper (2012))

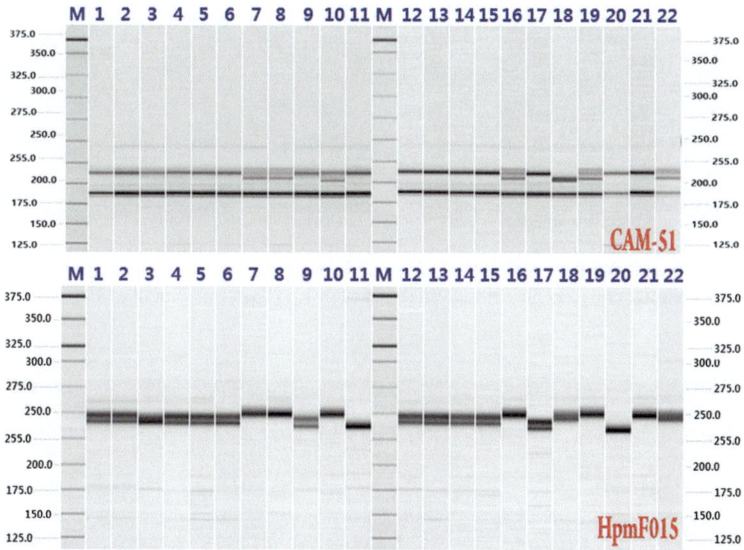

Fig. 4.16 Electrophoresis of PCR products amplified by SSR markers (CAM-51 and HpmF015) used for cultivar identification. Lanes 1, 2, and 12: 'Sunchaksoon'(sample from grower A); Lanes 3, 4, and 13: 'Sunchaksun' (sampled from grower B); Lane 5: 'Sunchaksoon' (seed sample from grower A); Lanes 6 and 15: 'Ochu' (seed samples from Korea Seed & Variety Service); Lanes 7 and 16: 'Sunchaksoon' (seed samples from Korea Seed & Variety Service); Lanes 8–11: varieties from other seed companies; Lane 14: 'Sunchaksoon' (seed sample from grower A); Lanes 17–22: varieties from other seed companies

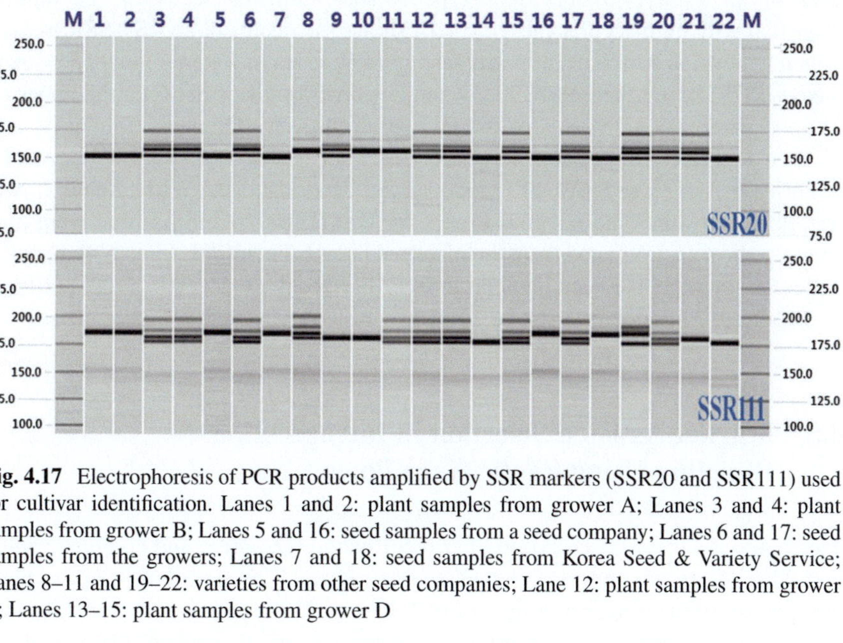

Fig. 4.17 Electrophoresis of PCR products amplified by SSR markers (SSR20 and SSR111) used for cultivar identification. Lanes 1 and 2: plant samples from grower A; Lanes 3 and 4: plant samples from grower B; Lanes 5 and 16: seed samples from a seed company; Lanes 6 and 17: seed samples from the growers; Lanes 7 and 18: seed samples from Korea Seed & Variety Service; Lanes 8–11 and 19–22: varieties from other seed companies; Lane 12: plant samples from grower C; Lanes 13–15: plant samples from grower D

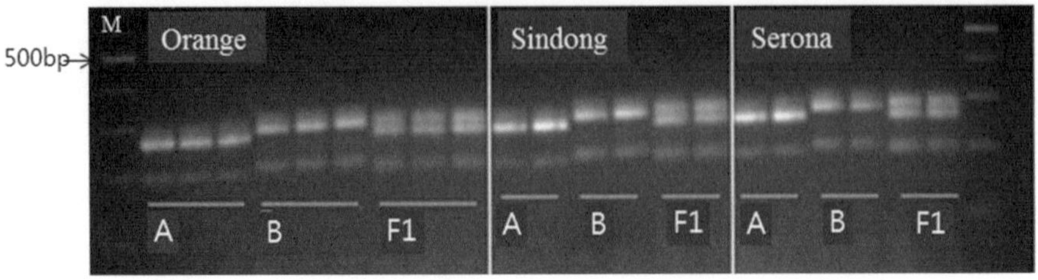

Fig. 4.18 EST-SSR marker WMU00569 developed for purity tests of the watermelon F_1 hybrids 'Orange', 'Sindong' and 'Serona'. *A* maternal parent of F_1, *B* paternal parent of F_1

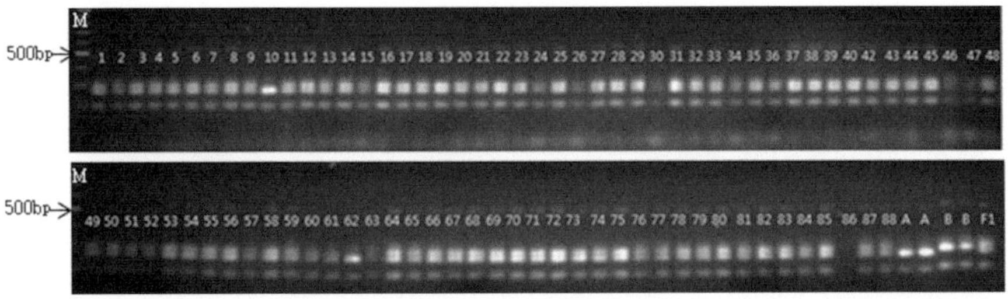

Fig. 4.19 Purity tests of the watermelon F_1 hybrid 'Sindong' using the EST-SSR marker WMU00569. *A* maternal parent of F_1, *B* paternal parent of F_1. Sample numbers 10 and 62 are suspected to be self-pollinated plants

lines of the F_1 hybrid, mixed off-type or self-pollinated seeds can be rapidly identified with high accuracy. Figure 4.18 shows the results of an EST-SSR marker screen to develop a marker for purity testing of F_1 seeds in watermelons. The watermelon EST-SSR marker WMU00569 shows polymorphisms between the maternal (A) and paternal (B) parents of the F_1 hybrid cultivars 'Orange', 'Sindong', and 'Serona' and a heterozygous genotype for the F_1 seeds. Figure 4.19 shows the results of purity tests of F1 seeds (1–88) produced for 'Sindong' using WMU00569. In this test, all seeds were heterozygous except for two (lanes 10 and 62) that were presumed to be self-pollinated from the maternal parent. Therefore, when polymorphic molecular markers between parental lines of F_1 hybrid are developed, the purity of the produced F_1 population can be efficiently tested.

4.5 Marker Assisted Selection in Introgression Breeding Using Exotic Gene Pools

The genus of *Oryza* is composed of two cultivated (*O. sativa* and *O. glaberrima*) and 22 wild species representing 11 genome types. These wild species are phenotypically inferior in agronomic traits, however, it serves as a reservoir of genes for tolerance to various biotic and abiotic stresses, and also possesses yield-enhancing loci. Advances in tissue culture, molecular technologies, genomics and in-situ hybridization opened the opportunity to exploit and utilize the genetic variability from the distant *Oryza* genome through interspecific hybridization. One of the strategies to facilitate the efficiency in utilization of new genetic variation from wild species is through marker – assisted selection (MAS). Genes for resistance to bacterial blight, blast, grassy stunt virus, tungro virus, brown planthopper, tolerance to soil toxicity, and cytoplasmic male sterility have been transferred from wild species. Some of the introgressed wild species genes with broad-spectrum of resistance to diseases have been mapped and used in MAS. This section presents the useful traits from the exotic genepools that can be exploited and how MAS facilitates the isolation of useful genes from the distant genome of *Oryza*.

4.5.1 Wild Species of **Oryza**

The genus *Oryza* is composed of two cultivated species and 22 wild species representing 11 genome types (AA, BB, CC, BBCC, CCDD, EE, FF, GG, HHJJ, HHKK and KKLL), which are distributed in different geographic locations worldwide (Vaughan 1989; Khush 1997). *O. sativa* is cultivated all over the world, whereas *O. glaberrima* is cultivated only in Africa. Within *O. sativa*, two subspecies are distinguished, the *indica* and *japonica* types. The subspecies *japonica* has narrow genetic resources compared to *indica* subspecies which has a wide genetic diversity. The *Oryza* wild species are classified into three genepools (Table 4.5). The primary genepool is composed of the two cultivated species and six wild species (*O. nivara, O. meridionalis, O. barthii, O. longistaminata, O. glumaepatula*, and *O. rufipogon*) with the AA genome (Khush 1997). Gene transfer from wild AA genome species into cultivated species can be accomplished through conventional cross-breeding and backcrossing, although various kinds of reproductive barriers may interfere. There are ten wild species belong to secondary genepool which have a wide geographical distribution. The species are either diploid or tetraploid with six different types of genomes: BB (*O. punctata*), CC (*O. officinalis*, *O. rhizomatis* and *O. eichingeri*), BBCC (*O. punctata* and *O. minuta*), CCDD (*O. latifolia*, *O. alta* and *O. grandiglumis*), EE (*O. australiensis*) and FF (*O. brachyantha*). These species show non-homologous chromosome pairing making gene transfer into cultivated rice difficult. Six wild species represent the tertiary genepool. There are two diploid wild species, *O. granulata* and *O. meyeriana* possess the GG genome, four tetraploid species with three different genomes: *O. longiglumis* and *O. ridleyi* (HHJJ), *O. coarctata* which was previously called *Porteresia coarctata* for KKLL genome and *O. schlechteri* possess HHKK genome (Ge et al. 1999; Khush 1997). These tertiary genepool are highly cross-incompatible with the cultivated species (*O. sativa*).

Table 4.5 Wild species of *Oryza* with chromosome number, genome composition and their origin

Wild species	Genome	Chromosome no.	Origin
Primary genepool			
O. rufipogon Griff.	AA	24	Tropical Asia
O. nivara Sharma et Shastry	AA	24	Tropical Asia
O. longistaminata Chev. et Roehr	AA	24	Africa
O. barthii Chev. et Roehr	AA	24	Africa
O. meridionalis Ng	AA	24	Tropical Australia
O. glumaepatula Steud.	AA	24	South and Central America
Secondary genepool			
O. punctata Kotschy ex Steud.	BB, BBCC	24,48	Africa
O. minuta J.S. Presl. ex C.B. Presl.	BBCC	48	Philippines and Papua New Guinea
O. officinalis Wall ex. Watt	CC	24	Tropical Asia
O. rhizomatis Vaughan	CC	24	Sri Lanka
O. eichingeri Peter	CC	24	South Asia and East Africa
O. latifolia Desv.	CCDD	48	South America
O. alta Swallen	CCDD	48	South America
O. grandiglumis Prod.	CCDD	48	South America
O. australiensis Domin.	EE	24	Tropical Australia
O. brachyantha Chev. et Roehr	FF	24	Africa
Tertiary genepool			
O. granulata Nees et Arn. ex. Watt	GG	24	Southeast Asia
O. meyeriana Baill	GG	24	Southeast Asia
O. longiglumis Jansen	HHJJ	48	Indonesia
O. ridleyi Hook	HHJJ	48	South Asia
O. schlechteri Pilger	HHKK	48	Papua New Guinea
O. coarctata Roxb.	KKLL	48	India

Adapted from Jena (2010)

4.5.2 Useful Traits of Wild Species

Wild species are an important reservoir of useful genes for rice improvement particularly for tolerance to various biotic and abiotic stresses, cytoplasmic diversification, nutritional traits and new yield enhancing loci (Table 4.6). Many useful genes from AA genome species have been transferred by interspecific hybridization and selection (Fig. 4.20). Among the classical examples is the introgression of a gene for grassy stunt virus resistance derived from the cross of *O. nivara* with rice cultivars (Khush 1977) and *O. spontanea*, a CMS source in developing CMS lines for hybrid rice production. Other useful traits for biotic stress (bacterial blight, sheath blight, blast, rice yellow mottle virus, brown planthopper, and white backed planthopper) and abiotic stress (heat and drought, soil problem such as iron toxicity, P-deficiency, and aluminum toxicity) were introgressed from AA genome species into

Table 4.6 Wild species of *Oryza* with useful traits

Wild species	Genome	Useful traits
O. glaberrima	AA	Tolerance to drought, resistance to nematode and RYMV
O. rufipogon Griff.	AA	Source of CMS, stem elongation ability, resistance to BB, sheath blight and tungro tolerance, drought tolerant, yield enhancing loci
O. nivara Sharma et Shastry	AA	Resistance to grassy stunt virus and BB
O. longistaminata Chev. et Roehr	AA	Resistance to BB, yield enhancing loci
O. barthii Chev. et Roehr	AA	Resistance to BB, GLH, drought avoidance, sheathblight and RYMV
O. meridionalis Ng	AA	Stem elongation ability, drought avoidance
O. glumaepatula Steud.	AA	Source of CMS, stem elongation ability, tolerance to heat
O. punctata Kotschy ex Steud.	BB, BBCC	Resistance to BPH and ZLH, tolerance to drought
O. minuta J.S. Presl. ex C.B. Presl.	BBCC	Resistance to sheath blight, blast, BB and BPH
O. officinalis Wall ex. Watt	CC	Resistance to BPH, WBPH and GLH
O. rhizomatis Vaughan	CC	drought avoidance, resistance to blast
O. eichingeri Peter	CC	Resistance to BPH, WBPH and GLH
O. latifolia Desv.	CCDD	Resistance to BPH, higher biomass for yield
O. alta Swallen	CCDD	Resistance to stem borer and high biomass
O. grandiglumis Prod.	CCDD	Higher biomass for yield, submergence tolerance
O. australiensis Domin.	EE	Resistance to BPH and blast
O. brachyantha Chev. et Roehr	FF	Resistance to YSB, BB, blast and BPH
O. granulata Nees et Arn. ex. Watt	GG	Adaptation to aerobic soil
O. meyeriana Baill	GG	Adaptation to aerobic soil
O. longiglumis Jansen	HHJJ	Resistance to blast and BB
O. ridleyi Hook	HHJJ	Resistance to blast, BB and stemborer
O. schlechteri Pilger	HHKK	Stoloniferous
O. coarctata Roxb.	KKLL	Salt tolerance, C4-like

Modified from Jena (2010)
CMS cytoplasmic male sterility, *BB* bacterial leaf blight, *BPH* brown planthopper, *WBPH* white backed planthopper, *GLH* green leaf, *ZLH* zigzag leafhopper, *YSB* yellow stem borer, *RYMV* rice yellow mottle virus

cultivated rice varieties. The *Sub1A* alleles were also identified in *O. rufipogon* (Li et al. 2011).

Wild species belonging to the secondary gene pool of *Oryza* are distantly related to *O. sativa* and this gene pool has a wealth of valuable genes needed for rice improvement (Zeigler 2013). The majority of the identified useful genes in the secondary genepool are resistance loci to major biotic stresses. Four BPH resistance

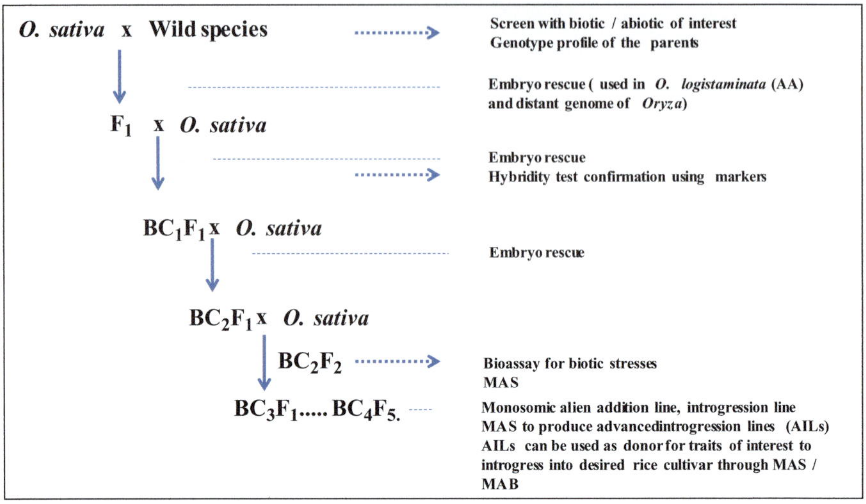

Fig. 4.20 Breeding scheme for the production of wild species introgression lines

genes (*Bph6, bph11, bph12*, and *Bph13*) and two QTLs (*QBph14* and *QBph15*) have been identified from *O. officinalis*, and two from *O. minuta* (*Bph20* and *Bph21*) and *O. australiensis* (*Bph10* and *Bph18*), which were transferred into cultivated rice varieties. Moreover, blast resistance genes derived from *O. minuta* and *O. australiensis* were similarly tagged with molecular markers (Liu et al. 2002; Jeung et al. 2007). The blast resistance gene *Pi40* from *O. australiensis* conferred broad spectrum and durable resistance to leaf and panicle blast isolates of different countries. The *Xa27* resistance gene for BB from *O.minuta* was mapped at the long arm of chromosome 6 within a genetic interval of 0.052 cM, and this gene conferred high level of resistance to 27 *Xoo* strains (Gu et al. 2004). The most distantly related wild species, *O. granulata, O. meyeriana, O. longiglumis, O. ridleyi and O. coarctata* have valuable genes such as adaptation to aerobic soil, water used efficiency, salinity tolerance, and resistance to bacterial blight, blast and stem borer (Brar and Khush 1997; Ge et al. 1999; Khush 1997).

4.5.3 MAS and Its Use in Exotic Genepool

Marker-assisted selection (MAS) involves selecting individuals based on the genotype profile of the trait of interest rather than the phenotypic profile (observable traits). The term 'marker assisted selection' was first used by Beckmann and Soller (1986). Since then, this method has gained interest by many plant breeders and geneticists and subsequently both the number of publications on MAS and QTL mapping has increased dramatically. The followings are the key methods for MAS; (1) Marker assisted introgression or marker assisted backcrossing (MABC), where one gene from a donor line is introgressed into the genetic background of a recipient parent by repeated backcrossing to the recipient parent. (2) Gene pyramiding schemes, where two or more parents line having one or more gene(s) of interest are crossed and then the offspring population is screened for individuals carrying both genes of interest (3) Marker-assisted recurrent selection where markers are used to improve the overall genetic value of a population with respect to some trait or suite of traits.

4.5.3.1 MAS for Biotic Stresses Resistance

Breeding for disease and pest resistance is dominating among publications since they are mainly controlled by major genes (Table 4.7). For disease resistance, two blast genes namely *Pi9* from *O. minuta* and *Pi40* from *O. australiensis* have been mapped and introgressed in cultivated varieties. The *Pi40* has been fine mapped using the *e*-landing approach. DNA marker 9871. T7E2b linked to *Pi40* gene at the 70-kb chromosomal region was obtained from NBS-LRR disease resistance motif sequences (Jeung et al. 2007; Fig. 4.21). This gene shows a broad-spectrum and durable blast resistance in rice (Suh et al. 2009). The *Pi40* gene has been incorporated into temperate rice cultivars of Turkey, Nepal, Russia and Bhutan through MAS and MABC and also used to pyramid with other blast genes (K.K. Jena unpublished). Recently, a novel blast resistance gene, *Pi54rh* identified from *O. rhizomatis* which belongs to the CC-NBS-LRR family of disease resistance genes with a unique Zinc finger (C_3H type) domain and confers broad spectrum resistance to blast (Das et al. 2012).

Out of 38 resistance genes for bacterial blight, seven genes (*Xa21, Xa23, Xa27, Xa29, Xa30, Xa33, Xa38*) are from wild species. The *Xa21* gene which is from *O. longistaminata* confers a broad spectrum of resistance to Philippine races (Ronald et al. 1992). Using the gene pyramiding approach, the *Xa21* gene was also combined with other BB resistance genes (*Xa4+Xa21, Xa4+xa5+Xa21)* in order to improved *indica* rice cultivars having broad-spectrum durable resistance to BB.

Three BPH resistance genes (*Bph14, Bph15, Bph18*) derived from the secondary genepool were fine-mapped. The gene, *Bph14* derived from *O. officinalis* conferred resistance to BPH biotype of China and the gene has been cloned (Du et al. 2009). The *Bph18* gene, which is derived from *O. australiensis,* has been used for MAS of BPH resistance in temperate japonica and tropical indica rice cultivars. The presence of *Bph18* in breeding lines in the japonica background provides enhanced resistance to the new biotype of BPH in Korea (Jena et al. 2006). A new cultivar 'Anmi' has been developed by incorporating the *Bph18* gene (Shu et al. 2011). Recently, one BPH gene (*Bph27*) from *O. rufipogon* was identified and fine-mapped at chromosome 4 (Huang et al. 2013).

One of the most damaging rice-infecting viruses in Africa is the rice yellow mottle virus (RYMV). So far, few sources of resistance for RYMV were identified in cultivated and one from wild species. Recessive genes (*rymv1-3* and rymv1-*4*) were found in two accessions (TOG 5681 and TOG5672) of O. *glaberrima* (Albar et al. 2003). Recently, one major recessive resistance gene (*RYMV*2) was identified *O. glaberrima* which confers high resistance to the RYMV (Orjuela et al. 2013).

Table 4.7 Useful gene of wild species of *Oryza* tagged with DNA markers transferred into cultivated rice

Wild species donor	Useful traits*	Identified genes/ QTLs	Genome	Reference
O. glaberrima	RYSV	*rymv 1–3, rymv 1–4; RYMV2*	AA	Albar et al. (2003); Thiémélé et al. (2010); Orjuela et al. (2013)
O. rufipogon	BPH	*Bph27*	AA	Huang et al. (2013)
	BB	*Xa23*		Zhang et al. (1998)
	Drought	*qSDT2-1, qSDT12-2*		Zhang et al. (2006b)
	Submergence	*OrSub1A-1, OrSub1A-2*		Li et al. (2011)
	Aluminum toxicity	QTL		Nguyen et al. (2003)
	Yield	*yld1, yld2*		Xiao et al. (1996)
O. nivara	Grassy stunt virus	GS	AA	Khush (1997)
	BB	*Xa30(t), Xa33, Xa38*		Cheema et al. (2008); Kumar et al. (2012)
O. logistaminata	BB	*Xa21*	AA	Ronald et al. (1992)
O. officinalis	BB	*Xa29(t)*	CC	Tan et al. (2004b)
	BPH	*Bph 6, bph11, Bph13, Bph14, Bph15*	CC	Jena et al. (2002); Hirabayashi et al. (2003); Renganayaki et al. (2002); Huang et al. (2001); Yang et al. (2002)
	WBPH	*Wbph7; Wbph8*		Tan et al. (2004a)
O. rhizomatis	Blast	*Pi54rh*	CC	Das et al. (2012)
O. minuta	BB	*Xa27*	BBCC	Gu et al. (2004)
	blast	*Pi9*		Liu et al. (2002)
	BPH	*Bph20, Bph21*		Rahman et al. (2009)
O. grandiglumis	Yield enhancing loci	QTL	CCDD	Yoon et al. (2006)
	Submergence	unknown		Okishio et al. (2015)
O. latifolia	Lodging resistance	Putative QTLs	CCDD	Angeles-Shim et al. (2014)
	BB	Putative QTLs		Angeles-Shim et al. (2014)
O. australiensis	Leaf and neck blast	*Pi40*	EE	Jeung et al. (2007)
	BPH	*Bph10, Bph18*		Ishii et al. (1994); Jena et al. (2006)

**BPH* brown planthopper, *BB* bacterial blight, *RYSV* rice yellow mottle virus, *QTL* Quantitative trait loci

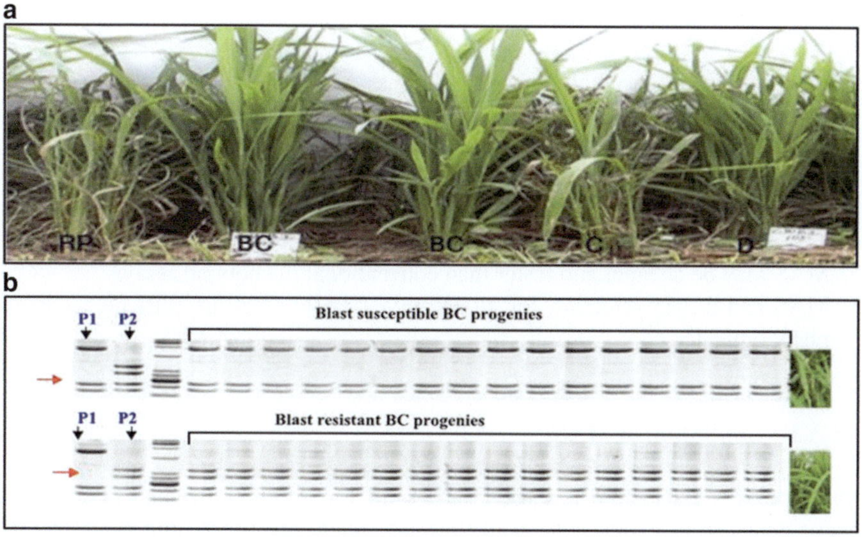

Fig. 4.21 **(a)** Blast resistance gene, *Pi40* transferred from *O. australiensis* derived introgression line, IR65482-4-136-2-2 using marker-assisted selection (MAS) into susceptible cultivated rice (RP) and produced blast resistant backcross progenies (BC), C = Susceptible check **(b)** Blast resistant BC lines tagged with DNA markers (→) which is absent in susceptible BC progenies (Jena 2010)

4.5.3.2 MAS for Abiotic Stress Resistance and Yield Related Traits

A few studies reported the successful application of MAS for abiotic stress tolerance and yield related traits. Among these studies include the introgression from wild species (*O. glumaepatula, O. rufipogon, O. grandiglumis*) in order to improve yield (Xiao et al. 1996; Yoon et al. 2006). Two QTLs (*qSDT2.1* and *qSDT12-2*) for drought tolerance, which were located in the chromosome 2 and 12, were identified in *O. rufipogon*, which increased the drought tolerance of the recipient rice cultivar (Zhang et al. 2006b). Another introgression line from *O. glaberrima* increased the yield of rice under drought condition (Bimpong et al. 2011). Recently, a strong and stable drought resistance QTL at the seedling stage was identified in introgression line IL395 of *O. rufipogon* and through whole genome marker analysis, these *O. rufipogon* segments were located in the chromosome 10 and 12 (Zhang et al. 2014). The *Sub1A*–like alleles (*OrSub1A-1* and *OrSub1A-2*) were identified in two accessions of *O. rufipogon* by the use of degenerate primers corresponding to the highly conserved regions of the *Sub1* locus (Li et al. 2011). Recently, two mechanisms of flooding ability were found in *O. grandiglumis*, one is the capacity of the internode to elongate when subjected to prolonged submergence and the other mechanism is like the *Sub1A,* which provides reduced shoot growth during flash flood at seedling stage and resumed growth after desubmergence (Okishio et al. 2015). Putative QTLs for strong stem located in chromosome 6 were identified from 19 disomic derivatives of *O. latifolia* (Angeles-Shim et al. 2014).

4.5.3.3 Advantage of MAS

1. It is not affected by environment conditions.
2. Recessive alleles can be determined by MAS, but not by phenotypic evaluation since the presence of recessive will be masked by the dominant allele.
3. Gene pyramiding or combining genes simultaneously (considered as one of the best MAS methods currently available).
4. Testing for specific traits where phenotypic screening is not feasible.
5. MAS may be cheaper and faster than conventional phenotypic assays.
6. MAS can aid in selecting for all the target alleles that are difficult to assay phenotypically.

4.5.3.4 Limitation of MAS

Marker assisted selection has been successful for introgressing and pyramiding major genes from exotic gene pool, however, there are some limitations of this breeding strategy:

1. It may be more expensive than conventional techniques especially for the start-up expenses and labor costs.
2. Recombination between the marker and the gene of interest may occur, leading to false positives.
3. Some markers that were detected must be converted to breeder-friendly markers that are more reliable and easier to use.
4. Markers developed for MAS in one population may not be suitable to other populations either due to lack of marker polymorphism or the absence of marker associations.

4.5.4 Conclusions

The cultivated rice that feeds more than half of the world population has been domesticated from ancestral wild species several thousand years ago. During the process of domestication only some selected traits/genes were fixed in the cultivated rice. However, many economically important genes to improve biotic and abiotic stress resistances that can defend the adverse effects of global warming are still present in the wild exotic germplasm of *Oryza*. Some of the high value genes from exotic species have been identified and transferred into cultivated rice through development of introgression lines, DNA marker development and MAS. At IRRI, we have produced MAALs and disomic introgression lines and transferred chromosome segments and genes in the background of elite cultivars. It is imperative to use the introgression lines having useful genes and transfer those genes into rice by using molecular genetics, biotechnology and bioinformatics approaches. The novel genes from exotic species of *Oryza* will enrich cultivated rice gene pool and make stable rice production and maintain food security.

4.6 High-Throughput Technologies in Marker-Assisted Selection

High-throughput molecular marker technologies have the potential to revolutionize marker-assisted breeding. The genomics revolution has led to a wealth of data on sequence variants, at the same time, more efficient techniques provide rapid and low-cost methods for genotyping large populations. With proper execution, these new advances promise to accelerate plant breeding efforts by enabling more precise selection of key alleles, while offering a faster and more efficient alternative to conventional selection methods. However, significant capacity building and training are required to develop the next generation of molecular breeders who can take advantage of these advances. The key steps towards using high-throughput marker technologies are presented below.

4.6.1 Marker Discovery

Instead of painstakingly developing a few markers at a time, now geneticists often have thousands, or even millions, of markers from which to select the most informative loci for various applications and sets of germplasm. Next-generation sequencing has led to high quality reference genomes for many crop species, and re-sequencing efforts have expanded the number of identified polymorphisms, including insertion/deletions (InDels), simple sequence repeats (SSRs), and single nucleotide polymorphisms (SNPs). For example, in rice the Oryza SNP project provided 160,000 high quality SNP markers across 20 varieties (McNally et al. 2009), while more recently the 3,000 Rice Genomes Project identified approximately 18.9 million SNP markers across 3,000 rice varieties (Li et al. 2014; Alexandrov et al. 2015; http://oryzasnp.org/iric-portal/). These SNP discovery pools provide a valuable resource for selecting markers for genome-wide scans and those that are associated with key breeding traits.

4.6.2 Selecting Markers for Genome-Wide Scans

Many applications in molecular breeding require sets of markers that are evenly spaced across the genome and informative for the germplasm being evaluated, including genetic diversity analysis, DNA fingerprinting, QTL and association mapping, and genomic selection. Previously, sets of SSR and InDel markers were used for genome-wide genotyping, but recently SNP markers have become the preferred type of marker due to their abundance across the genome, ease of data curation due to their biallelic nature, and the availability of rapid, cost-effective, and highly-multiplexed SNP genotyping systems. As most crop species have extensive SNP discovery pools in place, the key step for developing genome-wide SNP sets is to use selection criteria that will filter sets of candidate SNPs down to the more useful loci. The primary criteria for selection include: robust SNP markers with good performance using SNP genotyping systems (biallelic loci demonstrating high call rates/low missing data and expected rates of heterozygosity across diverse germplasm using the high-throughput SNP genotyping platform of choice), highly informative SNP loci (markers having a high minor allele frequency, which demonstrate high levels of polymorphism within and between germplasm groups of interest), and excellent genome coverage (evenly spaced SNP markers at the desired resolution for the application and genotyping platform being used). The required resolution of SNP markers across the genome depends on the application and the levels of linkage disequilibrium (LD) within the population being studied. For example, biparental populations for QTL mapping or background selection during MABC will have large LD blocks that can be detected with relatively few markers, such as 10–20 markers per chromosome or a total of 150–200 polymorphic SNP markers for a small genome such as rice, providing at least one marker per 10 cM. In contrast, higher marker resolution is required for applications such as whole genome association analysis, ranging from 1,000 to 100,000 markers for inbred species, but up to one million markers may be required for outcrossing species that have much smaller LD blocks.

4.6.3 Selecting and Validating Markers Associated with Key Traits

While some marker-assisted selection approaches, such as genomic selection, require genome-wide marker scans, other MAS methods take advantage of smaller sets of targeted trait-associated markers that are diagnostic for the desired alleles. For example, a set of 20–30 diagnostic SNP markers can be used to screen very large early-generation populations in a breeding program to eliminate a large number of individuals, so only the most promising lines are advanced to field trials. Ideally, these trait-specific markers will be located at the causal variants, also called the functional nucleotide polymorphism (FNP) or quantitative trait nucleotide (QTN) sites, but there are still very few confirmed functional SNPs that have been identified so far. While efforts to clone major genes and QTLs will continue to lead to the development of functional SNPs, most of the markers in use by breeders are flanking markers that are associated with the traits of interest through linkage. However, even in the case of closely linked markers, there is still a benefit to having "diagnostic" markers or haplotypes located at the gene of interest. When dealing with trait-specific markers associated with specific alleles of interest, two cases need to be carefully distinguished: that of flanking markers specific to a genetic donor using bi-parental crosses versus diagnostic markers that can be used to predict the desired allele across sets of diverse germplasm. When using bi-parental crosses to introgress a useful QTL from a specific donor, such as transferring the *Sub1* QTL for submergence tolerance from FR13A into other rice varieties, flanking markers will easily be able to track the FR13A introgression, even if they are a few hundred kilobases away, as long as they are polymorphic. However, once you apply these same flanking markers across diverse sets of germplasm, the trait association will likely be lost due to much smaller LD blocks across diverse accessions and potentially multiple alleles or other genetic loci leading to the same phenotypic effect. Thus for profiling of trait-enhancing QTL alleles across diverse germplasm, diagnostic haplotypes at the gene of interest are needed—since they are more likely to remain associated with the trait even if they do not come from the exact same genetic donor. For these purposes, the results from genome-wide association studies (GWAS) are also helpful, since they provide information on the frequency of trait-enhancing alleles across diverse accessions.

Once trait-specific SNP markers and haplotypes are identified, further validation is required to determine if they will be useful to deploy in breeding programs. The first step is to perform technical validation of the SNPs on the genotyping platform of choice to determine if the SNP marker has low rates of missing data and provides robust allele calls even across DNA of variable quality. This can be done simultaneously with validation of the SNP marker across diverse germplasm panels to determine the allele frequencies and to make sure the clustering performs as expected, with three genotype classes. Lastly, the SNP marker should be validated across additional segregating populations to confirm that the marker-trait associations remain strong even across different germplasm groups. The validated SNP markers can then be deployed across the breeding programs, providing additional feedback from the breeders if the marker is working as expected.

4.6.4 High-Throughput Genotyping Technologies

There are a number of high-throughput genotyping systems available that provide various price points and number of markers per sample to meet the demands of different breeding applications. For SSR and InDel markers, capillary electrophoresis provides more efficient methods than previous gel-based systems, but markers based on fragment sizing are still limited by their higher cost and constraints in multiplexing. Single SNP assays, such as TaqMan and KASPar assays, provide the flexibility of being able to run any number of SNPs and samples, and can be genotyped on a number of different systems. Small labs can set up relatively simple genotyping facilities using PCR machines and fluorescent plate readers, while lower costs and higher throughput can be achieved with the smaller reaction sizes provided by systems such as Fluidigm and the Douglas Scientific Array Tape. For running larger numbers of markers, highly multiplexed approaches, such as fixed microarrays or SNP chips, as provided by Illumina and Affymetrix, provide robust allele calling and low missing data rates, while genotyping by sequencing (GBS) approaches provide low cost alternatives, but with the challenges of more difficult data analysis and higher missing data rates.

With high-throughput genotyping systems now available, it has become feasible to fully integrate MAS into mainstream breeding programs (Thomson 2014). To handle very large breeding populations, however, requires industrial-scale DNA extraction facilities and marker production facilities that can efficiently deploy the desired set of markers across thousands of samples with a rapid turnaround time that allows for selection before crosses need to be made. Many of the large seed companies have set up large-scale production genotyping labs, and this trend will likely be taken up by public breeding programs due to the long-term benefits and cost-savings. Components of these systems often include seed chipping for DNA extraction from seeds or efficient methods for sampling leaf punches from the field, followed by automated DNA extraction systems, either crude preps or high quality purification methods, and finally to high-throughput genotyping and data analysis.

4.6.5 Bioinformatics and SNP Data Analysis for Breeding Applications

Now that high-throughput genotyping systems have become routine, the main challenge of the future will be to develop the bioinformatics tools, pipelines and databases to seamlessly link allele calling, data QC, data formatting, and downstream applications to enable breeders to use the marker data to make decisions for tracking introgressions and desired alleles, confirming seed identities and purity for quality control, and selecting parents and choosing the best progeny based on genomic estimated breeding values (GEBVs). There are a number of initiatives to develop modern breeding data management systems, such as the Breeding Management System from the Integrated Breeding Platform or the Breeding 4 Rice system being developed at IRRI, which aim to link digital phenotypic data entry, pedigree tracking systems, and breeding decision support tools. There are other initiatives that are focused on analyzing high-density marker data, including sequence analysis, imputation, tracking major alleles, and incorporating genome-wide predictions, such as the Genomic and Open-Source Breeding Informatics Initiative led by Cornell University. In the end, it will be the availability of integrated sequence and SNP databases and breeder-friendly software tools that will finally enable the widespread application of MAS into conventional breeding programs.

References

Albar L, Ndjiondjop M-N, Esshak Z et al (2003) Fine genetic mapping of a gene required for Rice yellow mottle virus cell-to-cell movement. Theor Appl Genet 107:371–378

Alexandrov N, Tai S, Wang W et al (2015) SNP-Seek database of SNPs derived from 3000 rice genomes. Nucleic Acids Res 43(D1):D1023–D1027

Angeles-Shim RB, Vinarao RB, Marathi B et al (2014) Molecular analysis of *Oryza latifolia* Desv. (CCDD genome)-derived introgression lines and identification of value-added traits for rice (*O. sativa* L.) improvement. J Hered 105(5):676–689

Beckmann JS, Soller M (1986) Restriction fragment length polymorphisms in plant genetic improvement. Oxford Surv Plant Mol Cell Biol 3:196–250

Bimpong IK, Serraj R, Chin JH et al (2011) Identification of QTLs for drought-related traits in alien Introgression lines derived from crosses of rice (*Oryza sativa* cv. IR64) × *O. glaberrima* under lowland moisture stress. J Plant Biol 54:237–250

Brar DS, Khush GS (1997) Alien introgression in rice. Plant Mol Biol 35:35–47

Cheema KK, Grewal NK, Vikal Y et al (2008) A novel bacterial blight resistance gene from *Oryzanivara* mapped to 38 Kbp region on chromosome 4 L and transferred to *O. sativa* L. Genet Res 90:397–407

Choudhary P, Khanna SM, Jain PK et al (2013) Molecular characterization of primary gene pool of chickpea based on ISSR markers. Biochem Genet 51:306–322

Collard BCY, Mackill DJ (2008) Marker-assisted selection: an approach for precision plant breeding in the twenty-first century. Philos Trans R Soc B 363:557–572

Das A, Soubam D, Singh PK et al (2012) A novel blast resistance gene, *Pi54rh* cloned from wild species of rice, *Oryza rhizomatis* confers broad spectrum resistance to *Magnaporthe oryzae*. Funct Integr Genomics 12:215–228

Davierwala AP, Reddy AP, Lagu MD et al (2001) Marker assisted selection of bacterial blight resistance genes in rice. Biochem Genet 39:261–278

Du B, Zhang W, Liu B et al (2009) Identification and characterization of *Bph14*, a gene conferring resistance to brown planthopper in rice. Proc Natl Acad Sci USA 106:22163–22168

Falke KC, Susic Z, Wilde P et al (2009) Testcross performance of rye introgression lines developed by marker-assisted backcrossing using an Iranian accession as donor. Theor Appl Genet 118:1225–1238

Fjellstrom R, McClung A, Shank A (2006) SSR markers closely linked to the Pi-z locus are useful for selection of blast resistance in a broad array of rice germplasm. Mol Breed 17:149–157

Ge S, Sang T, Lu BR et al (1999) Phylogeny of rice genomes with emphasis on origins of allohexaploid species. Proc Natl Acad Sci USA 96:14400–14405

Gu K, Tian D, Yang F et al (2004) High-resolution genetic mapping of Xa27(t), a new bacterial blight resistance gene in rice, *Oryza sativa* L. Theor Appl Genet 108:800–807

Havey MJ (1995) Identification of cytoplasms using the polymerase chain reaction to aid in the extraction of maintainer lines from open-pollinated populations of onion. Theor Appl Genet 90:263–268

He D, Liu Y, Cai M et al (2012) Genetic diversity of Lagerstroemia (Lythraceae) species assessed by simple sequence repeat markers. Genet Mol Res 11:3522–3533

Herzog E, Frisch M (2011) Selection strategies for marker-assisted backcrossing with high-throughput marker systems. Theor Appl Genet 123:251–260

Hirabayashi H, Kaji R, Okamoto M et al (2003) Mapping QTLs for brown planthopper (BPH) resistance introgressed from *O. officinalis* in rice. In: Khush GS, Brar DS, Hardy B (eds) Advances in rice genetics. International Rice Research Institute, Manila, pp 268–270

Hittalmani S, Parco A, Mew TV et al (2000) Fine mapping and DNA marker-assisted pyramiding of the three major genes for blast resistance in rice. Theor Appl Genet 100:1121–1128

Huang X, Feng Q, Qian Q et al (2009) High-throughput genotyping by whole-genome resequencing. Genome Res 19:1068–1076

Huang Z, He G, Shu L et al (2001) Identification and mapping of two brown planthopper resistance genes in rice. Theor Appl Genet 102:929–934
Huang D, Qiu Y, Zhang Y et al (2013) Fine mapping and characterization of *BPH27*, a brown planthopper resistance gene from wild rice (*Oryza rufipogon* Griff.). Theor Appl Genet 126:219–229
Ishii T, Brar DS, Multani DS et al (1994) Molecular tagging of genes for brown planthopper resistance and earliness introgressed from *Oryza australiensis* into cultivated rice, *O sativa*. Genome 37:217–221
Jena KK (2010) The species of the genus Oryza and transfer of useful genes from wild species into cultivated rice, O sativa. Breed Sci 60:518–523
Jena KK, Pasalu IC, Rao YK et al (2002) Molecular tagging of a gene for resistance to brown planthopper in rice (*Oryza sativa* L.). Euphytica 129:81–88
Jena KK, Jeung JU, Lee JH et al (2006) High-resolution mapping of a new brown planthopper (BPH) resistance gene, *Bph18*(t), and marker-assisted selection for BPH resistance in rice (*Oryza sativa* L.). Theor Appl Genet 112:288–297
Jeung JU, Kim BR, Cho YC et al (2007) A novel gene, *Pi40*(t) linked to the DNA markers derived from NBS-LRR motifs confers broad spectrum of blast resistance in rice. Theor Appl Genet 115:1163–1177
Jiang GH, Xu CG, Tu JM et al (2004) Pyramiding of insect- and disease-resistance genes into an elite indica, cytoplasm male sterile restorer line of rice, 'Minghui 63'. Plant Breed 123:112–116
Joseph M, Gopalakrishnan S, Sharma RK et al (2004) Combining bacterial blight resistance and Basmati quality characteristics by phenotypic and molecular marker-assisted selection in rice. Mol Breed 13:377–387
Kang WH, Hoang NH, Yang HB et al (2010) Molecular mapping and characterization of a single dominant gene controlling CMV resistance in peppers (*Capsicum annuum* L.). Theor Appl Genet 120:1587–1596
Khush GS (1977) Disease and insect resistance in rice. Adv Agron 29:265–341
Khush GS (1997) Origin, dispersal, cultivation and variation of rice. Plant Mol Biol 35:25–34
Kim S (2013) Identification of hyper variable chloroplast intergenic sequences in onion (*Allium cepa* L.) and their use to analyse the origins of male-sterile onion cytotypes. J Hortic Sci Biotechnol 88:187–194
Kim S (2014) A codominant molecular marker in linkage disequilibrium with a restorer-of-fertility gene (*Ms*) and its application in reevaluation of inheritance of fertility restoration in onions. Mol Breed. doi:10.1007/s11032-014-0073-8
Kim S, Yoo K, Pike LM (2005) Development of a PCR-based marker utilizing a deletion mutation in the DFR (dihydroflavonol 4-reductase) gene responsible for the lack of anthocyanin production in yellow onions (*Allium cepa*). Theor Appl Genet 110:588–595
Kim S, Lim H, Park S et al (2007) Identification of a novel mitochondrial genome type and development of molecular makers for cytoplasm classification in radish (*Raphanus sativus* L.). Theor Appl Genet 115:1137–1145
Kim HJ, Yang HB, Chung BN et al (2008) Survey and application of DNA makers linked to TSWV resistance. Korean J Hort Sci Technol 26:464–470
Kim S, Lee E, Cho DY et al (2009a) Identification of a novel chimeric gene, *orf725*, and its use in development of a molecular marker for distinguishing three cytoplasm types in onion (*Allium cepa*L.). Theor Appl Genet 118:433–441
Kim S, Lee Y, Lim H et al (2009b) Identification of highly variable chloroplast sequences and development of cpDNA-based molecular markers that distinguish four cytoplasm types in radish (*Raphanus sativus* L.). Theor Appl Genet 119:189–198
Kim HJ, Han JH, Kim S et al (2011) Trichome density of main stem is tightly linked to PepMoV resistance in chili pepper (*Capsicum annuum* L.). Theor Appl Genet 122:1051–1058
Kottapalli KR, Lakshmi Narasu M, Jena KK (2010) Effective strategy for pyramiding three bacterial blight resistance genes into fine grain rice cultivar, Samba Mahsuri, using sequence tagged site markers. Biotechnol Lett 32:989–996

Kumar PN, Sujatha K, Laha GS et al (2012) Identification and fine mapping of *Xa33*, a novel gene for resistance to *Xanthomonas oryzae* pv. *oryzae*. Phytopathology 102(2):222–228

Kwon SW, Cho YC, Lee JH et al (2011) Identification of quantitative trait loci associated with rice eating quality traits using a population of recombinant inbred lines derived from a cross between two temperate japonica cultivars. Mol Cells 31:437–445

Lee Y, Park S, Lim C et al (2008) Discovery of a novel cytoplasmic male-sterility and its restorer lines in radish (*Raphanus sativus* L.). Theor Appl Genet 117:905–913

Lee J, Do JW, Yoon JB (2011) Development of STS markers linked to the major QTLs for resistance to the pepper anthracnose caused by *Colletotrichum acutatum* and *C. capsici*. Hortic Environ Biotechnol 52:596–601

Lee J, Han JH, Yoon JB (2012a) A set of allele-specific markers linked to *L* locus resistant to Tobamovirus in *Capsicum* spp. Korean J Hortic Sci Technol 30:286–293

Lee WP, Lee J, Han JH et al (2012b) Validity test for molecular markers associated with resistance to *Phytophthora* root rot in chili pepper (*Capsicum annuum* L.). Korean J Hortic Sci Technol 30:64–72

Lestari P, Ham TH, Lee HH et al (2009) PCR marker-based evaluation of the eating quality of *Japonica* rice (*Oryza sativa* L.). J Agric Food Chem 57(7):2754–2762

Li ZX, Septiningsih EM, Quilloy-Mercado SM et al (2011) Identification of SUB1A alleles from wild rice *Oryza rufipogon* Griff. Genet Resour Crop Evol 58:1237–1242

Li JY, Wang J, Zeigler RS (2014) The 3,000 rice genomes project: new opportunities and challenges for future rice research. GigaScience 3:1–3

Liu G, Lu G, Zeng L et al (2002) Two broad spectrum blast resistance genes, *Pi9*(t) and *Pi2*(t) are physically linked on rice chromosome 6. Mol Genet Genomics 267:472–480

Liu SPLX, Wang CY, Li XH et al (2003) Improvement of resistance to rice blast in Zhenshan 97 by molecular marker-aided selection. Acta Botanica Sinica 45:1346–1350

McCouch SR, Kochert G, Yu ZH et al (1998) Molecular mapping of rice chromosomes. Theor Appl Genet 76:815–829

McNally KL, Childs KL, Bohnert R et al (2009) Genomewide SNP variation reveals relationships among landraces and modern varieties of rice. Proc Natl Acad Sci U S A 106:12273–12278

Myles S, Chia J, Hurwitz B et al (2010) Rapid genomic characterization of the Genus Vitis. PLoS One 5, e8219

Narayanan NNBN, Vera Cruz CM, Gnanamanickam SS et al (2002) Molecular breeding for the development of blast and bacterial blight resistance in rice cv. IR50. Crop Sci 42:2072–2079

Nas T, Sanchez D, Diaz G et al (2005) Pyramiding of thermosensitive genetic male sterility (TGMS) genes and identification of a candidate tms5 gene in rice. Euphytica 145:67–75

Neeraja C, Maghirang-Rodriguez R, Pamplona A et al (2007) A marker-assisted backcross approach for developing submergence-tolerant rice cultivars. Theor Appl Genet 115:767–776

Nguyen BD, Brar DS, Bui BC et al (2003) Identification and mapping of the QTL for aluminum tolerance introgressed from new source, *Oryza rufipogon* Griff. Into indica rice, (*Oryza sativa* L.). Theor Appl Genet 106:583–593

Ogawa T, Yamamoto T, Khush GS et al (1991) Breeding of near–isogenic lines of rice with ingle genes for resistance to bacterial blight pathogen (*Xanthomonas campestris* pv. oryzae). Jpn J Breed 41:523–529

Ogura H (1968) Studies on the new male sterility in Japanese radish, with special reference to the utilization of this sterility towards the practical raising of hybrid seeds. Mem Fac Agric Kagoshima Univ 6:39–78

Okishio T, Sasayama D, Hirano T et al (2015) Ethylene is not involved in adaptive responses to flooding in the Amazonian wild rice species *Oryza grandiglumis*. J Plant Physiol 174:49–54

Orjuela J, Deless EF, Kolade O et al (2013) A recessive resistance to rice yellow mottle virus is associated with a rice homolog of the CPR5 gene, a regulator of active defense mechanisms. Mol Plant Microbe Interact 12:1455–1463

Peleman JD, van der Voort JR (2003) Breeding by design. Trends Plant Sci 8:330–334

Perez L, Redoña E, Mendioro M et al (2008) Introgression of Xa4, Xa7 and Xa21 for resistance to bacterial blight in thermosensitive genetic male sterile rice (Oryza sativa L.) for the development of two-line hybrids. Euphytica 164:627–636

Prigge V, Maurer HP, Mackill DJ et al (2008) Comparison of the conserved with the simulated distributions of the parental genome contribution in two marker-assisted backcross program in rice. Theor Appl Genet 116:739–744

Rahman ML, Jiang W, Chu SH et al (2009) High resolution mapping of two rice brown planthopper resistance genes, *Bph20*(t) and *Bph21*(t), originating from *Oryza minuta*. Theor Appl Genet 119:1237–1244

Ramalingam J, Basharat HS, Zhang G (2002) STS and microsatellite marker-assisted selection for bacterial blight resistance and waxy genes in rice, Oryza sativa L. Euphytica 127:255–260

Renganayaki K, Feitz AK, Sadasivam S et al (2002) Mapping and progress toward map-based cloning of brown planthopper biotype-4 resistance gene introgressed from *Oryza officinalis* into cultivated rice, *O. sativa*. Crop Sci 42:2112–2117

Römer P, Hahn S, Jordan T et al (2007) Plant pathogen recognition mediated by promoter activation of the pepper *Bs3* resistance gene. Science 318:645–648

Ronald PC, Albano B, Tabien R et al (1992) Genetic and physical analysis of rice bacterial blight resistance locus, *Xa21*. Mol Gen Genet 236:113–120

Ruffel S, Gallois JL, Moury B et al (2006) Simultaneous mutations in translation initiation factors eIF4E and eIF(iso)4E are required to prevent pepper veinal mottle virus infection of pepper. J Gen Virol 87:2089–2098

Sanchez ACBD, Huang N, Li Z et al (2000) Sequence tagged site marker-assisted selection for three bacterial blight resistance genes in rice. Crop Sci 40:792–797

Sato Y (1998) PCR amplification of CMS-specific mitochondrial nucleotide sequences to identify cytoplasmic genotypes of onion (*Allium cepa*L.). Theor Appl Genet 96:367–370

Septiningsih EM, Pamplona AM, Sanchez DL et al (2009) Development of submergence-tolerant rice cultivars: the Sub1 locus and beyond. Ann Bot 103:151–160

Shin MS, Park SZ, Shin HT et al (1994) Breeding of near-isogenic lines for resistance to bacterial blight, *Xanthomons oryzae* pv. *oryzae*, in rice. Korean J Breed 26(3):238–242

Shin MS, Noh TH, Lee JK et al (2000) Breeding of japonica near-isogenic lines for resistance to bacterial blight in rice. Korean J Breed 32(3):291–295

Song S, Kim C, Moon JS et al (2014) At least nine independent natural mutations of the *DFR-A* gene are responsible for appearance of yellow onions (*Allium cepa* L.) from red progenitors. Mol Breed 33:173–186

Stellari GM, Mazourek M, Jahn MM (2010) Contrasting modes for loss of pungency between cultivated and wild species of *Capsicum*. Heredity 104:460–471

Stewart C Jr, Kang BC, Liuet K (2005) The *Pun1* gene for pungency in pepper encodes a putative acyltransferase. Plant J 42:675–688

Stewart C Jr, Mazourek M, Stellari GM et al (2007) Genetic control of pungency in *C. chinense* via the *Pun1* locus. J Exp Bot 58:979–991

Shu JP, Jeung JU, Kim YG et al (2011) A brown planthopper resistant and high grain quality rice variety 'Anmi' developed by molecular breeding method. Korean J. Breed Sci 46:152–159

Suh JP, Roh JH, Cho YC et al (2009) The *pi40* gene for durable resistance to rice blast and molecular analysis of *pi40*-advanced backcross breeding lines. Phytopathology 99:243–250

Suh JP, Jeung JU, Noh TH et al (2013) Development of breeding lines with three pyramided resistance genes that confer broad-spectrum bacterial blight resistance and their molecular analysis in rice. Rice 6:5

Swamy P, Panchbhai AN, Dodiya P et al (2006) Evaluation of bacterial blight resistance in rice lines carrying multiple resistance genes and Xa21 transgenic lines. Curr Sci 90:818–824

Singh S, Sidhu JS, Huang N et al (2001) Pyramiding three bacterial blight resistance genes (*xa5, xa13* and *Xa21*) using marker-assisted selection into indica rice cultivar PR106. Theor Appl Genet 102:1011–1015

Tai T, Dahlbeck D, Stall RE et al (1999) High-resolution genetic and physical mapping of the region containing the *Bs2* resistance gene of pepper. Theor Appl Genet 99:1201–1206

Takagi H, Abe A, Yoshida K et al (2013) QTL-seq: rapid mapping of quantitative trait loci in rice by whole genome resequencing of DNA from two bulked populations. Plant J 74(1):174–183

Takeuchi Y, Ebitani T, Yamamoto T et al (2006) Development of isogenic lines of rice cultivar Koshihikari with early and late heading by marker-assisted selection. Breed Sci 56:405–413

Tan GX, Weng QM, Ren X et al (2004a) Two whitebacked planthopper resistance genes in rice share the same loci with those for brown planthopper resistance. Heredity 92:212–217

Tan G, Ren X, Weng QM et al (2004b) Mapping of new resistance gene to bacterial blight in rice line introgressed from Oryza officinalis (In Chinese). Yi Chuan Xue Bao 31:724–729

Taran B, Wqarkentin TD, Vandenberg A (2013) Fast track genetic improvement of ascochyta blight resistance and double podding in chickpea by marker-assisted backcrossing. Theor Appl Genet 126:1639–1647

Thiémélé D, Boisnard A, Ndjiondjop M-N et al (2010) Identification of a second major resistance gene to Rice yellow mottle virus, RYMV2, in the African cultivated rice species, *O. glaberrima*. Theor Appl Genet 121:169–179

Thomson MJ (2014) High-throughput SNP genotyping to accelerate crop improvement. Plant Breed Biotechnol 2:195–212

Toenniessen GH, O'Toole JC, DeVries J (2003) Advances in plant biotechnology and its adoption in developing countries. Curr Opin Plant Biol 6:191–198

Toojinda T, Trangoorung S, Vanavichit A et al (2005) Molecular breeding for rainfed lowland rice in the Mekong Region. Plant Prod Sci 8:330–333

Vaughan DA (1989) The genus *Oryza* L.: current status of taxonomy, IRRI research paper series 138. International Rice Research Institute, Manila, p 21

Wang Z, Jia Y, Rutger JN et al (2007) Rapid survey for presence of a blast resistance gene Pi-ta in rice cultivars using the dominant DNA markers derived from portions of the Pi-ta gene. Plant Breed = Zeitschrift fur Pflanzenzuchtung 126:36–42

Wyatt LE, Eannetta NT, Stellari GM et al (2012) Development and application of a suite of non-pungency markers for the *Pun1* gene in pepper (*Capsicum* spp.). Mol Breed 30:1525–1529

Xiao J, Grandillo S, Ahn SN et al (1996) Genes from wild rice improve yield. Nature 384:223–224

Yan J, Yang X, Shah T et al (2010) High-throughput SNP genotyping with the GoldenGate assay in maize. Mol Breed 25:441–451

Yang HY, Ren X, Weng QM et al (2002) Molecular mapping and genetic analysis of a rice brown planthopper (*Nilaparvata lugens* Stal) resistance gene. Hereditas 136:39–43

Yang HB, Liu WY, Kang WH et al (2012) Development and validation of *L* allele-specific markers in *Capsicum*. Mol Breed 30:819–829

Yeam I, Kang BC, Lindeman W et al (2005) Allele-specific CAPS markers based on point mutations in resistance alleles at the *pvr1* locus encoding eIF4E in *Capsicum*. Theor Appl Genet 112:178–186

Yi G, Lee SK, Hong YK et al (2004) Use of Pi5(t) markers in marker-assisted selection to screen for cultivars with resistance to Magnaporthe grisea. Theor Appl Genet 109:978–985

Yoon DB, Kang KH, Kim KJ et al (2006) Mapping quantitative trait loci for yield components and morphological traits in an advanced backcross population between *Oryza grandiglumis* and the *O. sativa* japonica cultivar Hwaseongbyeo. Theor Appl Genet 112:1052–1062

Yoshimura S, Yoshimura A, Iwata N et al (1995) Tagging and combining bacterial blight resistance genes in rice using RAPD and RFLP markers. Mol Breed 1:375–387

Zeigler RS (2013) Food security, climate change and genetic resources. In: Jackson M, Ford-Lloyd B, Parry M (eds) Plant genetic resources and climate change. CAB International, Boston, pp 1–15

Zhang Q, Lin SC, Zhao BY et al (1998) Identification and tagging of a new gene for resistance to bacterial blight (*Xanthomonas oryza*e pv. *oryza*e) from *O. rufipogon*. Rice Genet Newsl 15:138–142

Zhang J, Jiang G, Xu Y et al (2006a) Pyramiding of Xa7 and Xa21 for the improvement of disease resistance to bacterial blight in hybrid rice. Plant Breed 125:600–605

Zhang X, Zhou S, Fu Y et al (2006b) Identification of a drought tolerant introgression line derived from Dongxiang common wild rice (*O. rufipogon* Griff.). Plant Mol Biol 62:247–259

Zhang F, Cui F, Zhang L et al (2014) Development and identification of a introgression line with strong drought resistance at seedling stage derived from *Oryza sativa* L. mating with *Oryza rufipogon* Griff. Euphytica 200:1–7

Zhou PH, Tan YF, He YQ et al (2003) Simultaneous improvement for four quality traits of Zhenshan 97, an elite parent of hybrid rice, by molecular marker-assisted selection. Theor Appl Genet 106:326–331

Chapter 5
Genomics-Assisted Breeding

Ik-Young Choi, Ho-Jun Joh, Gibum Yi, Jin Hoe Huh, and Tae-Jin Yang

Abstract Since the *Arabidopsis* genome was sequenced, hundreds of plant genomes have either been sequenced or are in sequencing progress. Reference genome sequences and large-scale genome sequencing technologies have initiated a new era in molecular breeding. The field of genomics is progressing rapidly and has already provided invaluable practical products for plant molecular breeding. Here, we review progress in genome sequencing technology and its application to plant breeding. We introduce various genomics tools and discuss how next-generation genome sequencing and genotyping technologies have been applied to high-throughput molecular breeding. We also describe the use of epigenome analysis to interpret phenotypic variations that cannot be explained by simple genetics based on the underlying DNA sequence alone, but rather by epigenetically-controlled mechanisms.

5.1 Introduction

5.1.1 Background

Ever since the discovery of the genetic material by Oswald Avery in 1944 (Avery et al. 1944), DNA sequences have been of fundamental interest to scientists who study biology and its applications. Scientists apply Mendel's law of heredity to crop

---

Author contributed equally with all other contributors.

I.-Y. Choi
Green-Bio Institute of Science and Technology, Seoul National University, Seoul, Republic of Korea
e-mail: choii@snu.ac.kr

H.-J. Joh • T.-J. Yang (✉)
Department of Plant Science, Seoul National University, Seoul, Republic of Korea
e-mail: zerosight@snu.ac.kr; tjyang@snu.ac.kr

G. Yi • J.H. Huh
Department of Plant Science and Plant Genomics and Breeding Institute, Seoul National University, Seoul, Republic of Korea
e-mail: gibumyi@gmail.com; hjujh@snu.ac.kr

H.-J. Koh et al. (eds.), *Current Technologies in Plant Molecular Breeding*,
DOI 10.1007/978-94-017-9996-6_5

breeding via cross-hybridization. The principle of hybridization breeding is to recombine genes from specific parents and select progeny with superior gene combinations. The process includes maintaining germplasm with allele diversity, evaluating traits, and selecting useful allele combinations to improve agricultural performance. Selecting individuals with the best combination of alleles for tens of thousands of genes is the most important step in successful breeding. Conventional breeding depended on phenotype alone, without allele information. Marker-assisted selection allows breeding for specific alleles that control qualitative traits. Marker-assisted backcross breeding improved genome-wide selection to incorporate a few alleles from a donor parent and to effectively recover other genome components in the recurrent parent. Genome-assisted breeding improved the genome-wide understanding of many agriculturally important genes and the simultaneous selection of many alleles related to valuable traits.

Early genomic analyses used genetic markers such as restriction fragment length polymorphism to construct linkage maps. Linkage, represented in centimorgans (cM), is calculated according to recombination values and reflect show often specific alleles occur between genetic markers in segregation populations, such as F_2 populations, recombinant inbred lines (RILs), and doubled haploid populations. The linkage map allows agriculturally important genes to be grouped, ordered, and positioned relative to DNA markers on the same chromosome. Better genetic maps are derived from many evenly-distributed high-quality markers, such as simple sequence repeats (SSR) and single nucleotide polymorphisms (SNPs). The number of large-scale genetic maps has rapidly increased via the development of genomics and the use of modern instruments for high-throughput SNP genotyping. Meanwhile, the construction of large insert genomics libraries using bacterial artificial chromosomes (BACs) allowed physical maps of whole plant genomes to be built. Comparing the physical and linkage maps made the first generation of plant genome projects with accurate chromosome level sequences possible; for example, the model plant *Arabidopsis thaliana* (Arabidopsis Genome Initiative 2000) or the model crop *Oryza sativa* (rice; International Rice Genome Sequencing Project 2005) projects. Even though the field of genomics is developing rapidly, many breeders have difficulty using genome-wide information and developing efficient genomics-assisted breeding programs.

5.1.2 Development of Genome Sequencing Technology

Genome sequencing methods were developed in 1977, and since then technology has improved very rapidly. Progress was revolutionized by the automation of sequencing, beginning with partial automation and proceeding through three generations of fully automated systems, each of which substantially improved sequencing capacity (Fig. 5.1) (Shendure 2013; Palermo et al. 2013).

Early genome projects were based on the construction of genetic and physical maps composed of BAC clones and complete BAC-to-BAC sequences of the clones that represented the minimum-tilting path (minimal set of overlapping clones

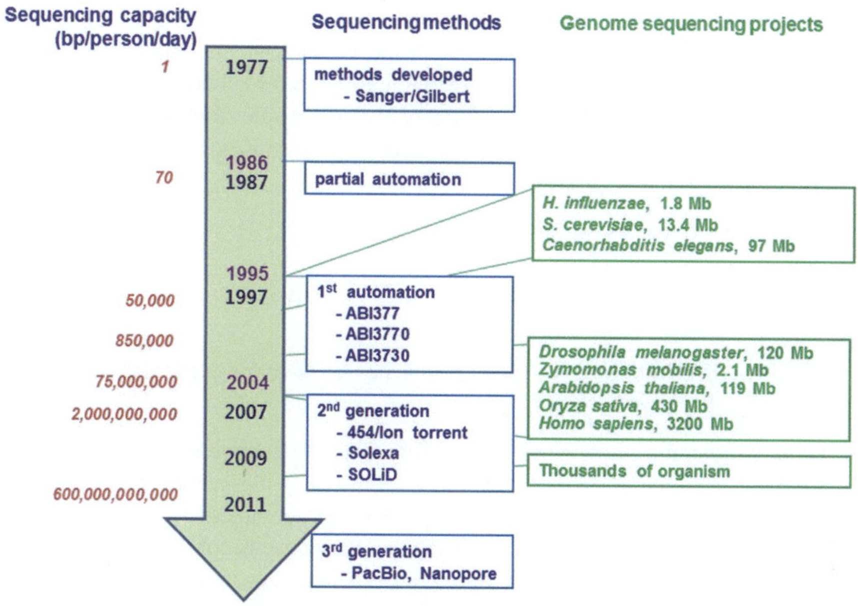

Fig. 5.1 Brief history of genome sequencing progress up to now

needed to provide complete coverage) for the whole genome of the target organism. The first-generation human and rice genome projects cost more than 3 billion US dollars to complete using automatic sequencers based on the Sanger method (International Human Genome Sequencing Consortium 2001). Next-generation sequencing (NGS) technology dramatically increased the genome-sequencing capacity with fast, accurate, and low cost results, which initiated a new era of fast, cheap genome projects for plants, including those with large genomes. NGS also promoted large and collective studies of expressed genes from various tissues or treatments via digital expression analysis with transcriptome data and this development expanded research into non-coding RNA.

The first NGS instrument, the GS FLX, was developed in 2006 by 454 Life Sciences (Branford, CT, USA) and could produce an average of 600 Mb of genetic information in 600-bp read lengths at a time (Rothberg and Leamon 2008). In 2007, the Illumina Corporation (San Diego, CA, USA) developed the HiSeq2000 platform, which could produce 600-Gbp raw reads at a time with average lengths of 100 bp (Bentley et al. 2008). Other novel NGS technologies such as Ion Proton of Life Technologies, part of Thermo Fisher Scientific Inc. (Waltham, MA USA), is also available. In addition, large numbers of long assembled reads can be generated using the Illumina technology (SanDiego, CA USA) by support of modified tools such as 'Moleculo' (Quail et al. 2012). Those NGS instruments and tools have been applied successfully in most of the recent genome sequencing and transcriptome profiling projects.

Third-generation instruments, such as the single molecule real time sequencing developed by Pacific Biosciences (Menlo Park, CA, USA) and nanopore sequencing developed by Oxford Nanopore Technologies (Oxford, UK) (Eid et al. 2009; Quail et al. 2012) can produce 20-kb long reads from single strands of DNA. These technologies have allowed full-length sequences of most genes to be obtained and have promoted complete genome assembly projects.

5.1.3 Recent Approaches to Genome Analysis

Here we briefly introduce various approaches to genome analysis using NGS technologies that have become popular.

- *De novo* genome sequencing and assembly: NGS technology has allowed the complete genome sequencing of many crops, even those with large genomes such as corn, pepper, and ginseng. As a result, complete reference genome sequences including all genes unique to the plant have become available, which provides useful information for comparative genomics with related species.
- Re-sequencing: For species with complete reference genome sequences, such as rice, cabbage, and maize, small-scale genome sequences of accessions with different genotypes (called resequencing) can be produced. Resequence data can be compared with the reference genome to elucidate allele diversity related to important characteristics. Millions of SNPs can be made available for association analysis, mapping, and allele mining from genetic resources.
- Transcriptome analysis: All expressed genes in specific tissues or at certain times can be compiled into reference datasets by mRNA sequencing, a process called transcriptomics or RNAseq. Transcriptomics can uncover all the expressed genes in plants without reference genomes and identify the functions of novel genes in the target plants.
- Digital expression profiling: Transcriptome sequences produced by NGS include all the genes expressed in a target tissue at specified time points. Digital expression levels can be determined by counting the reads for each gene. By comparing the expression levels from different tissues, times, or treatments, genes that are highly or rarely expressed or that have tissue-specific expression patterns can be identified. The expression patterns of genes that respond to specific environmental factors or that are related to phenotypes such as disease resistance can also be studied.
- Epigenetic analysis: NGS technology can be used to sequence CpG islands where DNA methylation and silencing occur. Small RNAs such as microRNAs or long non-coding RNAs also function in epigenetic control of many important genes and can be sequenced using NGS technology.
- Organelle genome sequencing and biodiversity study: NGS technology can generate data from small amounts of template DNA. Therefore, complete chloroplast and mitochondrial genome sequences can be obtained using low-coverage NGS.

5.1.4 Application of NGS Technology to Genome-Assisted Breeding

The goal of crop breeding is to produce individuals with excellent agricultural traits, and its success depends on how effectively favorable alleles from various breeding lines can be integrated into a superior line or cultivar. Despite advances in bioengineering, the best way to accumulate beneficial alleles in one plant is still through several rounds of cross-hybridization between different genotypes and proper selection of the progeny with the best allele combinations.

Conventional breeding depended on phenotype inspection without considering specific allele combinations. However, when genetic maps and other information became available for individual crops, massive numbers of alleles responsible for important qualitative and quantitative traits were revealed, and DNA marker-assisted selection was established. First-generation methods increased the efficiency of traditional breeding for individuals with good allele combinations without the need for time-consuming phenotypic evaluation (Fig. 5.2).

Genomics-assisted breeding has allowed the selection of beneficial alleles for multiple loci throughout the genome. NGS-based genomics data reveal the genotypes of all genes in a genome and allow individuals with the best combinations to be selected. Figure 5.3 depicts an example of genomics-assisted breeding used to select an individual with the optimum allele combination between two parental lines.

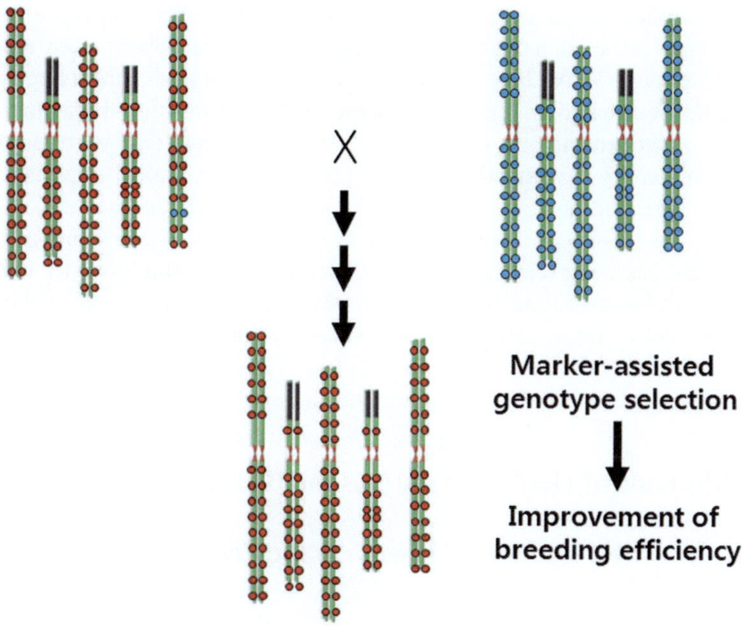

Fig. 5.2 Marker-assisted backcross breeding to introgress one good allele into an elite breeding line from a wild resource

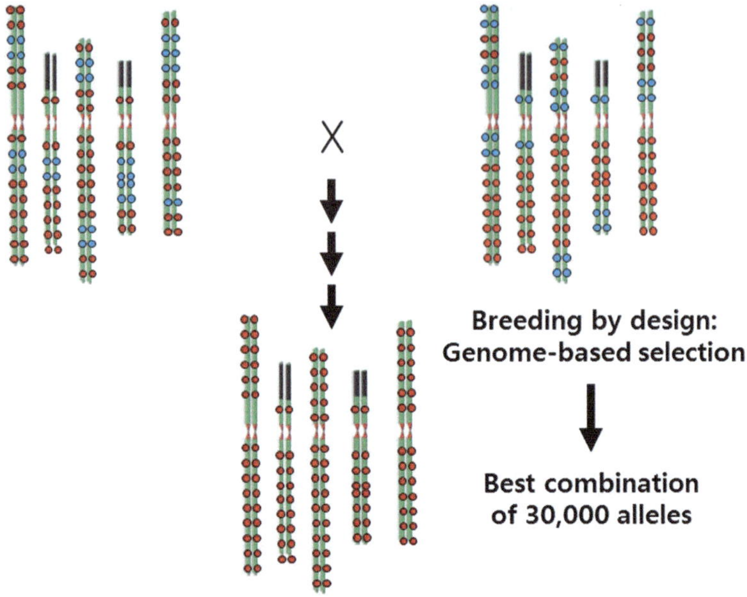

Fig. 5.3 Genomics-assisted breeding using genome-wide allele information

Reference genomes and genome-wide genotyping based on the resequencing data of many individuals with various agricultural traits led to the development of genome-wide association studies (GWASs) for important traits, even complex quantitative ones. GWAS has been used to identify novel alleles from large germplasm collections based on reference genome sequences and high-throughput technologies for sequencing and genotyping. Based on linkage disequilibrium (LD), GWAS can identify new functional alleles for many agriculturally important traits in diverse germplasm (Fig. 5.4). NGS-based genomics approaches, such as genotyping-by-sequencing (GBS) can also be used for ultra-high resolution genetic mapping and marker development. Overall, progress in genomics-assisted breeding has led to a concept called 'Breeding by Design', which allows all of the good alleles distributed among accessions to be integrated into one superior breeding line or cultivar (Fig. 5.4).

5.2 Methods of Genome Analyses for Breeding

5.2.1 Massive Parallel DNA Sequencing Technology

First-generation DNA sequencing was based on the Sanger method in which a labeled dideoxyribonucleotide was attached to terminate the polymerization reaction and the length of the sequence was detected using an automatic capillary laser detection system.

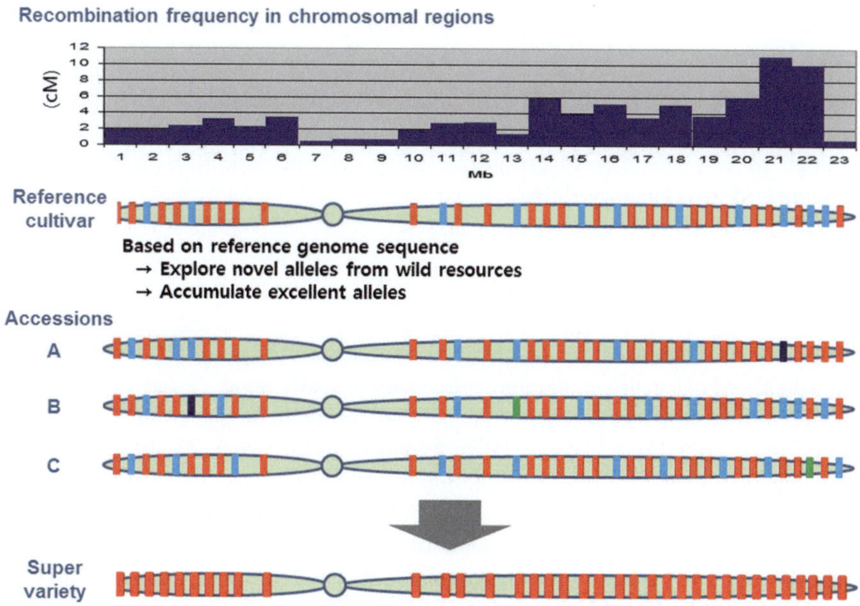

Fig. 5.4 Breeding by design using trait-related markers to incorporate good alleles into one cultivar

Second-generation systems used low-cost and highly accurate equipment that maintained the efficiency of the first generation systems while producing huge amounts of parallel data. The GS FLX (454 Life Sciences, now owned by Roche) and the Genome Analyzer (Solexa; now owned by Illumina and upgraded to the HiSeq platform) platforms are examples of these systems. The basic principle is to detect a signal produced during DNA synthesis.

Third-generation sequencers use a new method called single-molecule, real-time (SMRT) DNA sequencing, which can produce large numbers of long reads (up to about 20 kb from every DNA molecule). Unlike second-generation sequencers that require PCR amplifications to be performed before sequencing, SMRT sequencers can sequence the original DNA fragment, so PCR bias is not introduced. Although the PacBio RS system produces long reads with a relatively high number of sequencing errors, the reads can be used to form scaffolds that can be of help in completing the genome assembly process (English et al. 2012; Zhang et al. 2012). Based on the advance of the PacBio technology, the sequencing errors can be corrected by self-error correction system or application of hybrid genome assembly strategy by combining with other platform NGS reads (Bashir et al. 2012; Koren et al. 2012).

Nanopore sequencing technology developed by Oxford Nanopore Technology (Oxford, UK) can directly obtain large number of long reads by simultaneous sequencing of every long DNA molecule based on an electrical signal that is produced when each nucleotide base passes through a membrane pore; however, this technology it not yet commercially available.

5.2.2 *NGS-Based Strategies and Tools for Genome Study*

In the past, the primary object of many studies was the analysis of DNA mutations in single genes. NGS has made it possible to study mutation or nucleotide variation through entire genomes, expression (including genome-wide RNA transcription), and non-coding regulatory RNA.

5.2.2.1 Whole Genome Sequencing

Long genome sequences can be obtained by assembly of small (100–600 bp) but high depth whole-genome shotgun sequences. Usually short paired-end reads of about 500-bp fragments are *de novo* assembled to obtain unique contigs (Fig. 5.5). Mate- and pair-end reads from relatively long DNA fragments (5–40 kbp) can be used to build longer scaffolds by bridge-joining adjacent contigs (Fig. 5.6). Scaffolds can be ordered and directed into super scaffolds or pseudomolecules for each chromosome by genetic mapping. The pseudomolecules can then be used as a reference genome sequence for genotyping other related species, cultivars, strains, or lines by resequencing or GBS for genomics-assisted breeding.

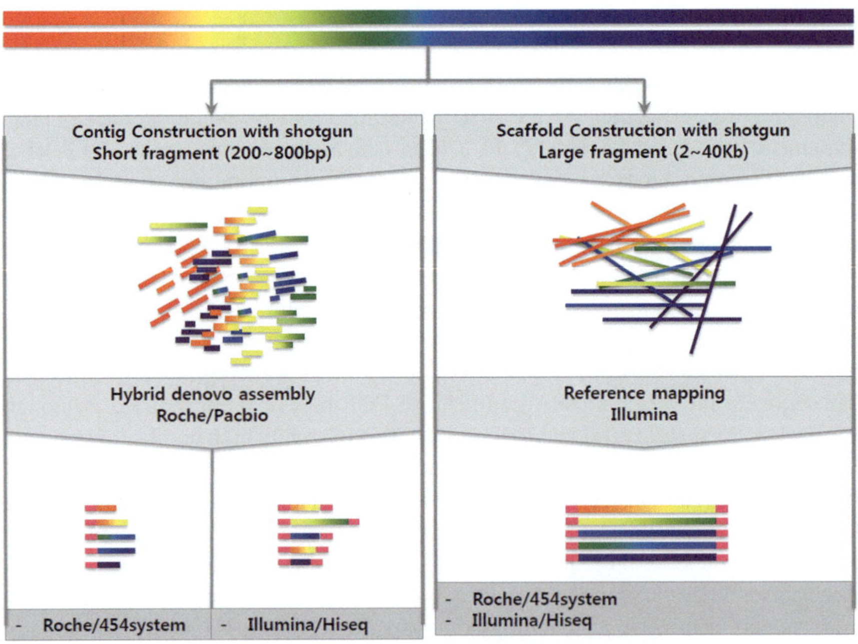

Fig. 5.5 Shotgun sequencing strategy to obtain whole genome sequences. Contigs are obtained by sequencing small pieces of DNA and scaffolds are constructed by matching overlapping ends

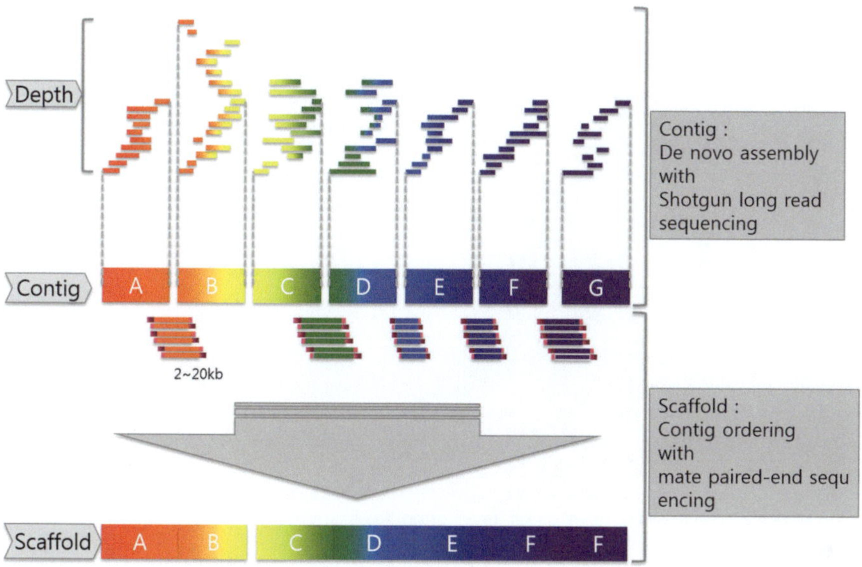

Fig. 5.6 Assembly of shotgun DNA sequences to build scaffolds. To make contigs, the genome should be sequenced more than 10 times; to make scaffold, the genome should be sequenced more than 20 times

Because NGS can produce huge amounts of short reads, complex and efficient computational methods are required to analyze the data. Various genome assemblers have been developed, such as SOAPdenovo (Luo et al. 2012), CLC assembly cell (www.clcbio.com), and All Path-LG (Butler et al. 2008). Many plant genomes have been assembled using SOAPdenovo (e.g., Wang et al. 2011; Kim et al. 2014; Liu et al. 2014a, b), while All Path-LG has been reported to produce better assemblies for microorganism and animal genomes (Ribeiro et al. 2012). Various hybrid assembly tools (e.g., Celera Assembler, Myers et al. 2000, http://wgs-assembler.sourceforge.net, or MIRA, http://mira-assembler.sourceforage.net) have also been applied to use sequences derived from different platforms, such as the short paired-end reads from the Illumina platform and long reads from the PacBio platform. Illumina data have high accuracy while PacBio data can help assemble longer sequences and offset the loss of data that can occur owing to PCR bias introduced in the other methods.

Recently, complete genomes were assembled successfully from only PacBio data using the hierarchical genome-assembly process (HGAP). HGAP creates a preassembled highly accurate genome sequence through an acyclic graph-based consensus process, which uses the longest read as a seed read and maps the short reads to it to produce a scaffold level assembly (Fig. 5.7) (English et al. 2012; Zhang et al. 2012). Application of this technology will be beneficial for the complete assembly of the complex plant genomes that have a highly duplicated genome structure.

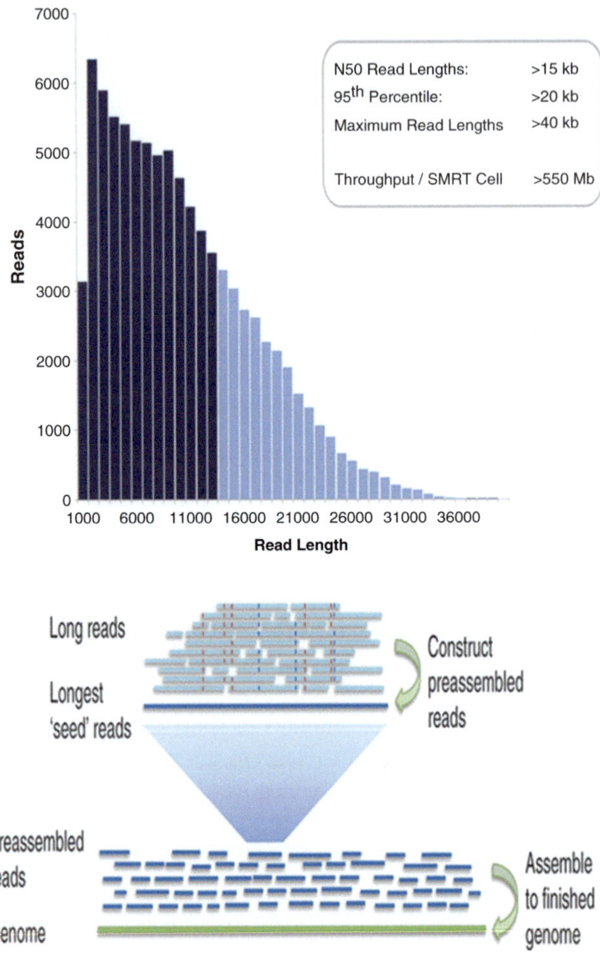

Fig. 5.7 Length distribution of PacBio long reads and a representation of the hierarchical genome-assembly process used to assemble the genome (http://www.pacificbiosciences.com/news_and_events/mediakit/)

5.2.2.2 Identification of Nucleotide Variation by Resequencing

An important part of future biological research will focus on how to effectively use the genomic information contained in the large collections of genotypes. Resequencing and genotyping by sequencing (GBS) have been applied to genotype various different accessions for comparison against a reference genome sequence (Fig. 5.8). Consider the reads as puzzle pieces that have to be matched with the corresponding sequence on the reference genome (Fig. 5.9).

De novo assembly: Sequencing without reference genes
Resequencing: Sequencing with reference genes

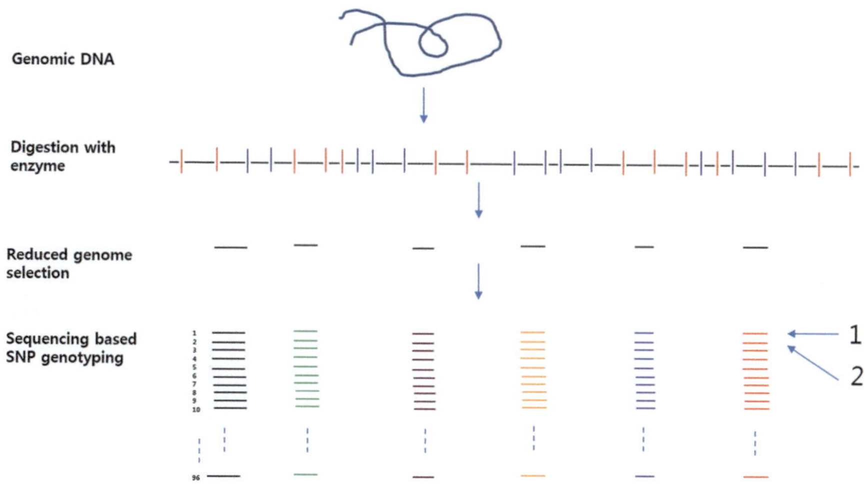

Fig. 5.8 Genotyping by sequencing strategies. The genome is fragmented using restriction enzymes and the fragments are sequenced and compared to discover SNPs

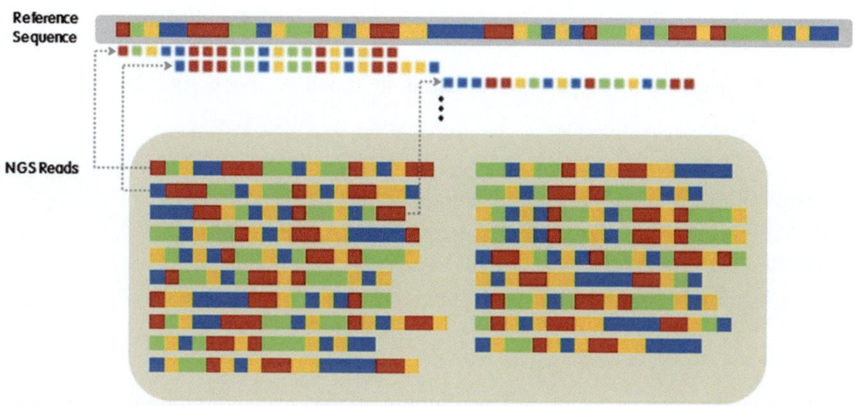

Fig. 5.9 Identification of nucleotide variations by resequencing. Representation of the puzzle pieces that are resequenced and matched to the reference genome

When resequenced reads are mapped onto the reference genome sequence, huge amounts of SNPs and insertion/deletions (InDels) or deletion/insertion polymorphisms (DIPs) can be detected among the different genotypes. The purpose of resequencing is to compare the genotype variation among different accessions (Fig. 5.10). Some caution is required because some of the detected SNPs may not be SNPs but differences that have arisen from sequencing errors. SNPs have to be authenticated by criteria of read depth of coverage or correct paired-end information to ensure accurate genotyping. Some SNP positions can be correlated with a trait, indicating that the regions may contain candidate genes or markers for the trait.

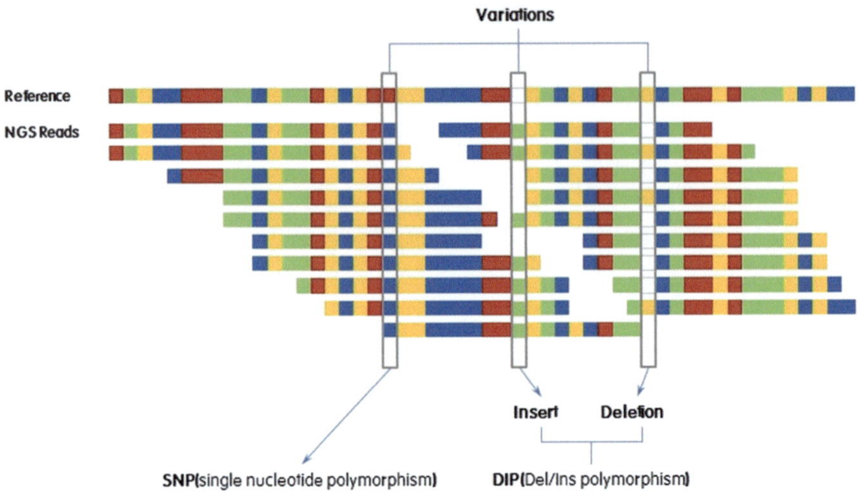

Fig. 5.10 Alignment of resequenced fragments to a reference genome to identify variable regions containing genotype variations

Reference-Guided Detection of Nucleotide Variation: Raw Read Mapping

Commonly used mapping tools are listed in the box. Each tool has a unique characteristic.

Bowtie: http://bowtie-bio.sourceforge.net/index.shtml
BWA: http://sourceforge.net/projects/bio-bwa/
SOAP: http://soap.genomics.org.cn/
Velvet: http://www.molecularevolution.org/software/genomics/velvet
CLC: http://www.clcbio.com/products/clc-assembly-cell/

Mapping data can be confirmed by aligning all the corresponding DNA sequencing reads under the target regions of a reference sequence using an assembly viewer program (Figs. 5.11 and 5.12).

Calling and Filtering Nucleotide Variations

Mapped reads can be compared with a reference sequence or resequenced samples can be compared to find candidate SNPs and DIPs using SAM tools and GATK (Fig. 5.13).

SAM Tools: http://samtools.sourceforge.net/
GATK: http://www.broadinstitute.org/gatk/

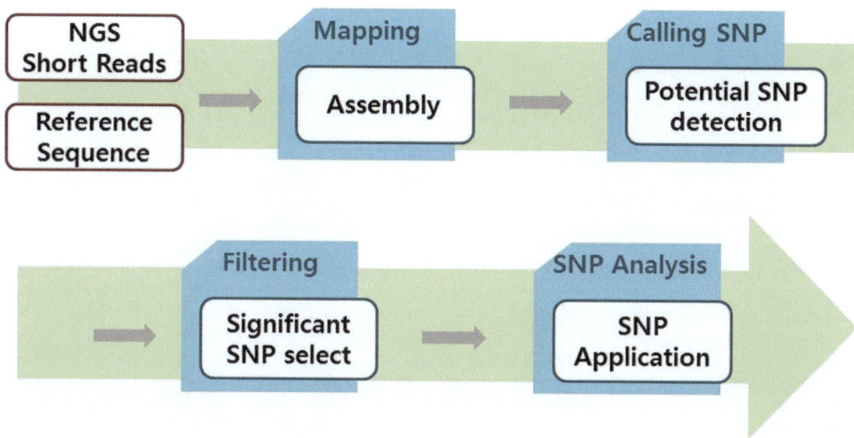

Fig. 5.11 Schematic representation of NGS data analysis through assembly and alignment to SNP discovery

Fig. 5.12 Representative alignment created using the CLC assembly view program

Figure 5.14 shows the VCF (Variant Call format) format that is obtained after variation calling using the SAM tools. VCF files contain reference name, chromosome, and nucleotide position of SNPs, InDels, sequence depth, and frequency. This information can be exported to user-friendly program files such as EXCEL (Fig. 5.15). Data filtering is necessary to remove sequencing errors and to identify genuine sequence variations.

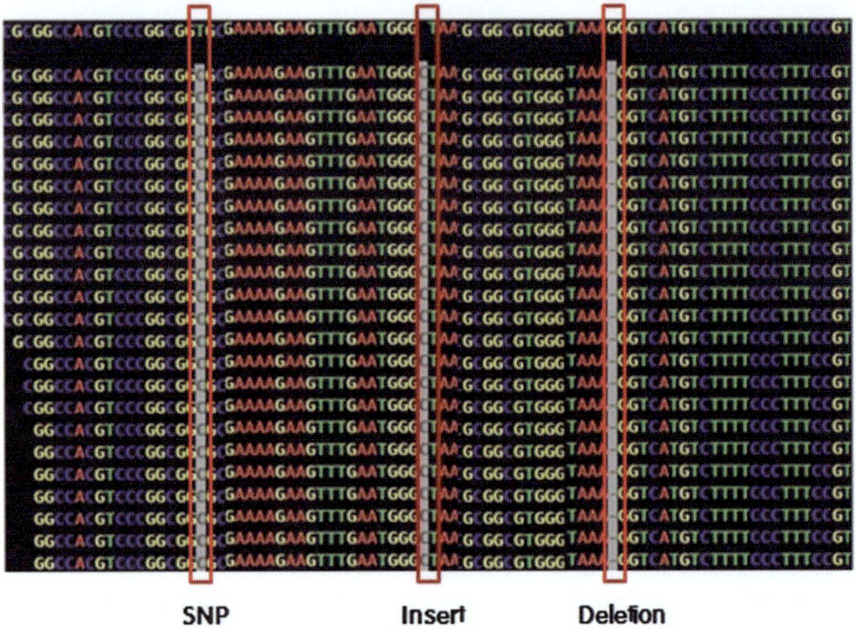

Fig. 5.13 View of a SNP, an insertion, and a deletion in a CLC generated alignment

#CHROM	POS	ID	REF	ALT	QUAL	FILTER	INFO	FORMAT	map.sort.bam
ref	70	.	T	A	13.9	.	DP=2;VDB=0.0084;AF1=1;AC1=2;DP4=0,0,2,0;MQ=25;FQ=-33	GT:PL:GQ	1/1:45,6,0:10
ref	73	.	C	T	13.9	.	DP=2;VDB=0.0084;AF1=1;AC1=2;DP4=0,0,2,0;MQ=25;FQ=-33	GT:PL:GQ	1/1:45,6,0:10
ref	358	.	G	A	7.59	.	DP=2;VDB=0.0198;AF1=1;AC1=2;DP4=0,0,0,2;MQ=25;FQ=-33	GT:PL:GQ	1/1:38,6,0:7
ref	361	.	T	A	13.9	.	DP=2;VDB=0.0198;AF1=1;AC1=2;DP4=0,0,0,2;MQ=25;FQ=-33	GT:PL:GQ	1/1:45,6,0:10
ref	1588	.	C	T	13.9	.	DP=2;VDB=0.0144;AF1=1;AC1=2;DP4=0,0,2,0;MQ=25;FQ=-33	GT:PL:GQ	1/1:45,6,0:10
ref	658733	.	C	T	58.5	.	DP=4;VDB=0.0225;AF1=1;AC1=2;DP4=0,0,2,2;MQ=25;FQ=-39	GT:PL:GQ	1/1:91,12,0:21
ref	659314	.	T	C	75.1	.	DP=7;VDB=0.0008;AF1=1;AC1=2;DP4=0,0,2,5;MQ=25;FQ=-48	GT:PL:GQ	1/1:108,21,0:39
ref	659350	.	T	C	137	.	DP=11;VDB=0.0248;AF1=1;AC1=2;DP4=0,0,5,5;MQ=25;FQ=-57	GT:PL:GQ	1/1:170,30,0:57
ref	3511	.	G	GG	12.9	.	INDEL;DP=12;VDB=0.0001;AF1=1;AC1=2;DP4=0,0,5,0;MQ=20;FQ=-49.5	GT:PL:GQ	1/1:53,15,0:23
ref	5340	.	G	A	87	.	DP=12;VDB=0.0404;AF1=0.5006;AC1=1;DP4=3,0,3,5;MQ=25;FQ=-3.53;PV4=0.18,0.12,1,0.36	GT:PL:GQ	0/1:117,0,26:29
ref	5349	.	A	T	51	.	DP=9;VDB=0.0069;AF1=0.5;AC1=1;DP4=3,0,1,5;MQ=25;FQ=11.3;PV4=0.048,0.0066,1,1	GT:PL:GQ	0/1:81,0,38:41
ref	5487	.	A	T	24	.	DP=6;VDB=0.0081;AF1=0.5;AC1=1;DP4=0,3,1,2;MQ=25;FQ=18.4;PV4=1,0.018,1,1	GT:PL:GQ	0/1:54,0,46:48
ref	28995	.	T	A	14.2	.	DP=7;VDB=0.0192;AF1=0.5;AC1=1;DP4=3,1,0,3;MQ=25;FQ=17.1;PV4=0.14,1,1,1	GT:PL:GQ	0/1:44,0,68:47
ref	29101	.	A	T	84.1	.	DP=8;VDB=0.0320;AF1=1;AC1=2;DP4=0,0,4,2;MQ=25;FQ=-45	GT:PL:GQ	1/1:117,18,0:33
ref	29191	.	G	A	46	.	DP=8;VDB=0.0404;AF1=0.5;AC1=1;DP4=2,1,3,2;MQ=25;FQ=20.1;PV4=1,0.031,1,1	GT:PL:GQ	0/1:76,0,47:50
ref	29590	.	T	A	92.1	.	DP=12;VDB=0.0394;AF1=1;AC1=2;DP4=1,0,3,6;MQ=25;FQ=-32;PV4=0.4,0.0031,1,1	GT:PL:GQ	1/1:123,5,0:9
ref	29596	.	A	T	125	.	DP=12;VDB=0.0179;AF1=1;AC1=2;DP4=0,0,3,6;MQ=25;FQ=-54	GT:PL:GQ	1/1:158,27,0:51
ref	29722	.	G	A	5.46	.	DP=12;VDB=0.0042;AF1=0.4999;AC1=1;DP4=4,4,0,4;MQ=25;FQ=7.8;PV4=0.21,5.2e-05,1,1	GT:PL:GQ	0/1:34,0,123:34
ref	30022	.	C	T	39	.	DP=3;VDB=0.0127;AF1=1;AC1=2;DP4=0,0,1,2;MQ=25;FQ=-36	GT:PL:GQ	1/1:71,9,0:16
ref	30061	.	G	A	39	.	DP=3;VDB=0.0127;AF1=1;AC1=2;DP4=0,0,1,2;MQ=25;FQ=-36	GT:PL:GQ	1/1:71,9,0:16
ref	36881	.	A	G	39	.	DP=3;VDB=0.0071;AF1=1;AC1=2;DP4=0,0,1,2;MQ=25;FQ=-36	GT:PL:GQ	1/1:71,9,0:16
ref	36892	.	A	G	39	.	DP=3;VDB=0.0071;AF1=1;AC1=2;DP4=0,0,1,2;MQ=25;FQ=-36	GT:PL:GQ	1/1:71,9,0:16

Fig. 5.14 Variant Call format (VCF) obtained after variation calling using SAM Tools (http://samtools.sourceforge.net/)

Significant SNPs must be inferred by filtering. Although no standard filtering procedure is available, depth and variation cell rate (frequency) of the mapped reads are generally regarded as an optimal standard in the filtering process. SNPs can be divided into two types, homozygous and heterozygous (Fig. 5.16). Homozygous

1. **CHROM** Chromosome name: an identifier from the reference FASTA file.
2. **POS** Position (1st base has position 1). For an indel, this is the position preceding the indel.
3. **ID** Variant identifier. A unique identifier where available, else '.'.
4. **REF** Reference base. Either A, C, G, T, or N.
5. **ALT** Comma separated list of alternate non-reference alleles.
6. **QUAL** Phred-scaled probability of all samples being homozygous reference.
7. **FILTER** Semicolon delimited list of filters that the variant fails to pass.
8. **INFO** Semicolon delimited list of variant information:

   INDEL Indicating the variant in an INDEL
   DP Row read depth **VDB** (Variant Distance Bias)
   AF1 Max-likelihood estimate of the first ALT allele frequency.AF1 between 0.2 and 0.8: variant is heterozygous
   AC1 Max-likelihood estimate of the first ALT allele count
   DP4 Number of (1) forward ref, (2) reverse ref, (3) forward non-ref, and (4) reverse non-ref alleles used in variant calling (high-quality). Sum can be smaller than DP because low-quality bases are not counted.
   MQ Root-mean-square (RMS) mapping quality of covering reads
   FQ Consensus quality (Phred probability). If positive, FQ equals the Phred-scaled probability of there being two or more different alleles. If negative, FQ equals the minus Phred-scaled probability of all chromosomes being identical.
   PV4P-values for (1) strand bias, (2) baseQ bias, (3) mapQ bias, and (4) tail distance bias

9. **FORMAT** Colon delimited list of the format of individual genotypes in the following fields: **GT** genotype (1/1: homozygous, 0/1: heterozygous), **PL** List of Phred-scaled genotype likelihood, and **GQ** Genotype Quality.
10. **Sample(s)** Individual genotype information defined by FORMAT.

SNPs are SNPs for which the variation in a certain mapping position is present in all reads, while heterozygous SNPs are SNPs for which the variation is present in only some of the reads. If the mapping read depth is low, it is difficult to confirm a SNP as significant. Therefore, an adequate number of reads is necessary for correct genotyping. If the genome size of the target plant is large, sequencing for reduced targets—for example using GBS technology, which sequences only restriction enzyme sites—can increase sequencing depth for the target sites and thus provide convincing genotyping data.

Reference Name	Position	Status	Reference	Query	A	C	G	T	N	-	Other
Ref	1399	Difference	T	C	0	60	0	2	0	0	
Ref	1562	Difference	A	G	27	0	36	0	0	0	
Ref	1705	Difference	T	A	6	0	0	0	0	0	
Ref	2019	Insert	-	A	4	0	0	0	0	3	
Ref	2032	Difference	G	a	5	0	1	0	0	0	
Ref	2045	Difference	A	G	0	0	8	0	0	0	
Ref	2075	Difference	T	C	0	5	0	0	0	0	
Ref	2092	Difference	c	A	51	46	0	5	0	0	
Ref	2107	Difference	T	C	0	145	0	65	0	0	
Ref	2108	Difference	A	G	62	0	100	0	0	0	
Ref	2120	Difference	C	T	0	0	0	17	0	0	
Ref	2121	Insert	-	AG	0	0	0	0	0	20	
Ref	2154	Difference	A	C	51	67	0	0	0	0	
Ref	2184	Difference	G	A	50	0	46	0	0	0	
Ref	2224	Difference	A	-	11	0	1	0	0	167	
Ref	2226	Difference	G	A	215	0	1	0	0	0	
Ref	2233	Difference	T	G	94	1	387	70	0	0	
Ref	2237	Difference	C	G	167	6	366	13	0	0	
Ref	2244	Difference	T	C	0	7	0	6	0	0	

Fig. 5.15 EXCEL file of the exported VCF data in Fig. 5.14

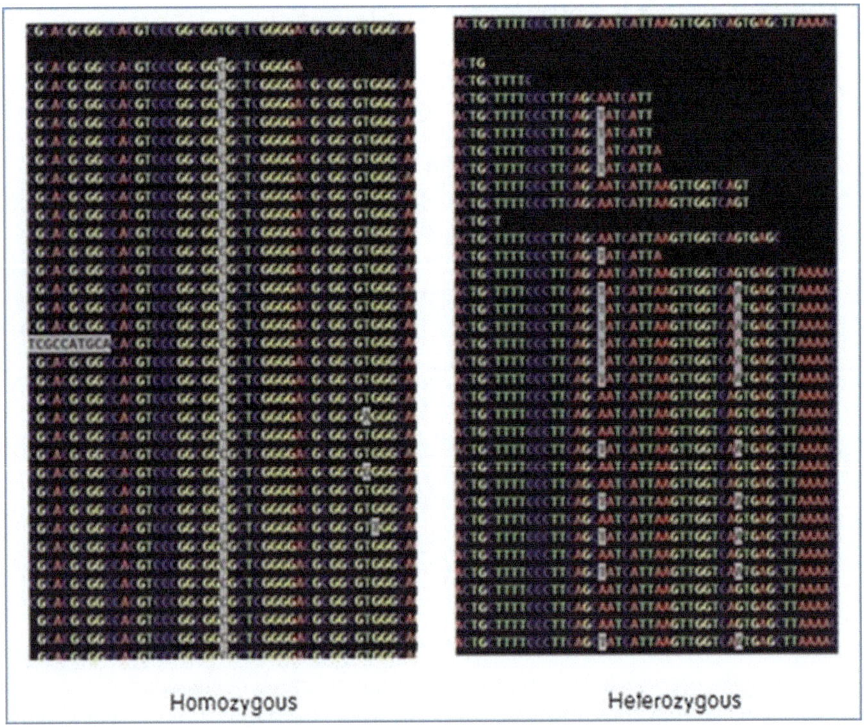

Fig. 5.16 View of a homozygous and a heterozygous site in a CLC generated alignment

SNPs and InDels in Coding Regions

Base changes in the coding regions of genes can lead to amino acid changes in the coded protein, which often affects the function of the protein (Fig. 5.17). SNPs that cause amino acid changes are called 'non-synonymous SNPs'. In DIPs or InDels,

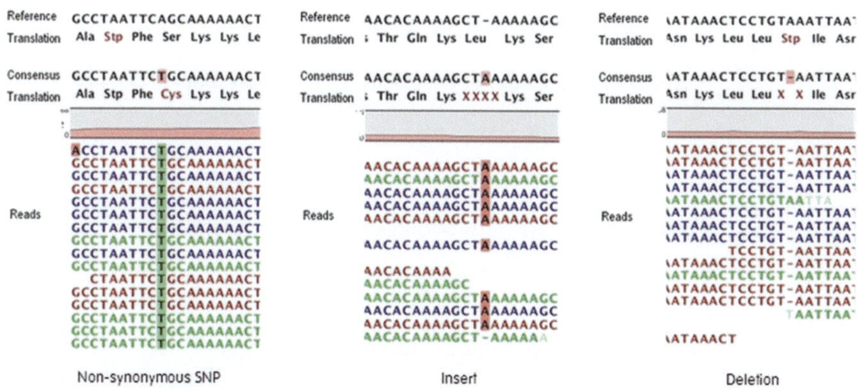

Fig. 5.17 Effects of a SNP, an insertion, and a deletion in the coding regions of a gene on the transcribed amino acid sequences

the insertion/deletion of one or two base pairs will change the reading frame, which will change the entire amino acid sequence. Thus, the identification of nucleotide variations in target genes among individuals in a large population will help characterize novel alleles.

Applications of SNP Data

Large numbers of SNP data can be identified by re-sequencing of many different genotypes. In general, SNPs occur at less than every 1,000 bp although the frequency will be different in different regions of the genome and in different species. SNP information is used mainly in comparative studies between species and between accessions within a species. For humans, SNPs could be used in advanced personalized medicine to create new medicines and customized medical treatments for each individual patient. For crops, SNPs can provide useful information to detect allele variance and to develop superior varieties by genome-assisted molecular breeding (Fig. 5.18).

5.2.2.3 Gene Expression Profiling

The *de novo* assembly of transcriptome sequences has provided almost complete gene datasets for the plants without a reference genome sequence. The assembled contig sequences can be annotated with gene functions using gene prediction methods and BLAST searches against public databases. When the short unassembled reads are mapped onto a reference gene set, the expression levels for each gene can be estimated (Fig. 5.19). Recently, large amounts of transcriptome data have been produced for many plant species. For important minor crops, resource plants, and medicinal plants that lack genomics data, transcriptome data have been used to identify novel genes related to important functions or complex metabolic pathways.

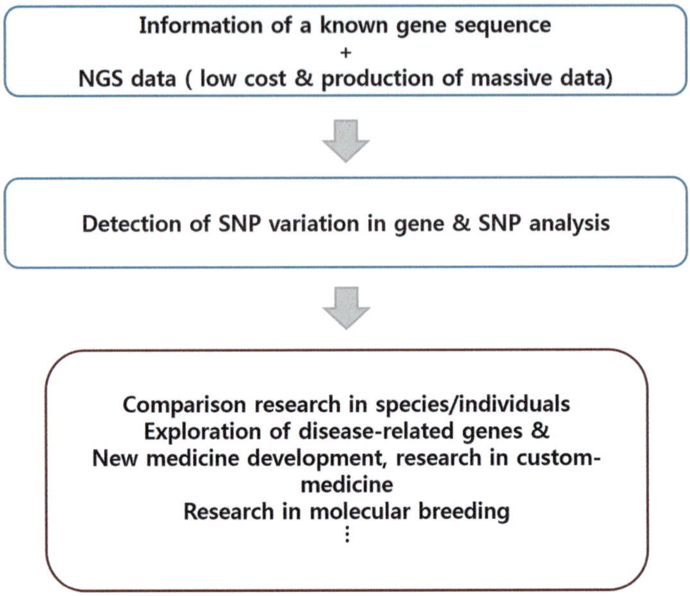

Fig. 5.18 SNPs identified by re-sequencing can be used in many diverse applications

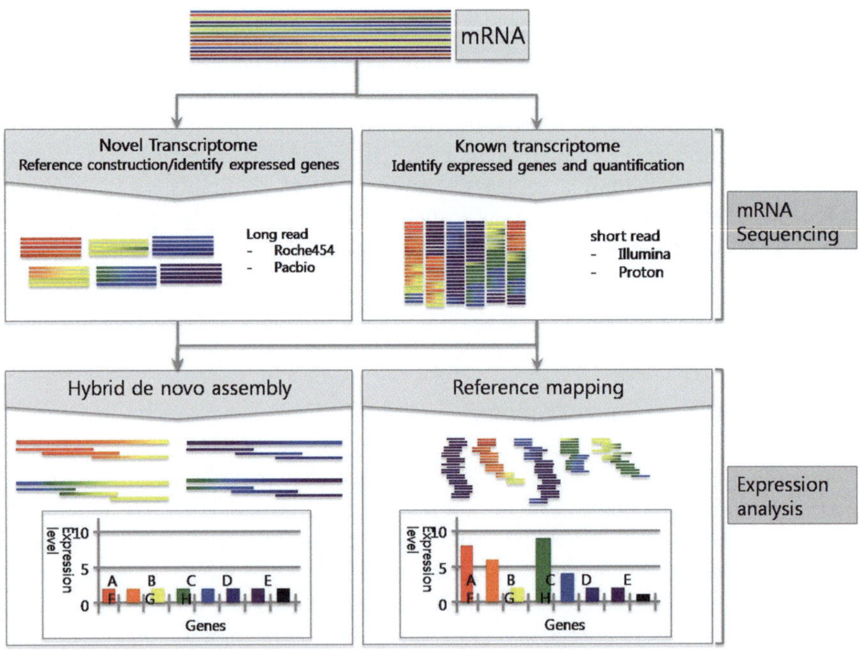

Fig. 5.19 Strategy for gene expression profiling of total RNA using short unassembled reads mapped to a known gene set

5.2.3 Genotyping Technology and Its Application

5.2.3.1 Genotyping by NGS

Quantitative characteristics are usually the most useful characteristics for crop improvement. Various alleles for any one genotype can be found among wild resources. High-throughput genotyping based on NGS technologies has opened up new opportunities for identifying huge amounts of novel alleles using segregating population between two accessions, or various wild accessions or germplasm. Genome-wide inspection of large collections of these resources together with phenotype data for each collection can help identify different alleles of novel genes. The availability of such allele information will promote the efficiency of molecular breeding for important traits, including agriculturally important quantitative traits (Fig. 5.20).

Resequencing and GBS have been used for the genotype mapping of populations to construct high-resolution bin maps. Resequencing data of 150 rice RILs with 0.02–0.05 × coverage were used for ultra-high resolution bin mapping (Huang et al. 2009). However, resequencing approaches cost too much for plants with large genome sizes, such as pepper (Kim et al. 2014) and ginseng (Choi et al. 2014). It is also difficult to obtain accurate genotype data using low coverage resequencing data for complex duplicated plant genomes such as those found in Brassicaceae crops (Yang et al. 2006; Liu et al. 2014a, b). GBS was developed to reduce the time and cost of obtaining sequences for accurate SNP calling. GBS reduces the genome complexity by creating a library using restriction enzymes and by tagging for different samples and applying multiplex sequencing. It has the advantage of creating high-quality polymorphism data for each sample data much lower cost than sequencing an entire genome. GBS detected thousands of SNPs that were used to build

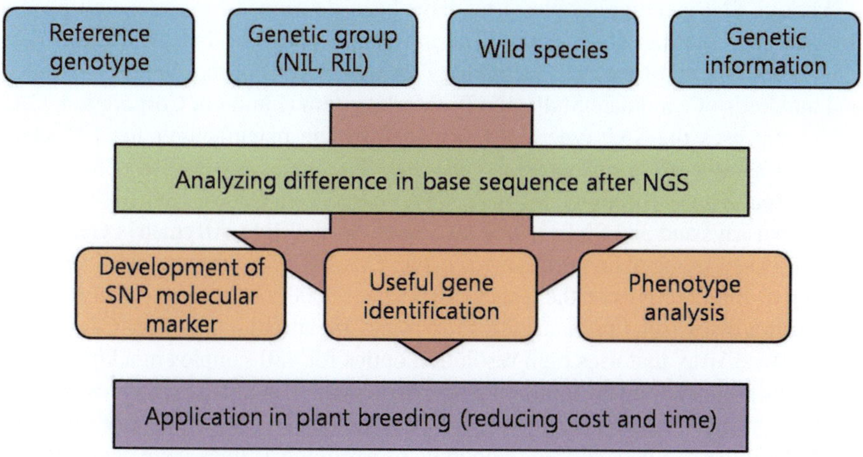

Fig. 5.20 High-throughput genotyping based on NGS technologies provides allele information for efficient molecular breeding

high-resolution genetic maps for barley and wheat, both of which have huge genome sizes (Poland et al. 2012).

Resequencing data and GBS data for large collections of wild resources or germplasm can be used to identify novel genes and quantitative trait loci (QTLs) by GWAS analysis. Many alleles and QTLs for 14 agriculturally important traits were identified by GWAS against 517 domestic accessions in China (Huang et al. 2009).

5.2.3.2 Genotyping by SNP Microarray

SNP array technology is a useful tool for genotyping large numbers of samples and large numbers of known variant regions at a time. High-density SNP chips have been used to find SNPs related to complex traits such as heading date, plant height, starch content, and grain weight. The SNPs that are represented on the chips can be selected from large-scale expressed sequence tag (EST) and genomic DNA datasets. Custom SNP genotyping panels can be designed for specific needs. High-density commercial SNP arrays have been developed for economically important plants including tomato (Sim et al. 2012), soybean (Song et al. 2013), and rice (Zhao et al. 2011). A total of 8,784 SNPs developed from NGS-derived sequences were tested for genotyping tomato germplasm and about 88 % of these were selected to study tomato genetic diversity. USDA/ARS/BARC/SGIL developed an Illumina Infinium BeadChip, the SoySNP50K Beadchip, that contained over 50,000 SNPs in soybean (*Glycine max* L. Merr.). When the SoySNP50K Beadchip was tested with 96 wild-type soybean accessions, 86 % (47,337 SNPs) of the represented genes were found to have polymorphic alleles (Song et al. 2013), confirming that the SoySNP50K Beadchip was a useful tool for soybean genotyping. A total of 413 diverse accessions of *O. sativa* were phenotyped for 34 traits using a 44K SNPs and selected for breeding based on the results (Zhao et al. 2011).

Various SNP genotyping platforms have been developed for various purposes; for example, TaqMan Open Array (Thermo Fisher Scientific Inc., Waltham, MA, USA) based on real-time PCR technology for the analysis of thousands of samples and hundreds of candidate SNPs, the BioMark system (Fluidigm Corporation, CA, USA) that uses the SNP-type assay provided by the manufacturer for 192 samples × 24 assays, 96 samples × 96 assays, and 48 samples × 48 assays (Chan et al. 2011), BeadXpress (Illumina, Inc. CA, USA)-based Vera Code assays for 480 samples × 96 assays and 384 SNPs in one kit (Trebbi et al. 2011), Affymetrix GeneChip System-based probe assays (Affymetrix, Inc. CA, USA) for thousands of samples with up to 900,000SNPs on the GeneChip Scanner 30007G (http://www.affymetrix.com/estore/index.jsp, Liu et al. 2014a, b), and iScan (Illumina, Inc. CA, USA) -based BeadArray that uses high resolution optics for 480 samples in a kit with up to one million SNPs on an Infinium chip (Westengen et al. 2012). The most appropriate platforms for a particular study can be selected based on the sample numbers or SNP numbers to be analyzed. Hundreds of *Arabidopsis* lines were analyzed for 250,000 SNP targets using an Affymetrix chip for GWAS (Atwell et al. 2010). In barley, major agricultural traits and QTLs were identified by GWAS for 957 SNP targets using Illumina GoldenGate technology (Pasam et al. 2012).

5.3 Plant Epigenome and Its Application to Breeding

Heritable phenotypic variations are explained primarily by genetic variations—that is, differences in DNA sequence. Therefore, all breeding programs have focused on identifying genetically distinct individuals as breeding material. However, some phenotypic variation cannot be explained by genetic variation alone but rather by a mechanism that goes beyond the DNA sequence. In eukaryotes, the expression of genetic information stored in the genome is affected by different states of the chromatin, a complex structure of DNA and proteins. Euchromatin is a lightly packed form of chromatin that is transcriptionally proficient, whereas heterochromatin is densely packed and often renders inefficient gene transcription. DNA methylation and histone modifications are two major epigenetic mechanisms that modulate chromatin structures in the absence of changes in the underlying DNA sequences. The presence or absence of such epigenetic modifications determines the transcriptional properties of the chromatin, and, therefore, a virtually unlimited number of different transcriptional programs may be possible by varying the 'epigenome' (the global landscape of epigenetic modifications throughout the genome; Bernstein et al. 2007), which can change dynamically during development. During specific developmental stages, one set of genes may be actively transcribed while another set of genes is transiently held in a repressed state depending on the chromatin structure. It is thus conceivable that epigenetic gene regulation involving DNA methylation and histone modification is responsible for diversifying the transcriptional profile in eukaryotes, thereby providing great flexibility in response to different developmental and environmental cues. Robust transcriptional regulation is required to maintain the characteristics and functions of a specific cell type, and a distinct epigenomic profile serves as a cellular memory associated with cell identity. Therefore, when this memory fails, aberrant changes in epigenome structure often result in disregulation of global gene transcription, which sometimes leads to heritable phenotypic changes. However, variations in heritable epigenetic information can be important for crop plants because they can result in stable phenotypic changes in agronomically important traits in the same way that genetic variations normally do. Here, we will describe examples of epigenetic variations in plants and the implications of epigenetic changes for phenotypic variation and plant breeding.

5.3.1 Plant Epigenome

The 'DNA methylome' (the genome-wide DNA methylation pattern) was first investigated to delineate the epigenome structure of an organism, mainly because DNA methylation is a universal epigenetic modification that is relatively easy to analyze using high-throughput technologies. Several studies have reported that gene expression patterns are strongly correlated with DNA methylation and other epigenetic factors such as histone modifications and the abundance of small RNAs (Zhang et al. 2006, 2007; Lister et al. 2008). Therefore, DNA methylome profiling is often the first step towards building a draft epigenome, which can help establish a general relationship between the global epigenome structure and gene expression patterns. Pericentromeric regions are heterochromatic and enriched with repeat sequences such as transposable elements, which have been found to be heavily methylated, along with a high abundance of small RNAs. This finding has led to the suggestion that the primary role of small interfering RNAs (siRNAs) is to trigger DNA methylation through the RNA-directed DNA methylation (RdDM) pathway (Chan et al. 2005; Zhang et al. 2006; Henderson and Jacobsen 2007). However, not all siRNAs are associated with DNA methylation, implying that it is more likely that an RdDM-independent mechanism is involved.

5.3.2 Epialleles and Epimutants

As mentioned, epigenetic alterations of genes may lead to changes in their expression. In some cases, both altered epigenetic states and expression patterns are mitotically/meiotically stable and inherited by the offspring. An individual with altered epigenetic modifications may produce offspring with different phenotypes, despite them having the same DNA sequence as individuals with no such modifications. Individuals with different phenotypes that are not caused by DNA mutations but by epigenetic alterations are called 'epimutants' and the corresponding epigenetic alleles that are responsible for the altered phenotypes are called an 'epialleles'. Transgenerational epigenetic inheritance involves the formation of epialleles and their consistent propagation in the next generations. Notably, most of the naturally occurring or artificially induced epialleles are caused by changes in DNA methylation (Zhang and Hsieh 2013).

Some of the first epimutants reported in *Arabidopsis* were induced in mutant backgrounds defective in DNA methylation. The METHYLTRANSFERASE1 (*MET1*), and DECREASE IN DNA METHYLATION1 (*DDM1*) genes encode a maintenance DNA methyltransferase and a SWI/SNF-like ATP-dependent chromatin remodeling factor, respectively (Vongs et al. 1993; Jeddeloh et al. 1999). The *met1* or *ddm1* mutants have a genome-wide decrease in DNA methylation levels and display severe developmental defects and abrupt activation of transposable elements (Kato et al. 2003; Lippman et al. 2004). Further, epimutations are induced at a high frequency against these mutant backgrounds, probably because of abrupt changes in the global DNA methylation patterns. Interestingly, some phenotypic changes in these mutants can be inherited by the offspring, even without an original causative mutation. For example, the late flowering phenotype of the *fwa* mutant was caused by a *ddm1* mutation, which induced a release of gene silencing by decreasing DNA methylation. After outcrossing to a wild-type, an ectopically expressed *fwa* epiallele was consistently transmitted as a dominant allele following Mendelian segregation (Kakutani 1997). A later study verified that a loss of DNA methylation in the repeat regions upstream of the gene caused transcriptional activation of *FWA* leading to late flowering (Soppe et al. 2000). An epigenetically silenced version of *SUPERMAN* (*SUP*) in *Arabidopsis* is another example of the formation of an epiallele caused by global DNA methylation changes (Jacobsen and Meyerowitz 1997).

In addition to the artificially induced epimutations in DNA methylation-defective mutants, several studies have reported epimutants and epialleles that occur spontaneously in nature. The first naturally occurring epiallele was reported in *Linaria vulgaris*, in which an asymmetric flower was transformed into a radially symmetric (peloric) flower (Cubas et al. 1999). The abnormal flower structure was found to be caused by DNA hypermethylation and silencing of the *Lcyc* gene, which encodes a *CYCLOIDEA* gene homolog that is known to regulate floral asymmetry in *Antirrhinum* (Cubas et al. 1999). Intriguingly, this mutant was first described more than 250 years ago by Linnaeus, suggesting that an epiallele can be stable enough for the maintenance of transgenerational epigenetic inheritance over many generations.

Several epimutations associated with agronomically important traits have also been found in garden plants. In tomato (*Solanum lycopersicum*), fruit ripening is inhibited in *colorless non-ripening* (*cnr*) mutants (Manning et al. 2006). The *Cnr* gene encodes an SBP-box transcription factor and its promoter in the *cnr* mutants was found to be hypermethylated leading to gene silencing (Manning et al. 2006). A follow-up study revealed that DNA methylation at the *Cnr* promoter region gradually decreased during fruit ripening in wild-type plants, indicating that DNA methylation was developmentally regulated (Zhong et al. 2013).

In many cases, epimutations are associated with the insertion of transposable elements. For example, the insertion of a hAT transposon near the *CmWIP1* gene in melon (*Cucumis melo*) was found to induce the spreading of DNA methylation to the promoter leading to gene silencing (Martin et al. 2009). The expression of *CmWIP1*was shown to promote carpel abortion and the development of male flowers, while its silencing suppressed anther development and resulted in the production of mostly female flowers (Martin et al. 2009). Therefore, CmWIP1 seemed to act as a sex-determination switch in melon flower development.

A recent study showed that expression of the tomato *VTE3* gene, which encodesthe2-metyl-6-phytylquinol methyltransferase that catalyzes the final steps of γ- and δ-tocopherol (vitamin E) biosynthesis, was modulated by the differential DNA methylation of a short interspersed nuclear element (SINE) located in the promoter region of the gene (Quadrana et al. 2014). DNA methylation of the *VTE3* promoter was found to be spontaneously reverted, which generated different epialleles with varying expression levels, indicating that naturally occurring epialleles may also be responsible for regulating metabolite contents. It is now recognized that repeat-induced DNA methylation and gene silencing is quite prevalent in plant epimutants. The wild-type *Arabidopsis FWA* gene, mentioned above, is also silenced by DNA methylation of a SINE in the promoter region, and, an *Arabidopsis bonsai* (*bns*) epiallele has also been associated with hypermethylation of long interspersed nuclear elements (LINEs) (Saze and Kakutani 2007).

Several epialleles associated with phenotypic variations have also been reported in rice. The rice *Epi-d1* allele is a spontaneously induced allele of the *DWARF1* (*D1*) gene whose expression is silenced by DNA methylation (Miura et al. 2009). The rice *Epi-d1* mutant is a dwarf and metastable, displaying chimeric features of both the normal and dwarf phenotypes. Another rice epimutation was found to occur in the *WEALTHY FARMER'S PANICLE* (*WFP*) gene encoding OsSPL14, which was identified by QTL analysis for panicle branching (Miura et al. 2010). *OsSPL14* was highly expressed during the reproductive stage to promote panicle branching and high grain yield, but its expression was limited in Nipponbare (the standard rice variety) because of DNA methylation. The introduction of a less methylated *OsSPL14*^{WFP} allele into Nipponbare resulted in increased rice production (Miura et al. 2010), suggesting that an epiallele associated with an agronomically important trait could be a valuable resource for crop breeding.

5.3.3 Epimutants as a Breeding Resource

Considering the genetic paucity of domesticated crop plants, the induction of epimutants with diverse phenotypic variations may be a promising strategy to diversify resources in crop breeding because a change in DNA methylation has been shown to be a major cause of epiallele generation. Although epimutations can be induced in DNA methylation-defective mutant backgrounds, such as *met1* and *ddm1* in *Arabidopsis*, most crop species lack equivalent mutant lines in which to generate epimutants. As an alternative, a DNA methyltransferase inhibitor 5-aza-2′-deoxycytidine (5-azadC) has been used as an epimutation inducer. For example, rice plants resistant to the bacterial pathogen *Xanthomonas oryzae* were obtained by 5-azadC treatment of seeds (Akimoto et al. 2007). The resistance resulted from hypomethylation of a disease resistance gene *Xa21G*. A hypomethylated population of *Brassica rapa* with large phenotypic diversity and *B. napus* with increased yield were also generated by 5-azadC treatment (Hauben et al. 2009; Amoah et al. 2012).

In *Arabidopsis*, epigenetic recombinant inbred lines (epi-RILs) were generated from a cross between wild-type and a *met1* mutant (Reinders et al. 2009). These lines were genetically identical but epigenetically distinct (Fig. 5.21), and displayed variable morphological or developmental characteristics for traits such as flowering time, height, weight, and pathogen resistance (Reinders et al. 2009). This study strongly suggested that a variety of phenotypic variations were generated by epigenetic alterations that were mostly caused by DNA methylation changes, and that the corresponding epialleles were stably heritable for many generations. Therefore, it is highly plausible that epigenetically induced heritable traits in crop plants can be used as a valuable resource in plant breeding.

5.3.4 Concluding Remarks and Future Directions

There is a huge demand on diverse genetic variations that can be used as breeding resources. Therefore, many scientists and breeders have looked for wild relatives carrying useful traits of agronomic importance. Diversifying epigenetic variations

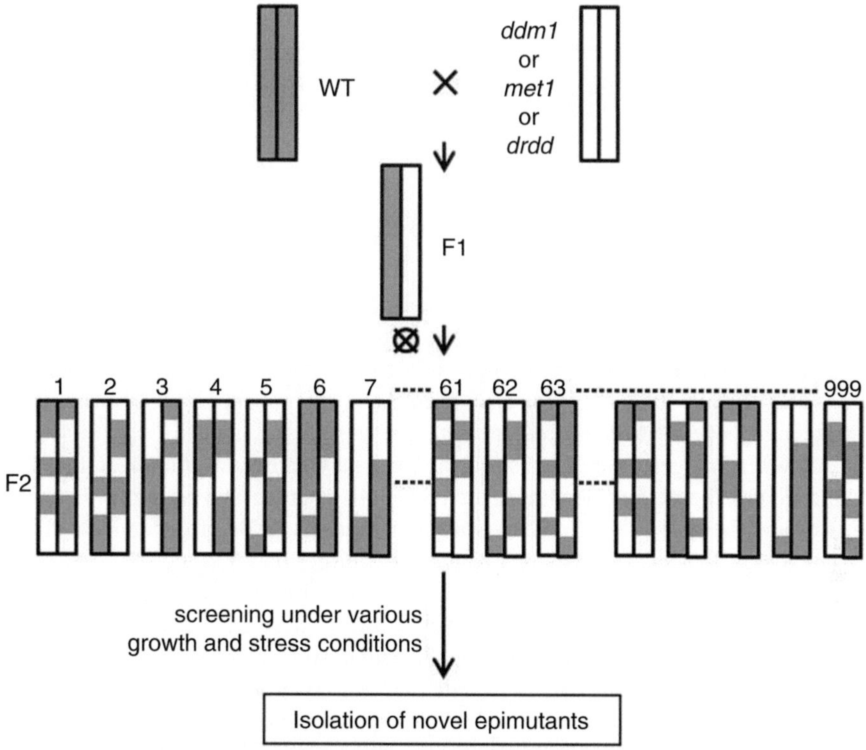

Fig. 5.21 Induction of epimutants in a DNA methylation-defective mutant background

in a limited pool of germplasm could be another way to increase the utility of current breeding materials by unleashing many hidden characters. A number of epialleles can arise either in a DNA methylation defective mutant background or by 5-azadC treatment, but epimutations stochastically can occur anywhere in the genome. Recent advances in genome editing technology have allowed the genome in many organisms to be tailored (Gaj et al. 2013). It has also been demonstrated that the plant genome can be arbitrarily manipulated (Chen and Gao 2013; Maeder et al. 2013; Mendenhall et al. 2013). Current genome editing technologies are based primarily on engineered nucleases in which DNA nuclease is fused to sequence-specific DNA binding modules such as ZF, TALE, and CRISPR/Cas (Gaj et al. 2013). This principle can be applied to epigenome engineering by fusing DNA binding modules to epigenetic modifiers such as DNA methyltransferase and DNA demethylase to induce epigenetic alterations at target loci. Recently, targeted DNA methylation was achieved in *Arabidopsis* using a ZF-SUVH2 fusion to recruit RdDM machineries to the target gene (Johnson et al. 2014). Therefore, in addition to the induction of novel epialleles in a random fashion, targeted epigenome editing may be another promising way to alter the expression of specific genes associated with a trait of interest. This "reverse epigenetics approach" should be highly feasible thanks to the development of genome editing tools and the increasing availability of genome and epigenome information for many crop plants. All scientists and breeders now belong to the post-genome era, and there is no doubt that the application of epigenetic principles to major crop species will become an important future direction of plant breeding.

5.4 Future of Genomics and Global Corporations

NGS technology needs expensive instruments and high-throughput bioinformatics pipelines; therefore individual researchers or breeders can no longer analyze huge amounts of genomics data within their own groups. As a result, genomics institutions and commercial companies have been established to support users who want to generate, access, and analyze large datasets.

A few leading genomics groups such as the Joint Genome Institution (JGI) in the United States, the Beijing Genome Institution (BGI) in China, and the Sanger Institute in England, have achieved amazing success in various fields of genomics and for thousands of organisms. Recently, many small bioinformatics groups or companies have been created in many countries to support researchers and breeders. Various user-friendly bioinformatics tools and web browsers have been made available to users to help them access and analyze public, and in some cases their own, datasets.

A recent example of a successful international collaboration, is the genomics-assisted breeding program established between the International Rice Research Institute (IRRI) and researchers in other countries (Bailey-Serres et al. 2010; Luo 2010) that developed superior rice cultivars by the accumulation of resistance genes for flooding, drought, potassium deficiency, as well as many important diseases. Many other successful applications of genomics-assisted breeding have been reported in many important crops and vegetables. Genomics information is also beginning to expand for many valuable wild-type resource plants, such as medicinal plants.

Good plant resources can be bottlenecks for expanding genomics research; therefore, collaborations between researchers will help in increasing the availability of these materials. The lack of highly efficient user-friendly bioinformatics techniques and pipelines is another limiting factor for the advancement of genomics-assisted breeding among breeders. Close collaboration between researchers and genomics and bioinformatics groups will help accelerate the 'breeding-by-design' approach for creating superior varieties of important plant species.

References

Akimoto K, Katakami H, Kim HJ et al (2007) Epigenetic inheritance in rice plants. Ann Bot 100(2):205–217

Amoah S, Kurup S, Rodriguez Lopez CM et al (2012) A hypomethylated population of Brassica rapa for forward and reverse epi-genetics. BMC Plant Biol 12:193

Atwell S, Huang YS, Vilhjálmsson BJ et al (2010) Genome-wide association study of 107 phenotypes in *Arabidopsis thaliana* inbred lines. Nature 465:627–631
Avery OT, Macleod CM, McCarty M (1944) Studies on the chemical nature of the substance inducing transformation of pneumococcal types: induction of transformation by a desoxyribonucleic acid fraction isolated from Pneumococcus type III. J Exp Med 79:137–158
Bailey-Serres J, Fukao T, Ronald P et al (2010) Submergence tolerant rice: *SUB1*'s journey from landrace to modern cultivar. Rice 3:138–147
Bashir A, Klammer AA, Robins WP et al (2012) A hybrid approach for the automated finishing of bacterial genomes. Nat Biotechnol 30:701–707
Bentley DR, Balasubramanian S, Swerdlow HP et al (2008) Accurate whole human genome sequencing using reversible terminator chemistry. Nature 456:53–59
Bernstein BE, Meissner A, Lander ES (2007) The mammalian epigenome. Cell 128(4):669–681
Butler J, MacCallum I, Kleber M et al (2008) ALLPATHS: de novo assembly of whole-genome shotgun microreads. Genome Res 18:810–820
Chan M, Chan MW, Loh TW et al (2011) Evaluation of nanofluidics technology for high-throughput SNP genotyping in a clinical setting. J Mol Diagn 13(3):305–312
Chan SW, Henderson IR, Jacobsen SE (2005) Gardening the genome: DNA methylation in *Arabidopsis thaliana*. Nat Rev Genet 6(5):351–360
Chen K, Gao C (2013) TALENs: customizable molecular DNA scissors for genome engineering of plants. J Genet Genomics 40(6):271–279
Choi SH, Lee BH, Kim HJ et al (2014) Ginseng gintonin activates the human cardiac delayed rectifier K+ channel: involvement of Ca2+/calmodulin binding sites. Mol Cells 37:656–663
CLC bio: CLC Assembly Cell user manual. http://www.clcbio.com/products/clc-genomics-workbench/
Cubas P, Vincent C, Coen E (1999) An epigenetic mutation responsible for natural variation in floral symmetry. Nature 401(6749):157–161
Eid J, Fehr A, Gray J et al (2009) Real-time DNA sequencing from single polymerase molecules. Science 323:133–138
English AC, Richards S, Han Y et al (2012) Mind the gap: upgrading genomes with pacific biosciences RS long-read sequencing technology. PLoS One 7(11):e47768
Gaj T, Gersbach CA, Barbas CF 3rd (2013) ZFN, TALEN, and CRISPR/Cas-based methods for genome engineering. Trends Biotechnol 31(7):397–405
Hauben M, Haesendonckx B, Standaert E et al (2009) Energy use efficiency is characterized by an epigenetic component that can be directed through artificial selection to increase yield. Proc Natl Acad Sci U S A 106(47):20109–20114
Henderson IR, Jacobsen SE (2007) Epigenetic inheritance in plants. Nature 447(7143):418–424
Huang X, Feng Q, Qian Q et al (2009) High-throughput genotyping by whole-genome resequencing. Genome Res 19:1068–1076
International Human Genome Sequencing Consortium (2001) Initial sequencing and analysis of the human genome. Nature 409:860–921
International Rice Genome Sequencing Project (2005) The map-based sequence of the rice genome. Nature 436:793–800
Jacobsen SE, Meyerowitz EM (1997) Hypermethylated SUPERMAN epigenetic alleles in Arabidopsis. Science 277(5329):1100–1103
Jeddeloh JA, Stokes TL, Richards EJ (1999) Maintenance of genomic methylation requires a SW12/SNF2-like protein. Nat Genet 22(1):94–97
Johnson LM, Du J, Hale CJ et al (2014) SRA- and SET-domain-containing proteins link RNA polymerase V occupancy to DNA methylation. Nature 507(7490):124–128
Kakutani T (1997) Genetic characterization of late-flowering traits induced by DNA hypomethylation mutation in Arabidopsis thaliana. Plant J 12(6):1447–1451
Kato M, Miura A, Bender J et al (2003) Role of CG and non-CG methylation in immobilization of transposons in arabidopsis. Curr Biol 13(5):421–426
Kim S, Park M, Yeom SI et al (2014) Genome sequence of the hot pepper provides insights into the evolution of pungency in Capsicum species. Nat Genet 46(3):270–278

Koren S, Schatz MC, Walenz BP et al (2012) Hybrid error correction and de novo assembly of single-molecule sequencing reads. Nat Biotechnol 30:693–700

Lippman Z, Gendrel AV, Black M et al (2004) Role of transposable elements in heterochromatin and epigenetic control. Nature 430(6998):471–476

Lister R, O'Malley RC, Tonti-Filippini J et al (2008) Highly integrated single-base resolution maps of the epigenome in Arabidopsis. Cell 133(3):523–536

Liu WY, Kang JH, Jeong HS et al (2014a) Combined use of bulked segregant analysis and microarrays reveals SNP markers pinpointing a major QTL for resistance to *Phytophthora capsici* in pepper. Theor Appl Genet 127(11):2503–2513

Liu S, Liu Y, Yang X et al (2014b) The *Brassica oleracea* genome reveals the asymmetrical evolution of polyploidy genomes. Nat Commun 5:3930

Luo LJ (2010) Breeding for water-saving and drought-resistance rice (WDR) in China. J Exp Bot 61:3509–33517

Luo R, Liu B, Xie Y et al (2012) SOAP*denovo*2: an empirically improved memory-efficient short-read de novo assembler. Giga Sci 1:18

Maeder ML, Angstman JF, Richardson ME et al (2013) Targeted DNA demethylation and activation of endogenous genes using programmable TALE-TET1 fusion proteins. Nat Biotechnol 31(12):1137–1142

Manning K, Tor M, Poole M et al (2006) A naturally occurring epigenetic mutation in a gene encoding an SBP-box transcription factor inhibits tomato fruit ripening. Nat Genet 38(8):948–952

Martin A, Troadec C, Boualem A et al (2009) A transposon-induced epigenetic change leads to sex determination in melon. Nature 461(7267):1135–1138

Mendenhall EM, Williamson KE, Reyon D et al (2013) Locus-specific editing of histone modifications at endogenous enhancers. Nat Biotechnol 31(12):1133–1136

Miura K, Agetsuma M, Kitano H et al (2009) A metastable DWARF1 epigenetic mutant affecting plant stature in rice. Proc Natl Acad Sci U S A 106(27):11218–11223

Miura K, Ikeda M, Matsubara A et al (2010) OsSPL14 promotes panicle branching and higher grain productivity in rice. Nat Genet 42(6):545–549

Myers EW, Sutton GG, Delcher AL et al (2000) A whole-genome assembly of Drosophila. Science 287(5461):2196–2204

Palermo RE, Tisonici-Go J, Korth MJ et al (2013) Old world monkeys and new age science: the evolution of nonhuman primate systems virology. ILAR J 54(2):166–180

Pasam RK, Sharma R, Malosetti M et al (2012) Genome-wide association studies for agronomical traits in a worldwide spring barley collection. BMC Plant Biol 12:16

Poland JA, Brown PJ, Sorrells ME et al (2012) Development of high-density genetic maps for barley and wheat using a novel two-enzyme genotyping-by-sequencing approach. PLoS One 7(2):e32253. doi:10.1371/journal.pone.0032253

Quadrana L, Almeida J, Asis R et al (2014) Natural occurring epialleles determine vitamin E accumulation in tomato fruits. Nat Commun 5:3027

Quail MA, Smith M, Coupland P et al (2012) A tale of three next generation sequencing platforms: comparison of Ion Torrent, Pacific Biosciences and Illumina MiSeq sequencers. BMC Genomics 13:341

Reinders J, Wulff BB, Mirouze M et al (2009) Compromised stability of DNA methylation and transposon immobilization in mosaic Arabidopsis epigenomes. Genes Dev 23(8):939–950

Ribeiro FJ, Przybylski D, Yin S et al (2012) Finished bacterial genomes from shotgun sequence data. Genome Res 22:2270–2277

Rothberg JM, Leamon JH (2008) The development and impact of 454 sequencing. Nat Biotechnol 26(10):1117–1124

Saze H, Kakutani T (2007) Heritable epigenetic mutation of a transposon-flanked Arabidopsis gene due to lack of the chromatin-remodeling factor DDM1. EMBO J 26(15):3641–3652

Shendure J (2013) 2012 Curt stern award address. Am J Hum Genet 92(3):340–344

Sim S-C, Durstewitz G, Plieske J et al (2012) Development of a large SNP genotyping array and generation of high-density genetic maps in tomato. PLoS One 7:e40563

Song Q, Hyten DL, Jia G et al (2013) Development and evaluation of SoySNP50K, a high-density genotyping array for soybean. PLoS One 8(1):e54985

Soppe WJJ, Jacobsen SE, Alonso-Blanco C et al (2000) The late flowering phenotype of fwa mutants is caused by gain-of-function epigenetic alleles of a homeodomain gene. Mol Cell 6(4):791–802

The Arabidopsis Genome Initiative (2000) Analysis of the genome sequence of the flowering plant Arabidopsis thaliana. Nature 408:796–815

Trebbi D, Maccaferri M, de Heer P et al (2011) High-throughput SNP discovery and genotyping in durum wheat (Triticum durum Desf.). Theor Appl Genet 123(4):555–569

Vongs A, Kakutani T, Martienssen RA et al (1993) Arabidopsis-thaliana DNA methylation mutants. Science 260(5116):1926–1928

Wang X, Wang H, Wang J et al (2011) The genome of the mesopolyploid crop species *Brassica rapa*. Nat Genet 43:1035–1039

Westengen OT, Berg PR, Kent MP et al (2012) Spatial structure and climatic adaptation in African maize revealed by surveying SNP diversity in relation to global breeding and landrace panels. PLoS One 7(10):e47832

Yang TJ, Kim JS, Kwon SJ (2006) Sequence-level analysis of the diploidization process in the triplicated FLOWERING LOCUS C region of Brassica rapa. Plant Cell 18:1339–1347

Zhang CQ, Hsieh T-F (2013) Heritable epigenetic variation and its potential applications for crop improvement. Plant Breed Biotechnol 1:307–319

Zhang K, Davenport KW, Gu W et al (2012) Improving genome assemblies by sequencing PCR products with PacBio. Biotechniques 53:61–62

Zhang X, Henderson IR, Lu C et al (2007) Role of RNA polymerase IV in plant small RNA metabolism. Proc Natl Acad Sci U S A 104:4536–4541

Zhang X, Yazaki J, Sundaresan A et al (2006) Genome-wide high-resolution mapping and functional analysis of DNA methylation in arabidopsis. Cell 126(6):1189–1201

Zhao K, Tung CW, Eizenga GC et al (2011) Genome-wide association mapping reveals a rich genetic architecture of complex traits in Oryza sativa. Nat Commun 2:467

Zhong S, Fei Z, Chen YR et al (2013) Single-base resolution methylomes of tomato fruit development reveal epigenome modifications associated with ripening. Nat Biotechnol 31(2):154–159

Chapter 6
Concept of Genome-Wide Association Studies

Chang-Yong Lee, Tae-Sung Kim, Sanghyeob Lee, and Yong-Jin Park

Abstract The human genome project, which ended in 2003, provided a human genome map with 99.99 % accuracy. This project stimulated association studies between traits and genes and allowed the discovery of new genes. In 2006, an innovational method known as genome-wide association study (GWAS) was developed. GWAS is different from the traditional method of studying associations between a few candidate genes and traits in that GWAS can handle about one million single-nucleotide polymorphisms (SNPs) simultaneously.

These days, GWAS is widely used to study genetic factors related to phenotype in many species, including plants. The advances in next generation sequencing (NGS) have allowed large amounts of genetic data to be obtained at a relatively low cost.

In this chapter, we will review GWAS including the associated statistical concepts and available software packages. We describe the use of GAPIT (Genomic Association and Prediction Integrated Tool) and demonstrate SNP calling using the widely used commercial CLC Genomic Workbench with re-sequenced cucumber genome data. Finally, we examine the current status of genomics research worldwide that requires whole-genome resequencing, one of the requirements of GWAS.

Author contributed equally with all other contributors.

C.-Y. Lee
Department of Industrial and Systems Engineering, Kongju National University, Cheonan, Republic of Korea
e-mail: clee@kongju.ac.kr

T.-S. Kim • Y.-J. Park (✉)
Department of Plant Resources, Kongju National University, Yesan, Republic of Korea
e-mail: yjpark@kongju.ac.kr

S. Lee
Department of Bio Resource Engineering & Plant Engineering Research Institute, Sejong University, Seoul, Republic of Korea
e-mail: sanglee@sejong.ac.kr

H.-J. Koh et al. (eds.), *Current Technologies in Plant Molecular Breeding*,
DOI 10.1007/978-94-017-9996-6_6

6.1 GWAS Methodology

GWAS has been used widely to investigate genetic variants associated with various diseases to identify markers for these diseases based on the hypothesis of "common disease, common variant" (Visscher et al. 2012). That is, if a common disease or trait is inherited, then its causal genetic variant is assumed to be common, even in unrelated subjects. The hypothesis implies that the genetic influence or cause of a common disease stems from allelic variants, such as SNPs.

Currently, analysis methods for genome-wide association studies (GWASs) rely mostly on statistical inference to estimate associations between traits (or diseases) and genetic variants, such as single-nucleotide polymorphisms (SNPs). In general, a GWAS infers these associations through a hypothesis test with pertinent test statistics. Association analyses for individual SNPs usually use Pearson's χ^2-test, Fisher's exact test, or the *F*-test under a null hypothesis of no association.

For a qualitative trait, the frequency of the different phenotype values is often used to infer associations. Generally, the trait is classified into two sample sets (test and control) and the genotype frequency at each SNP locus in each set is calculated and used to form a contingency table. This table can then be used to calculate the odds ratio or relative risk that is used in the hypothesis test. For a quantitative trait, a regression model is often used for an association study in which each SNP is an explanatory (or independent) variable and the trait is the response (or dependent) variable. In this case, the regression coefficient can be estimated using the least-squares method to test the null hypothesis of the coefficient being zero.

In GWAS, the population structure and the multiple tests must be considered. For population structure, population stratification must be eliminated to maintain the homogeneity of the population. Commonly used approaches include methods based on a statistical model (Pritchard et al. 2000) and principal component analysis (Price et al. 2006). The regression model can include population effects, such as age and sex, in addition to SNP loci. GWAS outputs can be visualized as Manhattan and quantile–quantile (Q-Q) plots (Gibson 2010).

From the perspective of statistical inference, GWAS is basically a multiple test problem because in general about one million SNPs are tested against several phenotypes at the same time. Thus, the significance level used for a single test may not be adequate for a multiple test. Consider n simultaneous tests with a significance level α as for the single test. In this case, the probability ζ that at least one test is false positive out of n is given as $\zeta = 1-(1-\alpha)^n$. For examples, when $\text{n} = 100$ and $\alpha = 0.05$ or 0.01, we obtain $\zeta = 0.99$ and $\zeta = 0.63$, respectively. This means that when the significance level for a single test is used for a multiple test, the chance of a false positive is either $\zeta = 0.99$ or $\zeta = 0.63$, which indicates the very high probability of a false positive. In Fig. 6.1, ζ versus α when $n = 100$ is plotted, which shows that ζ increases sharply as α varies. Thus, a pertinent significance level for the multiple test is required, and the family-wise error rate (FWER) and the false discovery rate (FDR) are the two methods that are most commonly used to handle the multiple test (Yekutieli and Benjamini 1999).

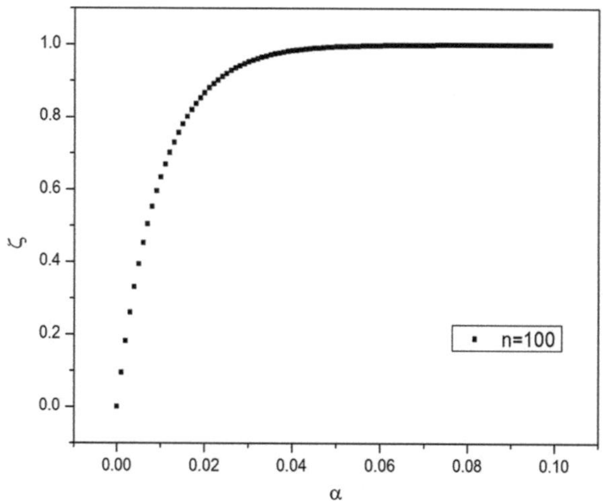

Fig. 6.1 Plot of ζ versus α when $n = 100$

The most commonly used is FWER, which was suggested by Bonferroni (Shaffer 1995), as the simplest and the most conservative method. To ensure that the probability of a false positive (or type I error) is less than –, the Bonferroni method assigns the significance level $\alpha' = \frac{\alpha}{n}$ for n multiple test. For example, when $\alpha = 0.01$ and $n = 1{,}000{,}000$, $\alpha' = \frac{0.01}{1{,}000{,}000} \approx 10^{-8}$. Thus, when the p-value of a test statistic satisfies $p < 10^{-8}$, it is concluded that the SNP is significantly associated with the trait. The Bonferroni method is easy to use and has the advantage of controlling the type I error to below – for the whole multiple test. The Bonferroni method is conservative in the sense that if a hypothesis test is found to be significant, the test will also be significant with other methods. Although the Bonferroni method is simple to use, the significance level tends to be so low that it loses its power; i.e., because the type I error is too conservative, the true negative problem may occur more often than one would expect. As a result, the null hypothesis may be accepted despite there being a significant association between a SNP and a trait.

FDR controls the rate of false decisions to less than a predetermined rate. That is, FDR can be set to tolerate a certain number of tests to be incorrectly decided; therefore, the FDR is the proportion of incorrect rejections among all rejections of the null hypothesis. The Benjamini–Hochberg method (Benjamini and Hochberg 1995) is the method most commonly used in FDR. Briefly, the Benjamini–Hochberg method ranks the p-values obtained from each test in ascending order and rejects all hypotheses for which the p-value is greater than a predetermined overall significance level q.

Suppose there are n p-values $p_1, p_2, \cdots, p_n$ from n multiple tests $H_1, H_2, \cdots, H_n$, the false discovery proportion q is calculated as follows: p-values are ranked in ascending order and denoted as $p_{(1)} \le p_{(2)} \le \cdots \le p_{(n)}$, and $H_{(j)}$ denotes the hypothesis with p-value $p_{(j)}$. Using the ranked p-values, k such that is determined as

$$k = \max_j \left\{ p_{(j)} \le \frac{j}{n} q \right\}. \tag{6.1}$$

Here, k is the maximum j that satisfies $p_{(j)} \le \frac{j}{n} q$; thus k depends on the significance level q. Based on q, all hypotheses $H_{(i)}\left(i = 1, 2, \cdots, k\right)$ are rejected. Note that Eq. 6.1 provides a critical value for accepting or rejecting a hypothesis with a significance level q. That is, by using $q_{(j)} = \frac{n}{j} p_{(j)}$, a minimum significance level $q_{(j)}$ is obtained from the p-value $p_{(j)}$ of the test. This $q_{(j)}$ value is called the FDR correction.

One problem with the FDR correction is that while $p_{(j)}$ is monotonic, $q_{(j)}$ is not. Consider, for example, a situation in which $p_{(a)} \ge p_{(b)}$, but $q_{(a)} \le q_{(b)}$. In this case, if $q_{(a)}$ is the critical value that causes $H_{(a)}$ to be rejected, then $H_{(b)}$ will also be rejected despite $p_{(b)} \ge p_{(a)}$. To resolve this problem, Yekutieli and Benjamini (2001) proposed an FDR adjustment method that calculated an adjusted FDR. While maintaining the basic concept of FDR, an FDR adjustment parameter $q^*_{(j)}$, (adjusted $q_{(j)}$) was defined as $q^*_{(j)} = \min\left\{q_{(k)}, k \ge j\right\}$. Thus, $q^*_{(j)}$ is a monotonic increasing function of $p_{(j)}$, which is called the FDR adjusted p-value.

6.2 Software Packages for GWAS

Various statistical analysis methods and genome preprocessing steps have been implemented in many software packages; among these, PLINK (Purcell et al. 2007) and GAPIT (Lipka et al. 2012) can be freely downloaded and are widely used. PLINK is an open-source whole genome association analysis toolset developed by Shaun Purcell and others at Harvard University and implemented using C/C++. The PLINK toolset consists of five parts: data management, summary statistics, population stratification, association analysis, inference ancestry. Recently, a Java-based software package, gPLINK, was developed to make it easier for biologists to use PLINK.

GAPIT, developed at the Institute for Genomic Diversity in Cornell University, is a freely available R package (http://www.r-project.org) that can perform GWAS and genome prediction (or selection). GAPIT has been widely used in GWAS of quantitative traits in plants. Because the main focus of this chapter is quantitative traits, here we explain, in some detail, the characteristics of GAPIT and how to use it.

The GAPIT project started in 2011 and is continuously being updated. GAPIT can be downloaded freely from http://www.maizegenetics.net/GAPIT. The GAPIT R package can handle many statistical methods for association tests between genetic variants and traits. TASSEL (Bradbury et al. 2007), written in JAVA program language, is a software package similar to GAPIT that can evaluate traits associations, evolutionary patterns, and linkage disequilibrium. However, unlike TASSEL, GAPIT can handle up to one million SNPs for more than 10,000 traits at the same time.

Various statistical methods are implemented in GAPIT, including the compressed mixed linear model, the efficient mixed model association (EMMA), and the population parameters previously determined approach. GAPIT also uses R libraries, such as multtest, gplots, LDheatmap, genetics, and a modified version of EMMA. Thus, before running GAPIT, it is necessary to first install these libraries. The EMMA R-package in particular provides an efficient way to estimate the variances that are need for the mixed liner model. At the time of writing (January 2015), the latest version of GAPIT was version 3.35.

GAPIT uses a mixed linear regression model for the association between phenotype and genotype (Bouchet et al. 2010). The mixed linear model includes a random effect on top of the usual linear model, and can be expressed using the Henderson's matrix notation as

$$y = X\beta + Zu + e, \tag{6.2}$$

where it is assumed that $u = N\left(0, K\sigma_g^2\right)$ and $e = N\left(0, I\sigma_e^2\right)$. In addition, $n \times 1$ vector of y represents phenotype (or trait), n is the number of phenotype values, which is the same as the number of samples. The right-hand side of Eq. 6.2 can be detailed as $n \times q$ design matrix X that represents fixed effects, such as the intercept, SNP, and the population structure. The $q \times 1$ vector of β represents the regression coefficients that need to be estimated for the fixed effect. If population structure is not considered, then q is nothing but the number of genotypes of a genetic variant (e.g., aa, aA, or AA) together with the intercept. The $n \times p$ matrix of Z is a design matrix that takes into account the kinship effect between samples, and the $p \times 1$ vector of u is another regression coefficient for the random effect. Here, p represents the number of clusters for the kinship matrix and satisfies $p \le n$; i.e., when the samples are not grouped, $p = n$. Last, the $n \times 1$ vector e represents the residual effect.

In addition, when the $p \times p$ kinship matrix K is used, the variance of regression coefficient u becomes $Var\left(u\right) = \sigma_g^2 K$, and the variance of residual is $Var\left(e\right) = \sigma_g^2 I$, where I is the $n \times n$ identity matrix. Thus, under the above conditions

$$E\left(y\right) = X\beta, Var\left(y\right) = V = \sigma_g^2 ZKZ^T + \sigma_e^2 I. \tag{6.3}$$

The best linear unbiased estimate, $\hat{\beta}$, of β for the fixed effect and the best linear unbiased prediction, $\hat{u}$, of u for the random effectcan be calculated as

$$\begin{pmatrix}\hat{\beta}\\ \hat{u}\end{pmatrix}=\begin{pmatrix}X^TX & X^TZ\\ Z^TX & Z^TZ+K^{-1}\dfrac{\sigma_e^2}{\sigma_g^2}\end{pmatrix}^{-1}\begin{pmatrix}X^Ty\\ Z^Ty\end{pmatrix}=\begin{pmatrix}C_{11} & C_{12}\\ C_{21} & C_{22}\end{pmatrix}\begin{pmatrix}X^Ty\\ Z^Ty\end{pmatrix}, \tag{6.4}$$

Where $\hat{\beta}$ can be expressed as

$$\hat{\beta}=\left(X^TV^{-1}X\right)^{-1}X^TV^{-1}y, V=\sigma_g^2ZKZ^T+\sigma_e^2I, \tag{6.5}$$

and the variance of $\hat{\beta}$ is $Var\left(\hat{\beta}\right)=C_{11}\sigma_e^2$. Here, the estimate of σ_e^2 is given as

$$\widehat{\sigma_e^2}=\frac{\left\{y^Ty-\left(\widehat{\beta^T}X^Ty+\widehat{u^T}Z^Ty\right)\right\}}{n-r(X)}, \tag{6.6}$$

where $r(X)$ is the rank of X, the number of genotypes. Additionally, the estimate of σ_g^2 is given as

$$\widehat{\sigma_g^2}=\frac{u^TK^{-1}u+\widehat{\sigma_e^2}\,\mathrm{trace}\left(K^{-1}C_{22}\right)}{n}. \tag{6.7}$$

To obtain $\hat{\beta}$, it is necessary to calculate V^{-1}, which needs $\hat{\beta}$. To resolve this problem, Henderson (1975) implemented an iterative method using Eqs. 6.6 and 6.7.

Once $\hat{\beta}$ is obtained, the hypothesis test can be performed. The null hypothesis that the genetic variants is not associated with the trait is used; i.e., $H_0: M\beta=0$, where M is a $s\times q$ matrix; here, s and q are the number of SNPs in the test and the number of genotypes, respectively. Under the null hypothesis, the test statistic F_0 follows the F distribution (Kang et al. 2008; Kennedy et al. 1992) as

$$F_0=\frac{\left(M\hat{\beta}\right)^T\left\{M\left(X^T\hat{V}^{-1}X\right)^{-1}M^T\right\}^{-1}\left(M\hat{\beta}\right)}{s\widehat{\sigma_e^2}}\sim F(s,n-q). \tag{6.8}$$

6.3 Using GAPIT

6.3.1 *Input for GAPIT*

The input files for GAPIT are as follows:

- Phenotype data
- Genotype data
- Kinship matrix calculated from genotype data
- Covariate variable (Q-matrix) or principal component (PC) for the population structure calculated from genotype data.

Note that each item in the input data is delimitated by a "Tab" and must be in text format. In addition, it is advisable to list the names of samples in alphabetical order starting with an upper case letter. Because the kinship and population structure matrices can be calculated from the genotype data, they are optional. Thus, the only essential input files are those containing the phenotype and genotype data.

6.3.1.1 Phenotype Data

GAPIT can carry out GWAS for multiple phenotypes sequentially if all information about a trait is included in the input file. The first column must contain the sample name, and subsequent columns contain the values of the phenotypes. If a value is unknown, it is denoted it as "NaN" or "NA". The first row in the phenotype data file contains the column header labels. An example of phenotype data for seven traits and ten samples are shown in Fig. 6.2.

6.3.1.2 Genotype Data

GAPIT accepts two types of genotype data files: HapMap format (Fig. 6.3) and numeric format. HapMap format is a general format that was designed to store the sequence data in the HapMap project. Each row contains SNP information

Taxa	Heading	Length	Width	LWR	Protein	Oil	Amylose
RWG-001	58	4.48	1.987	2.255	8	5.8	19.2
RWG-002	72	4.527	3.243	1.396	9.9	7	20.4
RWG-003	46	4.54	2.78	1.633	6.5	7.3	0
RWG-004	38	4.557	2.587	1.762	8.3	8.2	18.7
RWG-005	50	4.627	2.867	1.614	8.4	9.2	19.3
RWG-006	63	6.187	2.76	2.242	9.3	19	19.6
RWG-007	62	6.46	3.02	2.139	8.7	20.2	19
RWG-008	44	6.567	3.003	2.186	9.6	20.7	18.9
RWG-009	-999	0	0	1	0	0	-999
RWG-010	51	4.717	2.93	1.61	8.3	14.1	19.5

Fig. 6.2 Example of phenotype data input file for GAPIT

rs	allele	chrom	pos	strand	assembly	center	protLSID	assayLSID	panel	QCcode	RWG-001	RWG-002	RWG-003
chr07_1030	I/T	7	1030	NA	NA	NA	NA	NA	NA	NA	II	II	NN
chr07_1030	T/A	7	1030	NA	NA	NA	NA	NA	NA	NA	TT	TT	NN
chr07_1037	I/T	7	1037	NA	NA	NA	NA	NA	NA	NA	II	II	NN
chr07_1038	A/T	7	1038	NA	NA	NA	NA	NA	NA	NA	AA	AA	NN
chr07_1045	A/G	7	1045	NA	NA	NA	NA	NA	NA	NA	AA	AA	NN
chr07_1052	A/T	7	1052	NA	NA	NA	NA	NA	NA	NA	AA	AA	NN
chr07_1208	T/C	7	1208	NA	NA	NA	NA	NA	NA	NA	TT	TT	NN
chr07_1257	A/C	7	1257	NA	NA	NA	NA	NA	NA	NA	AA	AA	NN
chr07_1290	T/C	7	1290	NA	NA	NA	NA	NA	NA	NA	TT	NN	NN
chr07_1329	C/T	7	1329	NA	NA	NA	NA	NA	NA	NA	NN	NN	NN

Fig. 6.3 Example of HapMap format

Table 6.1 Relation between genotype and IUPAC code

Genotype	AA	CC	GG	TT	AG	CT	CG	AT	GT	AC
IUPAC code	A	C	G	T	R	Y	S	W	K	M

(chromosome and position), and each column contains sample information; thus, SNP and sample information can be stored in the same file. The first 11 columns display attributes of the SNPs and the remaining columns show the nucleotides observed at each SNP for each sample. GAPIT uses only three of the first 11 columns: "rs" for the SNP name, "chrom" for the chromosome containing the SNP, and "pos" for the base pair (bp) position of the SNP. The first row contains the column header labels and subsequent rows contain the information for each SNP. An example of genotype data in HapMap format is displayed in Fig. 6.3. Missing genotype data can be indicated by either "NN" or "N". GAPIT also accepts genotypes either in double bit or in the standard IUPAC code of a single bit as shown in Table 6.1.

GAPIT also accepts the numeric format used by EMMA. In the numeric format, each column represents a SNP and each row represents a sample. This format is inadequate when the number of SNPs is large, because the numbers of columns that are required becomes very large when there is a large number of SNPs. In the numeric format, genotype information is denoted by numbers; for example, homozygotes are denoted as "0" or "2", while heterozygotes are denoted as "1". The first row contains the header labels, which in this case are the SNP names, and the first column contains the sample name. It should be noted that the numeric format does not contain the chromosome or the positions of the SNPs. Therefore, two separate files for the numeric genotypic data (GD) and the position of each SNP (GM) along the genome must be provided as input to GAPIT. Examples of the GD and GM files are shown in Figs. 6.4 and 6.5.

6.3.2 *Visual Outputs of GAPIT*

GAPIT provides a series of output files for the GWAS results in two formats: csv (comma separate value) and pdf (printable document format). Because GAPIT can evaluate p-values for the association between about one million SNPs and a large

taxa	KNU0001.1	KNU0002.1	KNU0003.1	KNU0004.1	KNU0005.1
RWG-001	2	0	0	2	2
RWG-002	2	2	0	2	2
RWG-003	2	0	0	2	2
RWG-004	2	2	0	2	2
RWG-005	0	0	0	2	2

Fig. 6.4 Example of genotype data in numeric format

Fig. 6.5 Example of GM file showing chromosome number and SNP position

SNP	Chromosome	Position
KNU0001.1	1	157293
KNU0002.1	1	1283892
KNU0003.1	1	2414036
KNU0004.1	1	2829071
KNU0005.1	1	2912328

number of traits, it is almost impossible to examine the results for each SNP individually. Thus, visual outputs are provided so that candidate SNPs can be examined via numeric results. The most important visual output files are the Manhattan plot and the Q-Q plot, which are output as GAPIT.trait.Manhattan-Plot.Genomewise.pdf and GAPIT.trait.QQ-Plot.pdf files, respectively.

6.3.2.1 Manhattan Plot

The Manhattan plot is a scatter plot that displays p-values in the $-\log_{10}(p)$ scale versus the genomic position of the SNP and its chromosome number. In this case, $-\log_{10}(p)$ is the negative logarithm of the p-value obtained from the *F*-test of the null hypothesis of no association. Thus, large peaks correspond to small p-value, suggesting that the surrounding genomic region has a strong association with the trait, as shown in Fig. 6.6. The plot is called a Manhattan plot because it resembles the Manhattan skyscrapers in New York. GAPIT produces one Manhattan plot for the entire genome and individual Manhattan plots for each chromosome.

6.3.2.2 Quantile–Quantile Plot

The Q-Q plot is used to assess how well the model used in the GWAS accounts for population structure and familial relatedness (Fig. 6.7). If the null hypothesis is true for all SNPs, then the p-value should be distributed according to the uniform distribution over [0, 1]. If associations exist, then the corresponding p-values will deviate

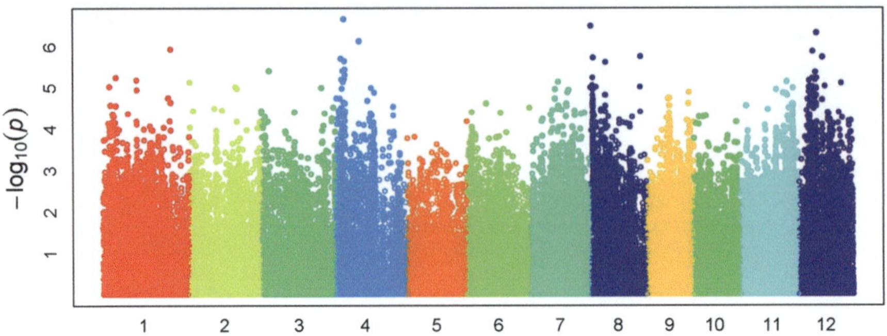

Fig. 6.6 Example of a Manhattan plot. The chromosome numbers are shown on the x-axis

Fig. 6.7 Example of a quantile–quantile (Q-Q) plot

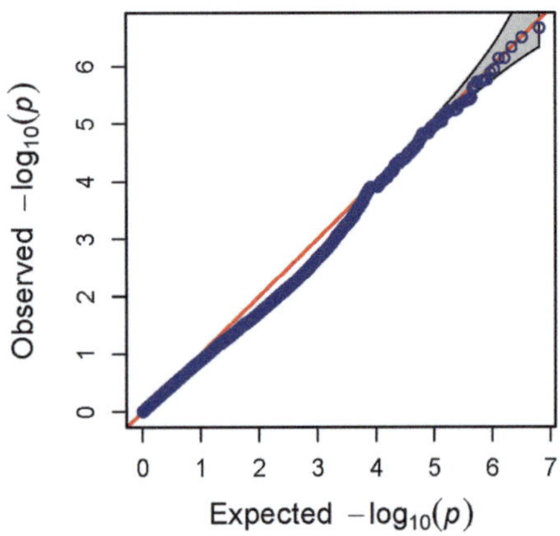

from the uniform distribution. The experimental p-values are plotted against the theoretical p-values; i.e., the Q-Q plot displays

$$\left(-\log_{10}\frac{i}{n}, -\log_{10} p(i)\right), i = 1, 2, \cdots, n, \tag{6.9}$$

Where n is the number of tested SNPs and $p(i)$ is the i-th smallest p-value. In the Q-Q plot, the negative logarithms of the p-values from the models fitted in the GWAS are plotted against the expected value under the null hypothesis of no association with the trait. Because most of the SNPs tested are probably not associated with the trait, the majority of the points in the QQ-plot should lie on the diagonal line. Deviations from this line suggest the presence of spurious associations owing to population structure and familial relatedness. Then, the SNPs on the upper right section of the graph deviate from the diagonal. These are the SNPs that are most likely associated with the trait under study.

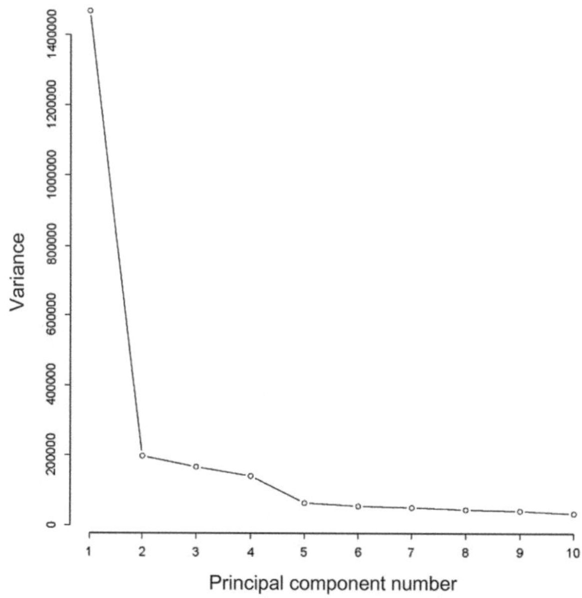

Fig. 6.8 Example of a principal components plot obtained by principal components analysis

6.3.2.3 Principal Component Plot

The principal component (PC) analysis estimates the effect of the population structure by analyzing multivariate data in terms of covariance structure of the data. The components that maximize the variance and reduce the multivariate data to a few important components, the principal components, are identified. For each PC included in the GWAS model, the observed PC values are plotted and output as a GAPIT.PCA.eigenValue.pdf file. Every possible pair of PCs is plotted against each other. A representative PC plot is shown in Fig. 6.8.

6.3.3 Numeric Outputs of GAPIT

6.3.3.1 GWAS Results

One of the most important GAPIT output files is the detailed summary of the GWAS results, which is output in the csv format and named *.GWAS. Results.csv. Each row contains the results for the tested SNPs with their p-values sorted in the ascending order. The columns contain the chromosome number (Chromosome), SNP position (Position), p-value (P.value), minor allele frequency (maf), sample size (nobs), R^2 value without considering the SNP (Rsquare.without.SNP), R^2 value

```
SNP,Chromosome,Position ,P.value,maf,nobs,Rsquare.of.Model.without.SNP,Rsquare.of.Model.with.SNP,FDR_Adjusted_P-values
chr07_13857471,7,13857471,2.83905134819353e-06,0.0357142857142857,84,-0.00198785207469943,0.316714868684625,0.717986415312488
chr07_26560346,7,26560346,6.14270994152183e-06,0.0714285714285714,84,-0.00198785207469943,0.292303551971943,0.717986415312488
chr07_12275431,7,12275431,7.20261942771912e-06,0.0476190476190476,84,-0.00198785207469943,0.287334101422252,0.717986415312488
chr07_12275435,7,12275435,7.20261942771994e-06,0.0476190476190477,84,-0.00198785207469943,0.287334101422252,0.717986415312488
chr07_27099256,7,27099256,8.68102895088106e-06,0.101190476190476,84,-0.00198785207469943,0.281533938955348,0.717986415312488
chr07_22054663,7,22054663,1.25659394775445e-05,0.0476190476190476,84,-0.00198785207469943,0.270133773357909,0.717986415312488
chr07_10050766,7,10050766,1.29390871175771e-05,0.0238095238095238,84,-0.00198785207469943,0.269236892375671,0.717986415312488
chr07_24094031,7,24094031,1.29390871175771e-05,0.0238095238095238,84,-0.00198785207469943,0.269236892375671,0.717986415312488
chr07_28816010,7,28816010,1.29390871175771e-05,0.0238095238095238,84,-0.00198785207469943,0.269236892375671,0.717986415312488
chr07_28934757,7,28934757,1.29390871175771e-05,0.0238095238095238,84,-0.00198785207469943,0.269236892375671,0.717986415312488
```

Fig. 6.9 Example of a GWAS.Results.csv output file

```
taxa,PC1,PC2,PC3,PC4,PC5,PC6,PC7,PC8,PC9,PC10,PC11,PC12,PC13,PC14,PC15,
RWG-001,-271.51122664515,-21.7525461464598,-71.0633079628977,-15.250518
RWG-002,-273.675786490748,-9.42651579024104,-65.4697278302784,-21.43106
RWG-003,433.000499776258,65.3229356222634,-52.720713604224,0.9786764249
RWG-004,-269.633188285851,-35.975154947833,-82.9562534915606,-16.358876
RWG-005,386.37827293756,61.2215568840685,-72.5255456458863,-24.77286920
RWG-006,410.915690084827,75.8805334342016,-74.8251472556878,227.3370525
RWG-007,33.9226281532374,-149.465646192077,-135.003157705196,70.7794810
RWG-008,194.377912377639,-291.658628531631,259.76459113213,-11.43406928
RWG-009,-262.51788909012,0.944079910420351,-51.8861217657203,-15.651749
RWG-010,390.952177181852,55.3763237267057,-43.3332403184925,20.39487560
```

Fig. 6.10 Example of a GAPIT.PCA.csv output file

considering the SNP (Rsquare.with.SNP), and the FDR adjusted p-value (FDR.adj.P.value). An example of a *.GWAS.Results.csv file is shown in Fig. 6.9.

6.3.3.2 Principal Component Analysis Results GAPIT.PCA.csv File

The results of the PC analysis are output in the *.GAPIT.PCA.csv file, which lists the PC values for each sample. Each row contains the PC of each sample. An example of a *.GAPIT.PCA.csv file is shown in Fig. 6.10.

6.3.3.3 Results of the Estimate of Allelic Effects

The results of the estimate of allelic effects are output in the *.Allelic_Effect_Estimates.csv file. As shown in Fig. 6.11, each column in the file contains information about SNP id, chromosome, bp position, and allelic effect estimate, and each row contains information on each SNP and the estimate of allelic effect. When genotype data in the HapMap format are used as input, the sign of the allelic effect is relative to the minor allele.

```
SNP,Chromosome,Position ,Allelic Effect Estimate
chr07_999,7,999,NA
chr07_1000,7,1000,NA
chr07_1024,7,1024,0.0780805305159046
chr07_1029,7,1029,-0.17523166029586
chr07_1036,7,1036,0.155730198531315
chr07_1045,7,1045,-0.0443036857591824
chr07_1208,7,1208,0.149711825092946
chr07_1257,7,1257,-0.025154167445476
chr07_1290,7,1290,0.0345966842021821
chr07_1329,7,1329,-0.0478462092466389
```

Fig. 6.11 Example of an Allelic_Effect_Estimates.csv output file

6.4 SNP Analysis Using NGS Data

Currently, most of NGS data comes from sequencing platforms market by Roche (CT, USA) (454 FLXsystem), Illumina (SD, USA) (Genome Analyzer, HiSeq, and MiSeq), Applied Biosystem® (SOLiD system), Life Technologies (CA, USA) (Ion Torrent), and Complete Genomics platforms (Van Dijk et al. 2014). Therefore, platform-specific software, like Cassava (Illumina) or Newbler (454 Life Sciences), are widely used to process NGS data (Miller et al. 2010). Because these software stand to be platform specific, other more general purpose software has been developed for SNP calling from NGS data from different sources. Galaxy is a popular Web- and cloud-based software for NGS data analysis (Blankenberg and Hillman-Jackson 2014). However, some of the general purpose software packages have the disadvantage of not being very user-friendly. Here, we present a widely used commercial software, the CLC Genomic Workbench v5 (http://www.clcbio.com/products/clc-genomics-workbench/), and describe its use for SNP calling from resequenced cucumber genome data (Unpublished).

6.4.1 *Downloading and Installing the CLC Genomic Workbench*

The CLC Genomics Workbench can be downloaded from (http://www.clcbio.com/products/clc-genomics-workbench/). The SNP identification analysis procedure that we describe here is based on the "CLC Genome Workbench" manual and appropriately modified for the resequenced cucumber genome data.

6.4.2 Importing Sequencing Data

As shown in Fig. 6.12, NGS data can be imported by clicking 'Import' in the Toolbar. The input file format can be selected in the 'Files of type' box (Fig. 6.13). There sequenced cucumber genome sequences used here were generated on an Illumina platform.

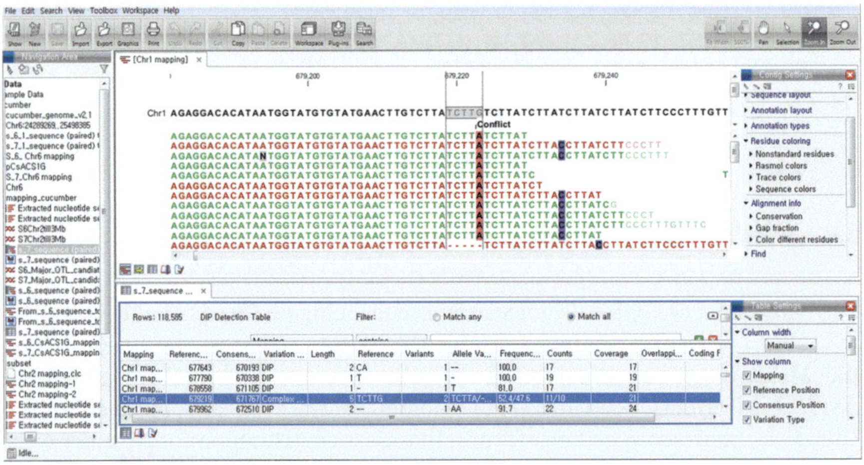

Fig. 6.12 Screenshot of CLC Genomics Workbench

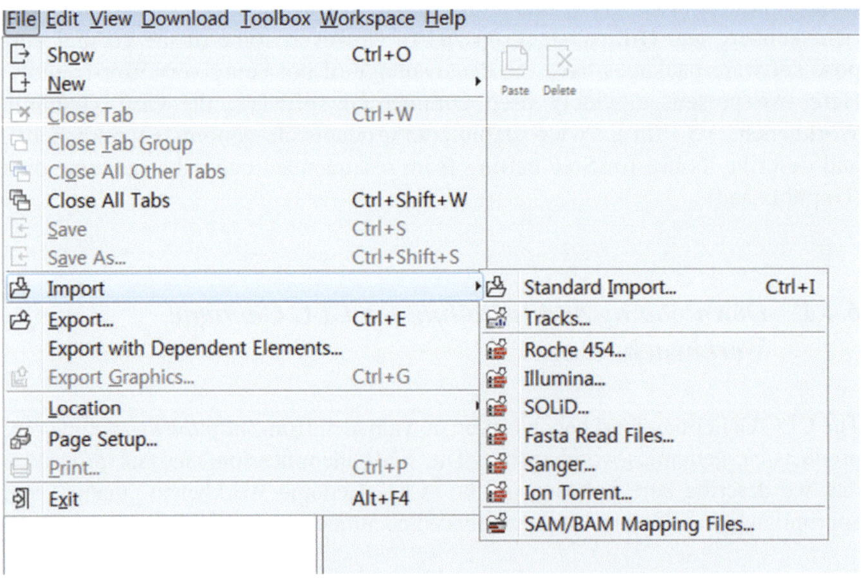

Fig. 6.13 Choosing the kind of data to be imported to the CLC Genomics Workbench

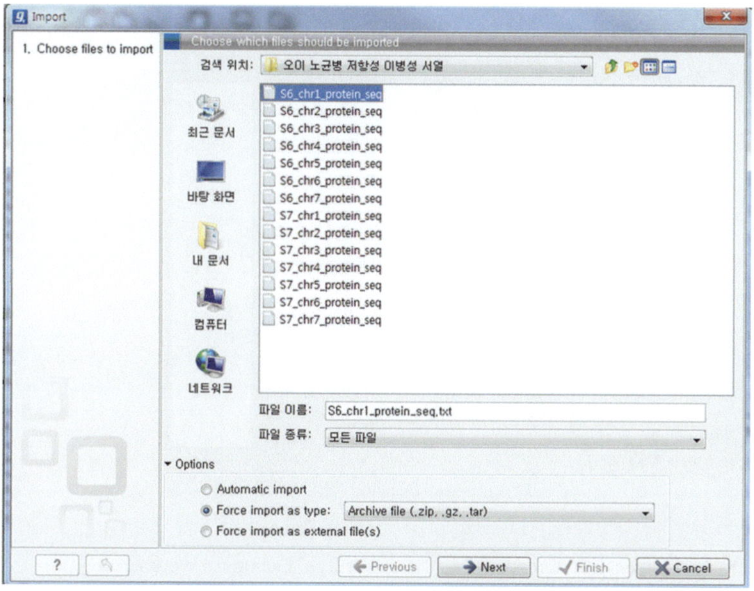

Fig. 6.14 Importing sequence data to the CLC Genomics Workbench

1. Choosing 'Illumina import' will open the dialog shown in Fig. 6.14.
2. Remove adaptor sequences, if needed. Choose either the single-end or paired-ends sequence, which ever corresponds to the NGS input data.
3. Click the 'Next' button.
4. Click the 'Save' button.
5. Click the 'Finish' button.
6. Go to
   File | Import | Standard Import| Locate '*file name for importing data*' | Select
7. Click the 'Next' button.

6.4.3 *Mapping Sequences Using Reference Mapping*

1. Go to
   After activating Toolbox | NGS Core Tools | Map Reads to Reference, as shown in Fig. 6.15.
2. Choose 'S6' and then move to right panel.
3. Click the 'Next' button.
4. Click the 'Reference' icon.
5. Click the '9930' as reference sequence. A popup window will open, as shown in Fig. 6.16. More than two sequences are shown as reference sequences.
6. Click the 'Next' button.

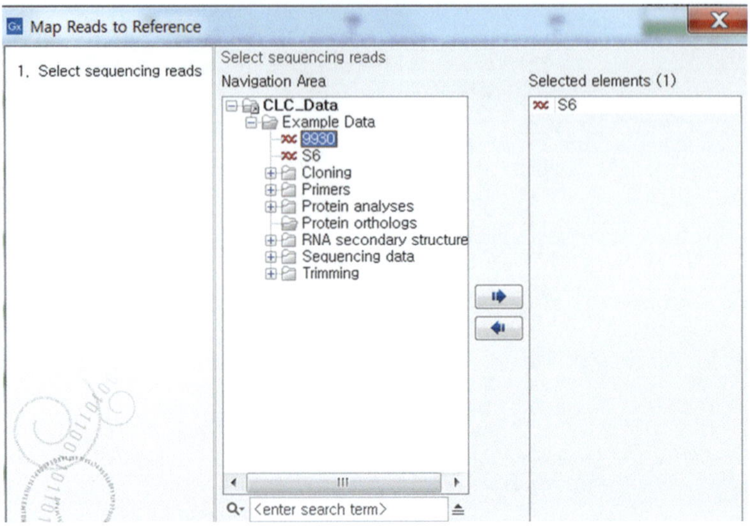

Fig. 6.15 Assignment of reference sequences in the CLC Genomics Workbench

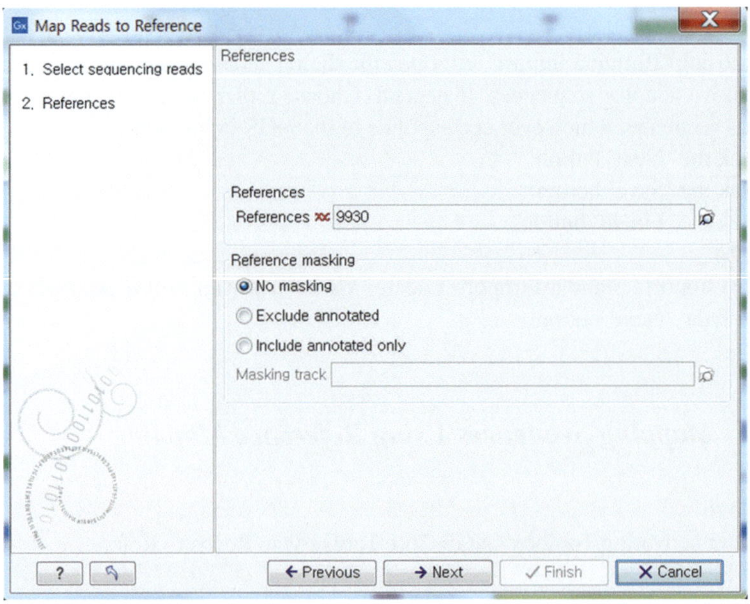

Fig. 6.16 Window in the CLC Genomics Work bench after assignment of a reference sequence

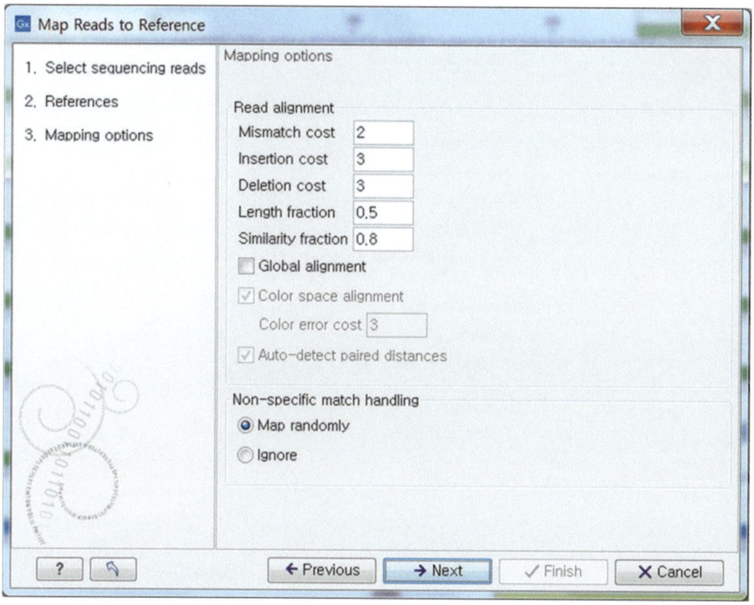

Fig. 6.17 Adjustment of the mapping parameter in the CLC Genomics Workbench

7. Choose the mapping option as shown in Fig. 6.17. Either the default mapping parameters can be chosen or users can adjust the parameters by clicking the 'reload' button and selecting the sequencing condition.
8. Click the 'Next' button. A window like the one shown in Fig. 6.18 will open.
9. Choose the output option by clicking the 'radio' button next to 'stand alone read mappings'.
10. Check the square box next to 'create report'.
11. Check the square box next to 'collect unmapped reads'.
12. Click the 'Save' and 'Finish' buttons.

6.4.4 *Results of the "Read Sequence Mapping" Analysis*

6.4.4.1 Analysis Results

After the mapping process is complete, the results are displayed in the 'Navigation area'.

1. 'List of non-mapped' reads displays all the sequences that were not been aligned to the reference sequence(s).
2. 'Report' displays all the mapping statistic information and the result(s) including the aligned sequences.
3. 'Mapping' displays all the alignment information.

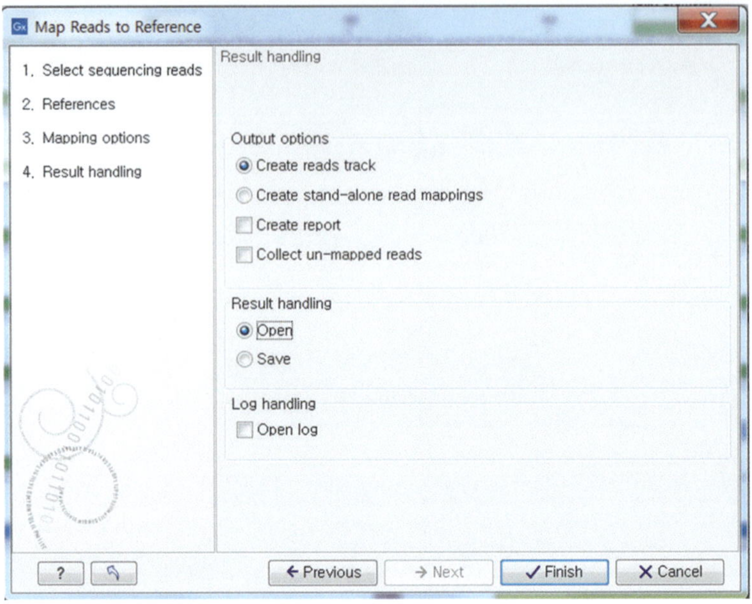

Fig. 6.18 Setting up the output file of mapping result in the CLC Genomics Workbench

6.4.4.2 Visualization of Results

The visualized alignment information can be obtained by double-clicking on the 'Mapping' object window. The viewing area will look like the example shown in Fig. 6.19.

1. Adjust 'compactness' located on the top of right pane (Read Mapping Settings).
2. On the default setting, red and green sequences represent the negative and positive strands, respectively. The sequence colors can be changed by clicking 'Residue coloring' in the right pane (Read Mapping Settings) and then clicking 'Sequence colors'.
3. The alignment result can be magnified by clicking either 'Zoom in' or 'Zoom out' on the topmost right position in 'Navigation area'.
4. The annotation type for the reference sequences can be changed by clicking 'Annotation type' in the right pane (Read Mapping Settings).
5. The sequence, annotation, sequence position, and gene name for a selected region of the reference sequence can be found by clicking 'Find' button. For example, for the region from 15,000 to 15,200 bp, first click the 'Find' button, then in the 'Find search box' input '15,000..15,200' and click the 'radio' button next to 'Position' (Fig. 6.20).
6. Then click the 'Find' button again. Coding sequence annotation information can also be found by going to 'Annotation' section and choosing 'CDS annotation type'.
7. To get the amino acid sequences, go to 'Nucleotide Info' section and click 'Translation' button.

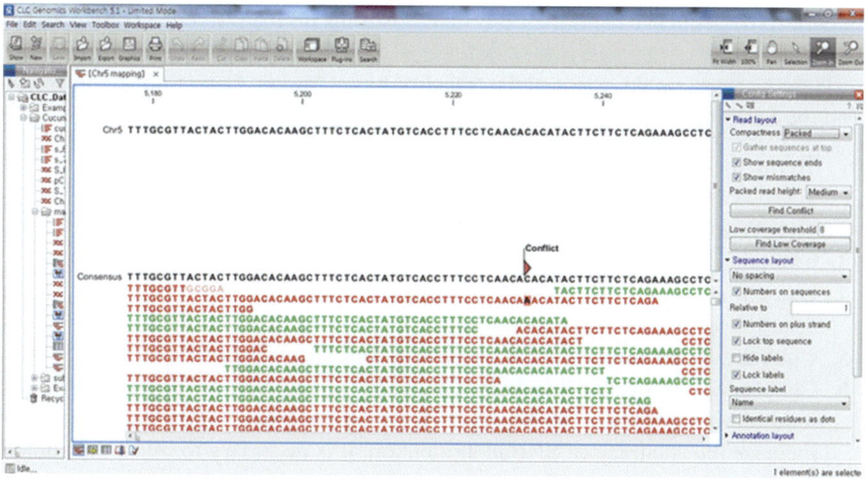

Fig. 6.19 Screenshot of the results after double-clicking the 'Mapping' object in the CLC Genomics Workbench. The default viewing type is 'packed view'

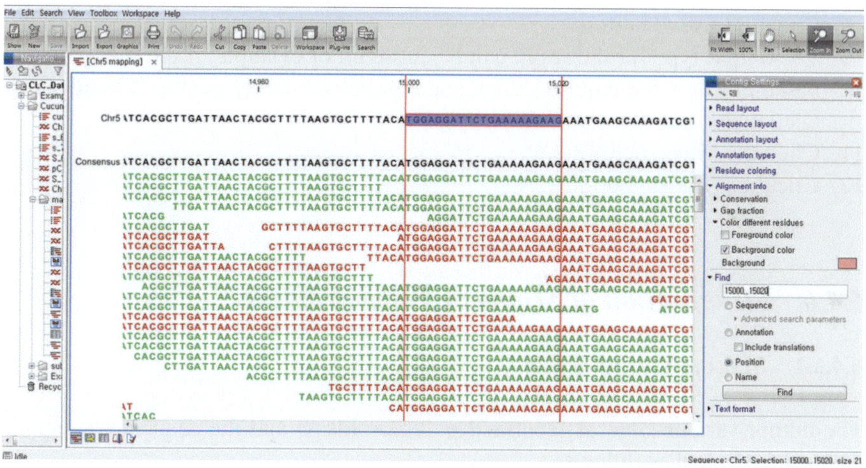

Fig. 6.20 Screenshot of the 'Find' function used to find a specific position in the sequence in the CLC Genomics Workbench

6.4.5 Finding Sequence Polymorphisms

The CLC Genomic Workbench uses two different types of algorithms, 'Quality-based Variant Detection' and 'Probabilistic Variant Detection', to find polymorphisms. After searching for polymorphisms, information for SNPs, multiple nucleotide variants, and insertions/deletions (InDels) can be obtained. The polymorphic sequences and their physical positions on the reference sequence can be obtained.

Here, as an example, we used the 'Quality-based Variant Detection' algorithm to find the SNP information.

1. To run the 'Quality-based Variant Detection' tool go to
   Toolbox | Resequencing Analysis | Quality-based Variant Detection.
2. The 'Wizard window' will open. Choose 'Mapping result object'.
3. Click the 'Next' button.
4. Use either the default parameter values or click the 'Help' button to find a description of each parameter. Choose appropriate parameter values.
5. Click the 'Next' button.
6. Choose the appropriate threshold. For example, set minimum coverage as 8 and minimum frequency as 35 %, and click the square box(es) of 'Variant filter'. To increase the accuracy of SNP identification, it is necessary to select the appropriate parameter values. The minimum frequency value is important because it should be set below 50 % to obtain heterozygote SNP identification.
7. Click the 'Next' button.
8. Set 'maximum expected allele' as 2 and 'genetic code' as standard. Detailed information can be obtained by clicking the 'Help' button.
9. Click the 'Next' button.
10. The output type for the result can be chosen. For example, if the follow track-based object is not required, click and un-click the boxes next to 'Create annotated table' and 'Create track', respectively.
11. Click the 'Save' button.
12. Click the 'Finish' button.

6.4.6 Visualization of Sequence Polymorphisms

6.4.6.1 Making Variant Table

The output variant table of identified polymorphisms and the sequence mapping data can be linked as follows.

1. Check the 'Mapping object' has popped-up in the viewing area.
2. Double click 'Variant table'.
3. Go to
   View | Split Horizontally
4. Double-click on a row in the 'Variant table'. The cursor will be moved to 'Sequence mapping result window'. Use the 'Zoom' button to adjust appropriate magnification to reveal the sequence polymorphism.

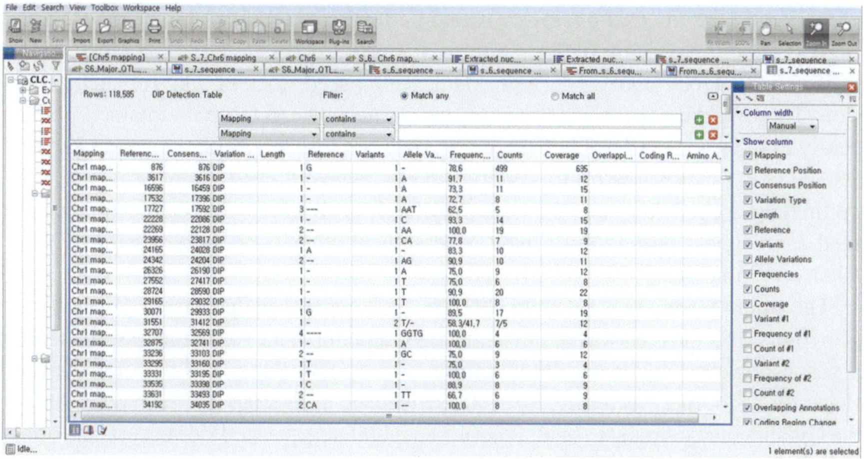

Fig. 6.21 Screenshot from the CLC Genomics Workbench showing detailed information for SNPs

6.4.6.2 Filtering Variant Table

Select a specific variant sequence using the 'filter' tool. For example, SNPs that lead to changes in the amino acid sequences of encoded proteins can be extracted from the variant table.

1. Set-up the filtering values as shown in Fig. 6.21.
2. Double-click on one of the row in the variant table and the variant information will be displayed as shown in Fig. 6.21.
3. 'Variant type' can be changed as shown in Fig. 6.21. One of the divergent variant types can be selected. For example, to extract all the information for In/Dels, 'In/Del' should be chosen as the variant type; to extract multiple nucleotide sequence variants, 'MNV' should be chosen.
4. Click 'Export' button and save in 'Text' or 'Excel' format, otherwise click 'Copy' and save the results as a specific spreadsheet.

6.5 Status of Genomics Research Related to Whole Genome Resequencing Worldwide

6.5.1 Rice Genetic Resources (IRRI/CAAS/BGI): An International Resequencing Project

To date, there are around 215,000 accessions of rice genetic resources worldwide. The Rice Genetic Resources research project was initiated to perform whole genome resequencing of 10,000 accessions that were representative of the genetic resources conserved in the International Rice Research Institute (IRRI) and the Chinese Seed Bank (Chinese Academy of Agricultural Science, CAAS,). The project also aimed to establish a DNA Bank and Tissue Bank with a seed stock of 5 kg for each accession. The original accessions used for resequencing need to be maintained permanently to bypass the heterozygosity problem, otherwise the results of whole genome resequencing may show up to 10–20 % heterozygosity even though rice is a self-pollinating crop. Thus, as is the case in IRRI, it is absolutely necessary to keep the seed stock intact and assign a new accession number once an accession is resequenced (Table 6.2).

The Beijing Genomics Institute (BGI) is a key partner in the project. When it first started in 2012, the plan was to finish the resequencing of all 10,000 accessions within 3 years; however, it is now clear that more time will be required to complete this. Recently, a summary of results for 3,000 sequenced accessions has been published in the GigaScience journal (Li et al. 2014). The 3,000 resequenced rice accessions originate from 89 countries. All 3,000 genomes had an average sequencing depth of 14×, with average genome coverage and mapping rates of 94.0 % and 92.5 %, respectively (Li et al. 2014). From the sequencing, approximately 18.9 million SNPs were discovered when aligned to the reference genome of the temperate japonica rice variety (Li et al. 2014). Furthermore, the raw sequencing data for the 3,000 genomes are publically available (Li et al. 2014). The global rice community will be able to use this data as a foundation for establishing a genetic/genomic database and information platform that will drive advances in rice breeding, genetics, and genomics (Table 6.2).

6.5.2 *Germplasm Resources (National Centre for Gene Research, CAS): A Chinese Resequencing Project*

The Bin Han group of researchers at the Chinese Academy of Sciences (CAS) National Research Center has performed whole genome resequencing of the largest number of genetic resources worldwide. The data analysis and results were reported at the Plant and Animal Genome (PAG) Symposium, San Diego, January 2007 and published in Nature Genetics in October, 2010 (Huang et al. 2010). From a collection of about 50,000 rice accessions that originated in China, 517 accessions were selected and genotyped with approximately one fold coverage (1×) genome sequencing depth. Approximately 3.6 million SNPs were identified (Huang et al. 2010). A GWAS was also conducted for 14 agronomical-related traits (Table 6.2).

Later, a new set of plant accessions, including an additional 100 Chinese japonica landraces and 330 diverse accessions from 33 countries, were added to the previous set of 520 Chinese landraces. A total of 950 accessions were resequenced on the Illumina Genome Analyzer IIx with the same approximately one-fold coverage as in the previous set. The resulting sequence dataset of 950 rice varieties consisted of 4.6 billion 73 bp paired-end reads. These results were published at Nature Genetics in December 2012 (Huang et al. 2012a). A total of 32 new loci

Table 6.2 Rice whole genome resequencing projects and GWAS

Country/Research institute	Main research contents	Reference
IRRI/CAAS/BGI	Targeting the key resources that are representative of world rice genetic resources (World rice genetic resources, ~215,000 accessions).	Rice Genome Symposium (2012)/Asia PAG (2013)/The 3000 project (2014) Gigascience
	Target number of accessions to be re-sequenced: approximately 10,000 accessions.	
	Step 1 (by March 2013): Resequencing of 3,042 accessions was completed with average 14× depth of whole genome sequencing	
	18.9 million SNPs were discovered.	
China	**<Han Bin group>**	Huang et al. (2010)/Huang et al. (2012a, b)/Xu et al. (2012)
	1st approach: 517 accessions of Chinese indica landrace were selected for whole genome resequencing with 1× depth in 2010	
	2nd approach: 100 additional Chinese japonica landrace accessions and 330 diverse cultivars selected from 33 countries were added to the previous set of 517 accessions and whole genome resequencing was performed with 1× depth in 2011.	
	3rd approach: whole genome resequencing of 1,083 *O.sativa* accessions, 446 *O.rufipogon* accessions, and 15 outgroup accessions of *O.sativa* resulted 2,621,077 SNPs, 619,132 small indels and 140,075 structural variants. GWAS for domestication-related genes of rice cultivars was performed.	
	<Wen Wang group>	
	Whole genome resequencing of 40 cultivated accessions and ten accessions of wild species (*O. rufipogon* and *O.nivara*) was conducted with 15× depth. GWAS for domesticated related genes (Kunming CAS/BGI).	

(continued)

Table 6.2 (continued)

Country/Research institute	Main research contents	Reference
Taiwan	China National Taiwan University (NTU) performed whole genome resequencing of seven recent Taiwan varieties.	Rice Genome Symposium (2012)
	5 Japonica: TC65, TK9, TNG67, TNG71, TNG72.	
	2 Indica: TCS 17, TNGS20.	
US	Cornell University of the United States selected 413 diverse accessions of *O. sativa* collected from 82 countries.	Rice Genome Symposium (2012)/Asia PAG (2013)/Zhao et al. (2011)
	Bred varieties and landraces were divided into five ecotype groups: Indica, Aus, Temperate japonica, Aromatic, Tropical japonica.	
	Initially, used SNP arrays for genotyping, now changed to whole genome resequencing (according to the report on March 2013).	
Japan	National Institute of Agricultural Science (NIAS) in Japan conducted 36× depth whole genome resequencing of 32 bred varieties and two landrace accessions.	Rice Genome Symposium (2012)
	Performing whole genome resequencing and developing RILs (recombinant inbred lines) and CSSLs (chromosome segment substitution lines).	
	Developing a Multiple Genome Viewer that will display sequence polymorphisms of each accessions and facilitate Forward/Reverse Genetics.	
	As a case study, functional analysis of flowering time gene (*Hd2*) was completed using GWAS.	
	Providing annotation of rice gene function: Update Q-TARO- QTLs annotation rice online database to OGRO (http://qtaro.abr.affrc.go.jp).	

(continued)

Table 6.2 (continued)

Country/Research institute	Main research contents	Reference
Korea	RDA and University have undertaken whole genome resequencing of major varieties including domestic and introduced accessions.	Rice Genome Symposium (2012)/Asia PAG (2013)
	National Food Research Institute performed 50× whole genome resequencing of 40 parental lines of NAM population.	
	Seoul National University research team is also conducting whole genome resequencing for Tongil-type and other major varieties for the research of grain quality related traits.	
	National Agricultural Research and Development Institute and Myongji University are conducting 100× depth whole genome resequencing of Dongjin, Il-Pum, Hwa-seong, Hwa-young, Mil-yang 23, Gi-ho, Il-mi.	
	Kongju National University are implementing 10× deep whole genome resequencing of 295 accessions including Korean authentic rice core set accessions and Korean bred varieties as a part of a GWAS.	
	Each university and research institute is conducting whole genome resequencing according to their needs.	

associated with flowering time and ten grain-related traits were identified (Huang et al. 2012a). Candidate genes for 18 associated loci were detected through detailed annotation (Huang et al. 2012a). Among the accessions used, 27 accessions of Korea-bred plants and landraces, and eight Tongil-type accessions were included (Huang et al. 2012a); however, the accession information turned out to be incorrect. Moreover, there were inconsistent records, even for the same accessions. Thus, careful examination and inspection will be required before the sequencing data for the Korea native accessions can be used (Table 6.2).

The Bin Han group also genotyped 1,083 accessions of *Oryza sativa*, 446 geographically diverse accessions of the wild rice species *O. rufipogon* and 15 outgroup accessions, which led to the identification of 2,621,077 SNPs, 619,132 small indels, and 140,075 structural variants. These results were published in Nature in October 2012 (Huang et al. 2012b). In that study, the sequence reads of 1,083 *O. sativa* accessions, 446 *O. rufipogon* accessions, and 15 outgroup accessions of *O. sativa* were aligned against the assembled genome sequence of wild rice accession W1943 using the same parameters as those used to align the sequences against the reference Nipponbare genome sequences. In this way, the genes related to evolution and domestication in the rice cultivars were identified (Huang et al. 2012b; Table 6.2).

6.5.3 Cultivated and Wild Species Research: Chinese Resequencing Project of the Kunming Institute (CAS) and BGI

The Kunming Institute (CAS) and BGI conducted whole genome resequencing of 40 cultivated accessions and ten accessions of wild rice species (*O.rufipogon* and *O.nivara*) with 15× depth. The results were published in Nature Biotechnology in 2012 (Xu et al. 2012). In this study, 6.5 million high-quality SNPs were obtained, and many domestication-related genes and genome-wide variation patterns were investigated (Xu et al. 2012; Table 6.2).

6.5.4 Rice Resequencing Project: National Taiwan University (NTU), Taiwan

NTU performed whole genome resequencing of seven modern rice varieties to a 7–33× depth. Among the varieties used, TC65, TK9, TNG67, TNG71, and TNG72 belonged to the Japonica ecotype, while TCS17 and TNGS20 belonged to the Indica ecotype (unpublished; Table 6.2).

6.5.5 Rice Resequencing and Genetic Diversity Project: Cornell University, USA

The Rice Genetic Resources research team led by Susan McCouch at Cornell University selected 413 accessions representing diverse rice genetic resources and genotyped them using a 44,100 SNP array and performed GWAS for 34 agronomical related traits. The 413 rice accessions fell into five distinct ecotypes, 87 were indica, 57 were aus, 96 were temperate japonica, 97 were tropical japonica, 14 were aromatic, and 62 were admixtures. This study provided huge amount of genetic information that can be readily implemented into GWAS and molecular breeding (Zhao et al. 2011). There search team led by Susan McCouch has also implemented a 10–60× depth whole genome resequencing project of selected rice accessions. In March 2013, the team reported at the PAG (Asia) Symposium in March 2013 that resequencing of 120 accessions was already completed through the Rice SNP Consortium (unpublished). These researchers have also implemented a high-resolution SNP array analysis to enable more powerful GWAS analyses (Table 6.2).

6.5.6 *Rice Resequencing and Genetic Resources Diversity Project: National Institute of Agricultural Science (NIAS), Japan*

Japan's Rice Genetic Resources research team, led by Dr. Yano, has selected 32 cultivated and two wild accessions to be resequenced to a 36× depth. The team also selected 18 useful varieties to be used to establish RILs (recombinant inbred lines) and CSSLs (chromosome segment substitution lines). In addition, a multiple genome viewer that can visualize genetic information and sequence variations for each accession is being developed, thus providing a basis for the facilitation of forward and reverse genetics.

The team announced the isolation and identification of a gene (*Hd2*) responsible for flowering time at the Chiang Mai Rice Functional Genomics International Symposium in November, 2012. To reinforce the existing sequence resources, the team has also performed the whole genome resequencing (44×) of Japanese bred varieties, and the upgraded IRGSP version with many corrections and supplements to the existing sequence base of IRGSP1.0 was released in late 2012 (http://rapdb.dna.affrc.go.jp/).

To annotate the 700 rice genes with known functions, the Japanese researchers have replaced the existing on-line QTL annotation database Q-TARO (http://qtaro.abr.affrc.go.jp/) with OGRO (http://qtaro.abr.affrc.go.jp/ogro). RiceXPro (http://ricexpro.dna.affrc.go.jp/) and RiceFREND (http://ricefrend.dna.affrc.go.jp/) are Japanese databases for gene expression information obtained by microarray analysis (Table 6.2).

6.5.7 *Rice Genetic Resources Diversity and Resequencing Project: Korea*

Recent studies of the diversity of rice genetic resources and whole genome resequencing projects are being implemented individually in Korea by the Seoul National University, Kangwon National University, National Food Research Institute, National Academy of Agricultural Science, and Kongju National University. The National Institute of Crop Science is implementing the whole genome resequencing (60× or higher) of 40 founder rice varieties as a part of the research project of the Molecular Breeding Agency. The project includes the establishment of a nested association mapping population and whole genome resequencing. The Seoul National University and Kangwon National University are conducting the whole genome resequencing of some elite lines and Tongil-type varieties (Table 6.2).

The National Academy of Agricultural Science and Myong-ji University have performed whole genome resequencing (100× or more) of rice varieties such as Dongjin, Il-Pum, Hwa-seong, Hwa-young, Mil-yang 23, and Gi-ho, Il-mi as part of their next-generation genomics research project. The Kongju National University has announced plans to establish a genetic information system of GWAS with Korea authentic core set accessions as a part of their next-generation genomics research

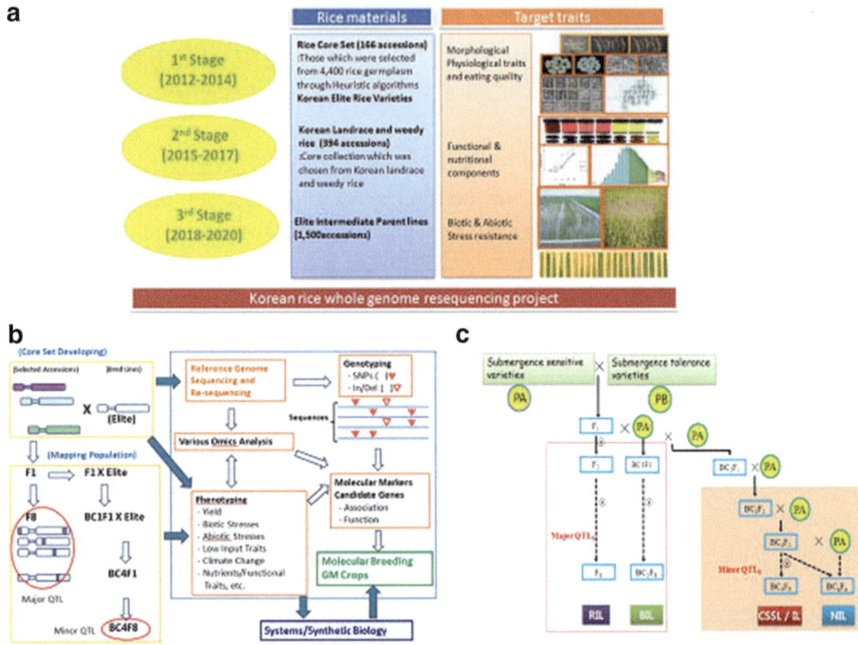

Fig. 6.22 Summary of the Korean rice whole genome resequencing project and related future research prospects. (**a**) Rice materials and target traits for each research stage of the resequencing project. (**b**) Schematic diagram of genomics-assisted breeding and genetics study. GWAS analyses followed by rice resequencing and relevant phenotyping pinpoint large numbers of agronomically useful alleles, leading to the detection of major and minor QTLs at F_8 and BC_4F_8, respectively. (**c**) Follow-up dissection of major and minor QTLs for related traits. These are exemplary mapping populations that were designed to find genes or to obtain useful intermediate parent lines underlying major and minor QTLs

project by implementing 10× depth whole genome resequencing. In 2013, the team was aiming to analyze 137 accessions of a heuristic core set and 159 Korean-bred cultivars (Table 6.2, Fig. 6.22).

Because rice did not originate in Korea, no wild resources are available; therefore, landraces (394 selected accessions) and weed-type resources are nationally important. In addition to establishing a molecular breeding system, it is also necessary to establish whole genome resequencing and an information system for old rice generations developed during the last 100 years in Korea (Table 6.2, Fig. 6.22).

6.5.8 Future Prospects

Plants are sessile organisms that have accumulated genetic-based natural variations from the diverse environments over time, creating many phenotypic variations. The recent advances in DNA sequencing technology have allowed for in-depth characterization of the genetic and genomic variation present in plant germplasm and in natural populations. New genomics tool, such as GWAS, can exploit these resources to associate the phenotypic variations with relevant genes or functional polymorphisms, thereby providing numerous insights on complex traits across many taxa. In particular, recent advances in the genome-scale measurement of molecular quantitative omics phenotypes (transcriptome, metabolome, and proteome) (Matsuda et al. 2015) and high-throughput phenotyping systems (Yang et al. 2014) are beginning to yield unprecedented insights for the associated genes underlying agronomically important genes, expediting genomics assisted breeding. Thus, in this environment, we consider that efforts to develop more efficient computational algorithms for GWASs would be highly desirable.

References

Benjamini Y, Hochberg Y (1995) Controlling the false discovery rate: a practical and powerful approach to multiple testing. J R Stat Soc Ser B Methodol 57:289–300
Benjamini Y, Yekutieli D (2001) The control of the false discovery rate in multiple testing under dependency. Ann Stat 29:1165–1188
Blankenberg D, Hillman-Jackson J (2014) Analysis of next-generation sequencing data using Galaxy. Methods Mol Biol 1150:21–43
Bouchet S, Servin B, Bertin P et al (2010) Adaptation of mixed linear model for genome-wide association studies. Nat Genet 242:355–360
Bradbury PJ, Zhang Z, Kroon DE et al (2007) TASSEL: software for association mapping of complex traits in diverse samples. Bioinformatics 23:2633–2635
Gibson G (2010) Hints of hidden heritability in GWAS. Nat Genet 42:558–560
Henderson CR (1975) Best linear unbiased estimation and prediction under a selection model. Biometrics 31:423–447
Huang X, Wei X, Sang T et al (2010) Genome-wide association studies of 14 agronomic traits in rice landraces. Nat Genet 42(11):961–967
Huang X, Zhao Y, Wei X et al (2012a) Genome-wide association study of flowering time and grain yield traits in a worldwide collection of rice germplasm. Nat Genet 44(1):32–39
Huang X, Kurata N, Wei X et al (2012b) A map of rice genome variation reveals the origin of cultivated rice. Nature 490(7421):497–501
Kang HM, Zaitlen NA, Wade CM et al (2008) Efficient control of population structure in model organism association mapping. Genetics 178:1709–1723
Kennedy B, Quinton M, Van Arendonk J (1992) Estimation of effects of single genes on quantitative traits. J Anim Sci 70:2000–2012
Li J-Y, Wang J, Zeigler R (2014) The 3,000 rice genomes project: new opportunities and challenges for future rice research. GigaScience 3(1):8
Lipka AE, Tian F, Wang Q et al (2012) GAPIT: genome association and prediction integrated tool. Bioinformatics 28(18):2397–2399
Matsuda F, Nakabayashi R, Yang Z et al (2015) Metabolome-genome-wide association study (mGWAS) dissects genetic architecture for generating natural variation in rice secondary metabolism. Plant J81(1):13–23
Miller JR, Koren S, Sutton G (2010) Assembly algorithms for next sequencing data. Genomics 95:315–327
Price A, Patterson N, Plenge R et al (2006) Principal components analysis corrects for stratification in genome-wide association studies. Nat Genet 38:904–909

Pritchard JK, Stephens M, Donnelly P (2000) Inference of population structure using multilocus genotype data. Genetics 155:945–959

Purcell S, Neale B, Todd-Brown K et al (2007) PLINK: a toolset for whole-genome association and population-based linkage analysis. Am J Hum Genet 81(3):550–575

Shaffer JP (1995) Multiple hypothesis testing. Annu Rev Psychol 46:561–584

Van Dijk EL, Auger H, Jaszczyszyn Y et al (2014) Ten year of next-generation sequencing technology. Trends Genet 30:418–426

Visscher PM, Brown MA, McCarthy MI et al (2012) Five years of GWAS discovery. Am J Hum Genet 90:7–24

Xu X, Liu X, Ge S et al (2012) Resequencing 50 accessions of cultivated and wild rice yields markers for identifying agronomically important genes. Nat Biotechnol 30(1):105–111

Yang W, Guo Z, Huang C et al (2014) Combining high-throughput phenotyping and genome-wide association studies to reveal natural genetic variation in rice. Nat Commun 5:5087

Yekutieli D, Benjamini Y (1999) Resampling-based false discovery rate controlling multiple test procedures for correlated test statistics. J Stat Plann Inference 82:171–196

Zhao K, Tung CW, Eizenga GC et al (2011) Genome-wide association mapping reveals a rich genetic architecture of complex traits in Oryza sativa. Nat Commun 2:467

Chapter 7
Identification of Mutagenized Plant Populations

Geung-Joo Lee, Dong-Gwan Kim, Soon-Jae Kwon, Hong-Il Choi, and Dong Sub Kim

Abstract The recent availability of large-scale genomic data has allowed researchers to begin deciphering plant gene function. Plant mutagenesis is a powerful tool for the identification and characterization of the function of specific genes linked to phenotypes. TILLING (Targeting Induced Local Lesions IN Genomes) is a general reverse-genetics tool that combines traditional mutagenesis with high-throughput methods of mutation discovery among mutagenized populations. The aim of TILLING is to identify mutagenized genotypes that affect specific phenotypes. Securing genetic diversity and selecting efficient progeny are the most important factors in plant breeding. During the evolutionary process, the gene pool has skewed towards a direction that is favorable to humans and some essential alleles may have been lost during the selection processes. Genome editing using an engineered nuclease is a target-directed, controlled, and predictable approach that can provide a genetically diverse gene pool and shorten cultivar development time. Depending on engineered nucleases and plant species types, the mutagenesis rate and outcomes show significant differences, indicating that nuclease types and characteristics, efficient target mutagenesis, construction and delivery of the nuclease, selection and verification of the mutants, and off-target mutagenesis need to be considered. Genomic sequence variations generated by chemical and/or physical mutagenesis can be strongly related to changes in phenotype. Haplotype analysis allows plant breeding knowledge to be expanded and can help improve understanding of diversity

---

Author contributed equally with all other contributors.

G.-J. Lee
Department of Horticultural Science, Chungnam National University, Daejeon, Republic of Korea
e-mail: gjlee@cnu.ac.kr

D.-G. Kim
Department of Plant Science, Seoul National University, Seoul, Republic of Korea
e-mail: kimdg@snu.ac.kr

S.-J. Kwon • H.-I. Choi • D.S. Kim (✉)
Advanced Radiation Technology Institute, Korea Atomic Energy Research Institute, Jeongeup, Republic of Korea
e-mail: soonjaekwon@kaeri.re.kr; hichoi@kaeri.re.kr; bioplant@kaeri.re.kr

H.-J. Koh et al. (eds.), *Current Technologies in Plant Molecular Breeding*,
DOI 10.1007/978-94-017-9996-6_7

at the genomic sequence level. Experiments using SNP haplotypes can provide insights into plant evolution, and can contribute to phenotypic analysis, and expectation and characterization of mutagenized plant phenotypes.

7.1 High-Throughput Genome Analysis Using Mutant Populations

7.1.1 TILLING (Targeting Induced Local Lesions IN Genomes)

Plant mutation breeding is widely used for crop improvement and is also a powerful tool for characterizing biological processes, specific gene functions, and linkages between mutations and phenotypes in the mutated plant. The forward-genetics approach (from phenotype to gene function) requires large mutated populations to screen for alterations of phenotypes and biological processes of interest in the target plants (Kurowska et al. 2011). The Food Agriculture and Organization (FAO)/International Atomic Energy Agency (IAEA) programme's Database of Mutation Enhanced Technologies for Agriculture (META, http://mvgs.iaea.org/) contains information about mutant varieties and mutant genetic stocks along with related publications. Mutated plants that show morphological changes have been studied to identify the mutated positions in their genomes. The altered phenotype can be isolated using classical methods such as map-based cloning or appropriate tagging methods. However, the forward-genetics approach is both time consuming and labor intensive (Krattinger et al. 2009; Zhang et al. 2009). With the past decade's rapid acquisition of large-scale genomic data, the reverse-genetics approach (from gene information to phenotype) has largely replaced the forward-genetics approach. Although the traditional reverse-genetics approach using T-DNA-inserted mutants is a very useful method for identifying gene function, its applications are limited because it can be applied to only model plants.

TILLING is a general reverse-genetics strategy that merges traditional mutagenesis with high-throughput methods of mutation discovery in plant and animal species (McCallum et al. 2000; Till et al. 2004b, 2007; Cooper et al. 2008) (Fig. 7.1). In 2000, McCallum et al. used automated denaturing high-performance liquid chromatography (dHPLC) to detect heteroduplexes of DNA from rare single nucleotide polymorphisms (SNPs) in an alkylating agent ethyl methanesulfonate mutagenized *Arabidopsis thaliana* population and used the term "TILLING" to describe their approach. Later, Colbert et al. (2001) introduced the use of CEL I endonuclease, which was isolated from bulk celery, to detect nucleotide polymorphisms in a same ethyl methanesulfonate mutagenized *A. thaliana* population. CEL I is a plant-specific extracellular glycoprotein that cleaves the 3′ end of single-stranded DNA heteroduplexes containing single nucleotide mismatches (Oleykowski et al. 1998). Using fluorescently-labeled primers (IRDye-700 and IRDye-800) to detect amplicons cleaved by CEL I on a Li-Cor (automatic acrylamide gel image system), Colbert et al. (2001) showed that CEL I detected a heterozygous SNP in pools of

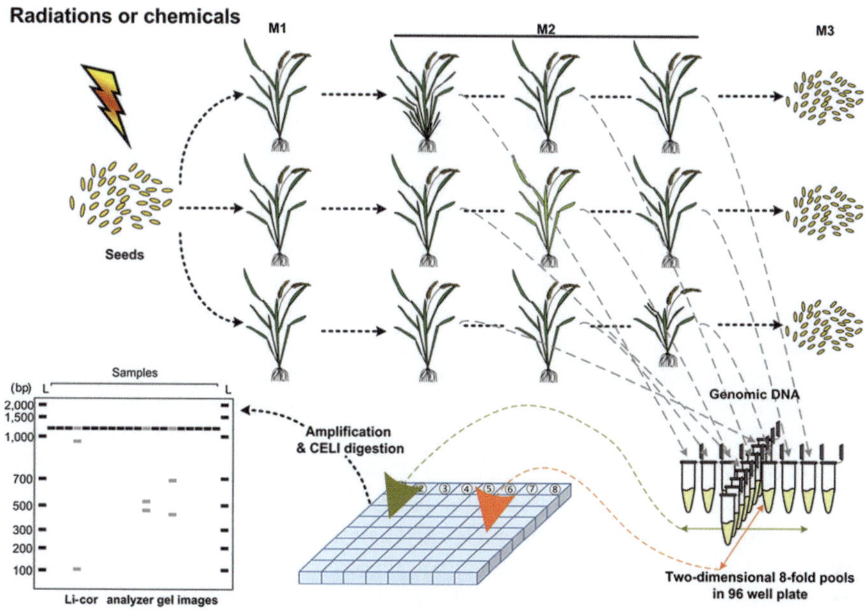

Fig. 7.1 TILLING procedure for mutation-treated rice. After seed mutagenesis using radiation or a chemical, several M1 plants were self-pollinated and numerous single M2 plants were grown for mutation detection. Eight-fold pools in a 96-well plate included the DNAs extracted from each M2 plant. Amplification and digestion were performed. After digestion, the DNA fragments were separated by denaturing polyacrylamide gel electrophoresis, such as Li-C or for two-channel detection

mutagenized plants. The pooling strategy allowed a 1-kb fragment to be screened on 3×96-well plates in a library of 3,000 mutagenized plants. Thus, these two approaches, using CEL I and pooling, have allowed TILLING to be applied to all plant species, regardless of the genome size, the ploidy level, or propagation method. TILLING also produces an allelic series of mutations, including hypomorphic alleles that are useful for genetic analysis. TILLING which was first used in the model plant *A. thaliana*, has now been applied in a variety of plant species including rice, maize, barley, lotus, and wheat to detect mutated locus/loci in their genomes (Perry et al. 2003; Caldwell et al. 2004; Slade et al. 2004; Till et al. 2004b, 2007).

7.1.2 Application of TILLING

To improve the high-throughput and efficiency of mutation detection by TILLING, a diverse set of old and new technologies have been used in different plant species (Wang et al. 2012). The aim of TILLING is to identify genotypes that affect specific phenotypes. The technique can also distinguish homozygous and heterozygous genotypes for a phenotype and can provide information on the position of the mutation in a nucleotide sequence.

7.1.2.1 De-TILLING

Various mutagens for introducing genomic variations are available; they include chemicals, radiation, and transgenic mutagenesis of plant genomes. Each of these mutagen sources can lead to structural variations in plant genomes. In particular, fast neutrons, a form of high-energy radiation can induce a broad range of DNA deletions that range in size from a few bases to megabases. Fast neutrons can also cause various random chromosome re-arrangements in plant genomes (Li et al. 2001; Men et al. 2002). Therefore, fast neutron radiation can change plant phenotypes through knockouts and disruptions of genes in plant genomes. Fast neutron mutagenesis has been applied in many plant species, including *A. thaliana*, *Medicago truncatula*, *Glycine soja*, *Hordeum vulgare*, and *Lotus japonicus* (Alonso et al. 2003; Oldroyd and Long 2003; Searle et al. 2003; Zhang et al. 2006; Hoffmann et al. 2007), and many phenotype-related genes have been identified and cloned as a result of this approach. The Deleteagene (Delete-a-gene) system (Li et al. 2002), which is based on neutron mutagenesis and high-throughput PCR screening, has been developed and named deletion-TILLING, or de-TILLING (Rogers et al. 2009) (Fig. 7.2). Currently, fast neutron mutagenesis screening has been developed with comparative genome hybridization and high-throughput next-generation sequencing (NGS) platforms which are powerful tools for the screening of copy number variation and structural genome variation. With the invention of techniques for preferentially amplifying mutant loci from large DNA sample pools, a novel strategy for

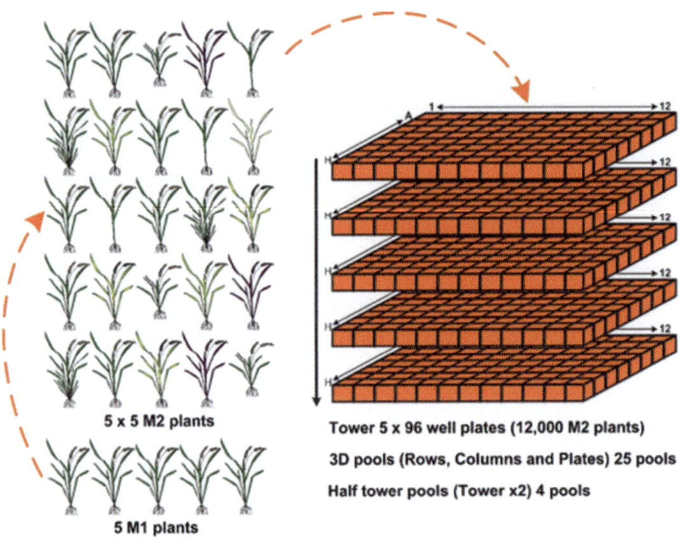

Fig. 7.2 De-TILLING tower for mutagenized rice. The tower of five plates consists of 480 pools. DNA from 25 seedlings was taken from the pooled DNA from the M2 plants of five mutagenized M1 rice plants. Each tower is initially pooled into 25 pools (row, column, and plate). The De-TILLING population is initially screened using the 54 half tower pools representing the 13 towers twice. After the deletion locus/loci are detected, 25 3D pools are screened to locate the mutant to an individual well

deeper pooling has been designed (Rogers et al. 2009). In this design, the population is segregated into a towered structure consisting of 96-well plates of DNA extractions and each tower is pooled to create a three-dimensional pool matrix of rows, columns, and plates (Fig. 7.2).

7.1.2.2 High Resolution Melting (HRM)-Based TILLING

HRM analysis is a high-throughput method that does not rely on mismatch cleavage by specific endonucleases but detects sequence variants by distinct patterns in DNA melting curves derived from the release of fluorescent dyes intercalated in double-stranded DNA by thermal denaturation (Gady et al. 2009). Pools possessing mutations form heteroduplexes that can be identified by differential melting curves compared with homoduplexes (Gady et al. 2009). HRM has been previously used to detect mutations in mutagenized plant populations including tomato (Gady et al. 2009), wheat (Botticella et al. 2011), and *Brassica rapa* (Lochlainn et al. 2011), with applications ranging from qualitative SNP detection to semi-quantitative analysis of DNA methylation. Real-time PCR is carried out using a touchdown method for thermal recycling with regents such as the LightCycler® 480 system. HRM curve data can be obtained at a rate of 25 acquisitions per temperature degree. Lochlainn et al. (2011) demonstrated that HRM can be used to identify mutations at multiple amplicons within a PCR fragment from ethyl methanesulfonate mutagenized populations of *Brassica rapa* using a single PCR primer pair. Small numbers of primer combinations lead to savings of both cost and time which allows rapid and efficient high-throughput genotyping of multiple alleles for some genes. This feature is particularly important in complex crop species in which paralogous genes may be present and the accumulation of mutations may be required to generate a new phenotype.

7.1.2.3 Conformation-Sensitive Capillary Electrophoresis (CSCE)-Based TILLING

CSCE methods are based on the fact that non-denatured heteroduplex DNA molecules form a different secondary structure to homoduplexes. Heteroduplex analysis uses non-degenerating polyacrylamide gels to show differences in running speeds of a semi-denaturing polymer, allowing the identification of pools containing a mutation within the target fragments. Thus, the CSCE assay uses different migration speeds between homoduplexes and heteroduplexes on capillary electrophoresis (Gady et al. 2009). The obtained peaks of heteroduplexes have a different shape than homoduplex peaks as shown in Fig. 7.3.

7.1.2.4 Eco-TILLING

Eco-TILLING (ecotype TILLING) uses the same principle as TILLING to discover variations in natural populations and has been shown to be efficient in detecting SNPs in large populations (Comai et al. 2004) (Fig. 7.4). No mutant population is

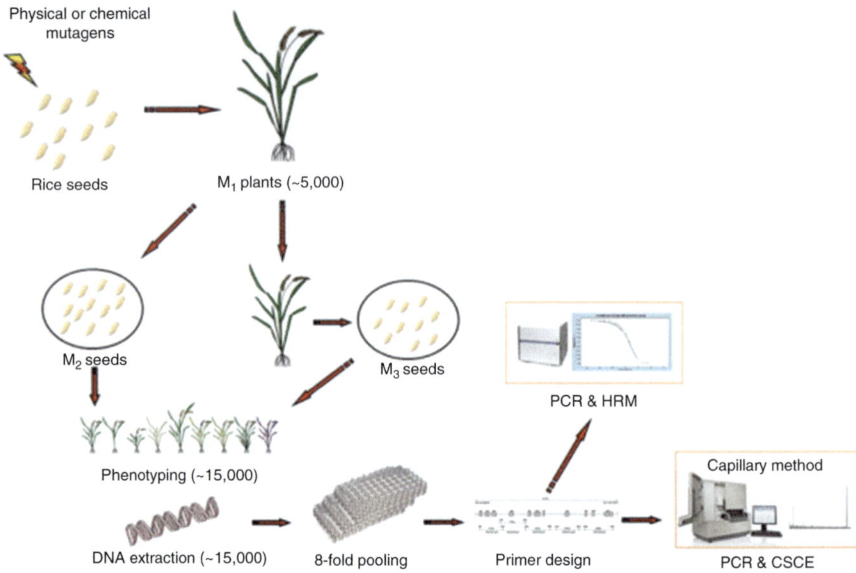

Fig. 7.3 Construction of TILLING population in rice and identification of mutants. After the general TILLING processes including treating mutagens, generating a mutation population, phenotyping, extracting DNA, and pooling are complete, mutations can be detected either by high-resolution melting (HRM)-based TILLING or conformation-sensitive capillary electrophoresis (CSCE)-based TILLING steps related to heteroduplex formation and enzyme digestion are not required using these two methods

required for Eco-TILLING; instead, different existing genotypes representing the genetic variation of a species or a certain group within a species are collected. This approach allows large numbers of samples to be rapidly screened to identify naturally occurring SNPs and/or small insertions and deletions (indels). This method has been used successfully to detect DNA size variations as well as the number of microsatellite repeats in Arabidopsis (Till et al. 2006). The TILLING process for mutation screening can be adapted to Eco-TILLING. DNA from only two individuals is pooled because, although a higher pooling depth could be applied, the expected high diversity in natural populations used for Eco-TILLING makes this approach inadvisable in most cases. Eco-TILLING was first modified for TILLING in a study to detect natural polymorphisms in *A. thaliana* (Comai et al. 2004). In this study, the genomic DNA of each ecotype was mixed with the reference genomic DNA in a 1:1 ratio and CEL I endonuclease was used to detect various variation in the cleaved amplicons (SNPs, indels, size variations, and microsatellite repeat numbers). Overall, Eco-TILLING proved to be an efficient method to haplotype individuals without needing to sequence the genomes of all individuals. Eco-TILLING has been used in plants such as common bean, melon, poplar, wheat, and rice (Gilchrist et al. 2006; Kadaru et al. 2006; Nieto et al. 2007; Wang et al. 2008; Galeano et al. 2009), but this approach is still not as widely used as TILLING.

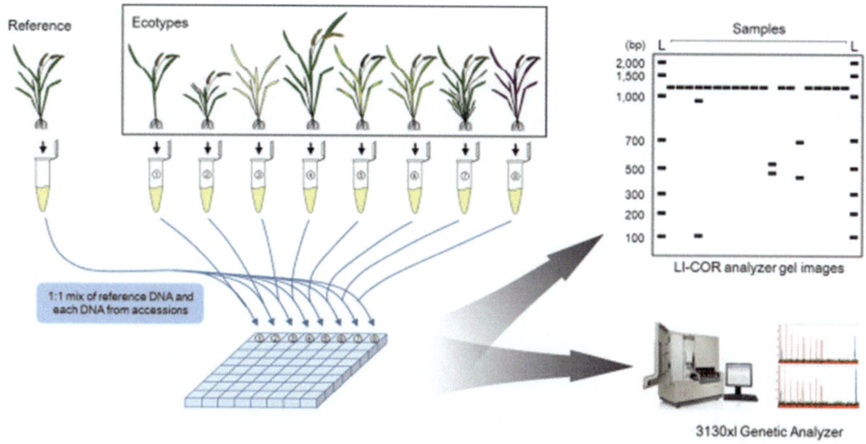

Fig. 7.4 Schematic representation of the Eco-TILLING process. Each of the DNA ecotypes is mixed with the DNA of a reference individual in the ratio of 1:1. The rest of the processes for mutation detection are the same as the other TILLING methods

7.1.3 TILLING Procedures and Considerations

7.1.3.1 DNA Extraction from Mutagenized Populations

Genomic DNA is extracted from each individual in a mutagenized plant population. For TILLING, the quality of DNA matters more than the quantity because in the pool, the small amount of DNA from an individual should represent itself in the series of experiments that follow. The quantity and quality of the DNA can be estimated by agarose gel electrophoresis and spectrophotometry. Low-quality DNA shows up as a smear in agarose gels and as a low A260/A230 ratio in spectrophotometry (Till et al. 2006). Although low-quality DNA samples sometimes can be solved by increasing concentration, high-quality DNA is recommended for further experiments (Till et al. 2006). After measuring the quantity and quality of all the DNA samples, their concentrations are normalized for DNA pooling and PCR analysis. The concentration of the individual DNA samples, which depends on the size of the genome, should be determined before pooling. For large genomes, high concentrations of DNA are required.

7.1.3.2 DNA Pooling

TILLING can be regarded as a high-throughput and cost-effective method because the number of samples that can be processed in each experiment is increased by pooling (Colbert et al. 2001; Till et al. 2003). It is recommended that each of the samples is first individually arrayed into multi-well plates to prevent pipetting errors and save the time consumption for pooling samples (Till et al. 2006). A 96-well plate is generally used because these plates are compatible with 8- or 12-channel pipettes.

One- to three-dimensional pooling strategies for TILLING have been reported (Till et al. 2006; Tsai et al. 2011). A one-dimensional (1D) pool consists of 8–12 samples that correspond to the number of samples in one row on a 96-well plate. When 8×8 grids are used, eight samples in one row of the plate are pooled for 1D pooling (Fig. 7.5). Two-dimensional (2D) pools are constructed by pooling arrayed samples in rows and columns (Fig. 7.5). For three-dimensional pooling, samples are pooled according to the x-, y-, and z- axes. A 3D pool usually contains samples from the xy-, yz-, and zx- coordinate planes (Fig. 7.5).

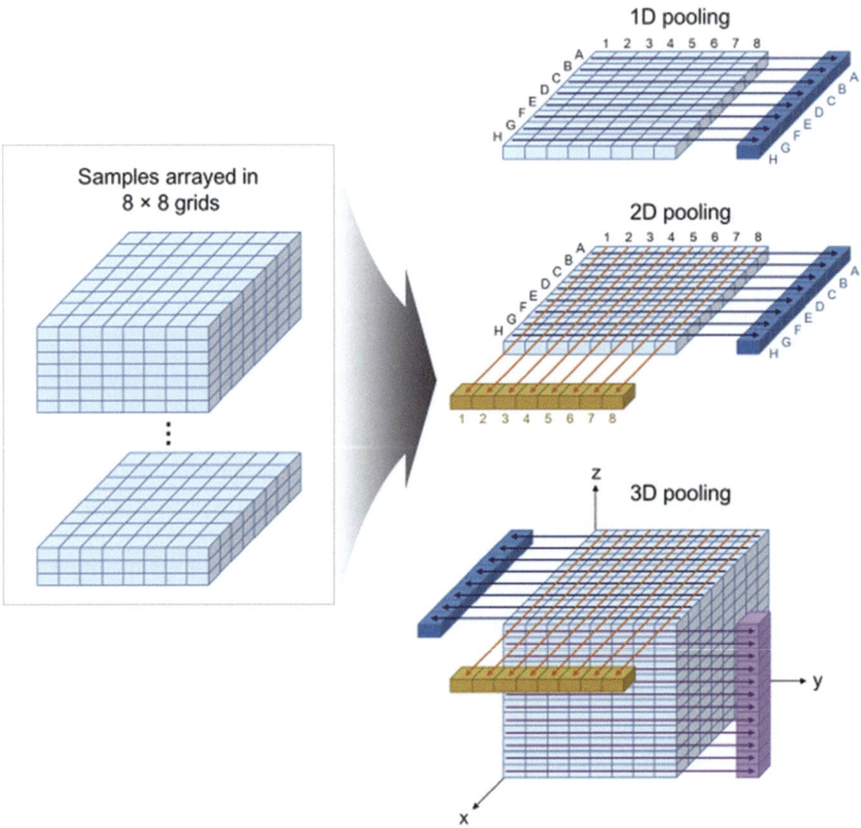

Fig. 7.5 Multi-dimensional pooling strategy for TILLING. From the samples arrayed in 8×8 grids, 1D, 2D, and 3D pools can be prepared as depicted in the right panel. The 1D pooling strategy is as follows: 64 samples are arrayed into an 8×8 grid and the samples are pooled using an eight-channel pipette to combine samples from each of the eight rows on the plate into a single column of the pooled plate (column A–H on the right of the 1D array), so that position A of the pool plate contains the A1–A8 samples on a grid. The 2D pooling strategy combines the samples on an 8×8 grid from each of the eight rows and columns so that positions A and 1 contain the samples A1–A8 and A1–H1, respectively. The 3D pooling strategy combines samples on an 8×8×8 grid stack. Samples are first pooled according to the x-, y-, and z-axes. Generally, the final 3D pools are constructed by combining the first pool generated along each of the axes again so that one pool contains samples from the xy-, yz-, and zx-coordinate planes

An adequate level of pooling is important to make TILLING high-throughput, fast, and reliable. The pooling depth needs to be determined empirically by preliminary testing pooled samples at different levels, because only a small number of individuals are likely to possess mutant alleles at a given locus in mutagenized populations (Till et al. 2006).

7.1.3.3 PCR Amplification and Heteroduplex Formation

Primer pairs should be specific to the unique target region to be amplified (Till et al. 2006) to reduce the probability of false-positive bands resulting from mispriming during screening steps. Therefore, the availability of the genome and target gene sequences of the species is an important part of the TILLING method. Carefully designed primer pairs for a unique gene in the species can then be used in touchdown PCR to amplify the genomic DNA and avoid producing non-specific targets. A preliminary test such as agarose gel electrophoresis is recommended to test the quality of the PCR amplicons before full-scale screening. After the success of the PCR is confirmed, amplicons are denatured and annealed by heating and cooling repeatedly to form heteroduplexes (Till et al. 2003, 2006).

7.1.3.4 Mutation Detection

Various methods for mutation detection have been reported and are constantly being improved. In the early days of TILLING, a dHPLC method was used for mutation detection (McCallum et al. 2000). This method has been replaced by heteroduplex-specific endonuclease-based methods using CEL I endonuclease, which is relatively cheaper and faster (Colbert et al. 2001; Till et al. 2006; Kurowska et al. 2011). The cleavage activity of celery juice extract (CJE) is comparable to that of CEL I; CJE has been used as a substitute for commercial nucleases because it is cheap and easy to obtain (Till et al. 2004a, 2006).

The LI-COR gel analyzer system (LI-COR Biosciences) with fluorescently labeled primers (IRDye-700 and IRDye-800) is a well-established platform for endonuclease-based TILLING screening that has shown high sensitivity (Colbert et al. 2001; Till et al. 2003, 2006) (Fig. 7.6). To reduce the cost of reagents, non-denaturing polyacrylamide gel electrophoresis was set up with ethidium bromide staining, which was shown to have comparable sensitivity to the LI-COR system (Uauy et al. 2009). Agarose gel-based screening systems have also been established although these systems often have low sensitivity (Sato et al. 2006; Dong et al. 2009; Rogers et al. 2009).

Two alternative platforms for high-throughput TILLING screening – HRM and CSCE (see Sects. 7.1.2.2 and 7.1.2.3) – that do not use endonucleases have been reported (Gady et al. 2009). These two platforms can save time because the heteroduplex digestion and gel electrophoresis steps can be skipped (Gady et al. 2009; Kurowska et al. 2011).

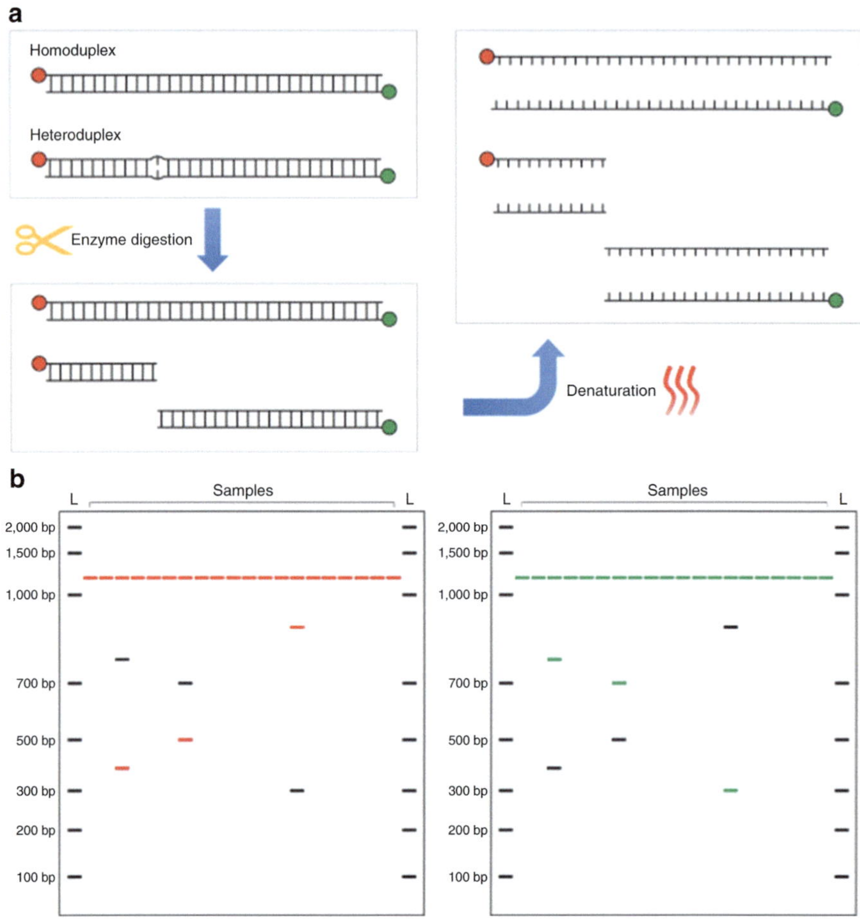

Fig. 7.6 Mutation detection using heteroduplex-specific endonuclease in TILLING. (**a**) PCR amplicons during enzyme digestion and denaturation. *Red* and *green circles* represent fluorescently labeled primers. (**b**) Mutation detection using denaturing polyacrylamide gel electrophoresis and an LI-COR DNA analyzer. The samples can be considered to have true mutations when the cleaved fragments appear in both two channels and the sum of the length of the cleaved fragments in the respective channels is equal to the full-length fragment (Redrawn from Till et al. 2006)

NGS technology opened a new era in genomics research and TILLING by sequencing was devised on the strength of the high throughput, low cost, and multiplex sequencing on NGS platforms (Tsai et al. 2011). Samples are pooled along the 3D axis in planes and each pool is used as a template for PCR. Individual target regions are amplified, normalized, and pooled together in accordance with their template pools. Then each amplicon pool is sheared, tagged with a unique index adapter, and sequenced on an Illumina platform. Sequenced reads are pre-processed and aligned to the reference target sequences through a bioinformatics pipeline for mutation calling (for details see Tsai et al. 2011).

7.1.4 *Case Study: TILLING Approach to Detect Mutations in High-Tryptophan-Accumulating Rice Mutants*

Tryptophan is an essential amino acid in rice and is a precursor of various secondary metabolites including vitamin B3, serotonin, and indole alkaloids (Saika et al. 2011). In humans, tryptophan deficiency can induce pellagra and cause low levels of serotonin, which have been related to depression, anxiety, aggression, and overeating (Lucki 1998; Hegyi et al. 2004). Therefore, the development of rice with increased amounts of tryptophan will likely provide great benefits for humans and animals.

In this study, to identify the mutations that resulted in high levels of tryptophan in rice, a total of 1,350 mutant lines among the M8 generation were used in a TILLING analysis (Kim et al. 2004; Chun et al. 2012). Calli formed from wild-type rice cv. Dongan were irradiated with various doses of gamma rays (30–120 Gy) and then the mutated calli were cultured in a medium that contained a tryptophan analogue, 5-methyltryptophan (5MT). Over-synthesis of tryptophan was observed in 5MT-resistant mutants with altered feedback inhibitions of anthranilate synthase, therefore, the anthranilate synthase encoding genes were selected as targets in the mutant line with high tryptophan content using the TILLING approach (Kisaka et al. 1996).

The rice anthranilate synthase alpha 1 subunit (*OASA1*) gene (LOC_Os03g61120) consists of 11 exons spanning 5.6 kb. To amplify *OASA1* coding regions, five primer sets were designed; the amplicons were approximately 770–1300 bp in length (Fig. 7.7). The 5′ ends of the forward and reverse primers were fluorescently labeled with FAM and HEX, respectively.

Normalized samples were allocated to 96-well plates for 2D pooling. After PCR amplification, heteroduplex formation was performed and the re-annealed PCR products were digested using CEL I. The CELI cleavage samples were subjected to capillary electrophoresis and analyzed using an ABI 3130 XL DNA sequencer (Applied Biosystems). When a SNP is located in a DNA fragment, two peaks are detected under cleavage PCR amplification through capillary electrophoresis

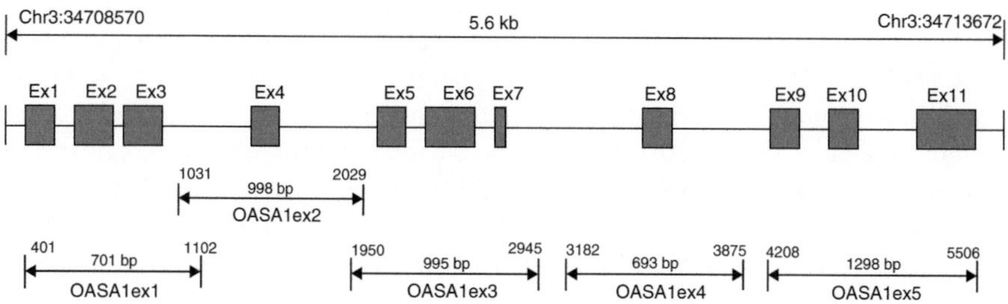

Fig. 7.7 Structure of the *OASA1* gene and design of PCR primer pairs for TILLING. The primer pairs were designed to amplify all 11 exons for polymorphism detection (Chun et al. 2012)

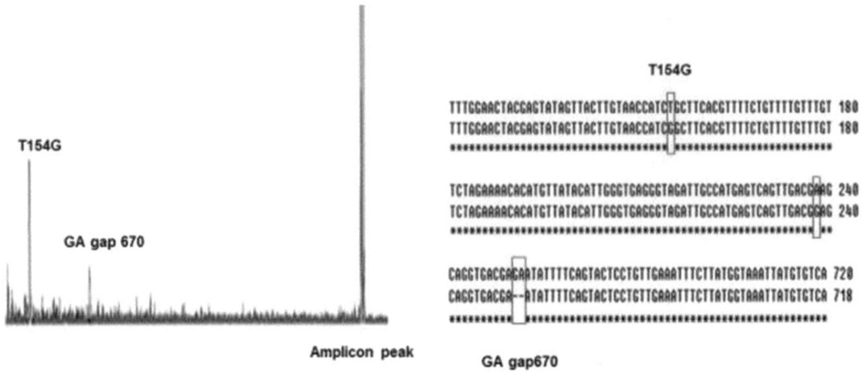

Fig. 7.8 Polymorphism detection of amplicons with OASA1ex3 from the mutant rice line MRVII-9 Capillary electrophoresis results (*left panel*) correspond to the sequencing results (*right panel*) (Chun et al. 2012)

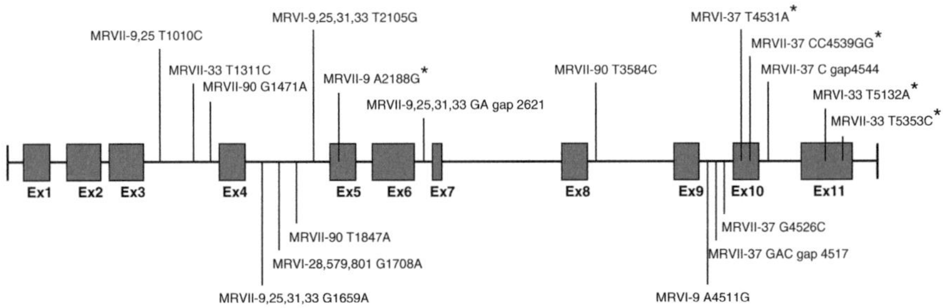

Fig. 7.9 Distribution of mutations in the *OASA1* gene in nine mutant rice lines. A total of 18 mutation sites were discovered by sequencing. *Asterisks* (*) indicate SNPs in the coding regions (Chun et al. 2012)

(Fig. 7.8). The CELI cleaved PCR products were subjected to further sequencing and analysis.

Among the 1,350 mutant lines screened using the TILLING method, nine lines produced one or two cut PCR fragments of the target region of *OASA1*. Sequencing of the genes in the nine mutant lines using five primer sets showed that a total of 31 SNPs were located in the exons and introns of *OASA1*. In particular, three mutant lines, MRVII-9, MRVII-33, and MRVII-37 contained five SNPs (Fig. 7.9), some of which were nonsynonymous SNPs that led to amino acid changes in the encoded protein sequences (Fig. 7.10). In the MRVII-9 mutant line, the SNP changed the amino acid at position 241 from glutamic acid (E) to glycine (G); in MRV33-33, the SNP changed alanine (A) into glycine (G) at position 560 and in MRVII-37, two amino acids valine (V) and serine (S) were changed to phenylalanine (F) at positions 495 and 496, respectively (Fig. 7.10).

The three mutant lines (MRVII-9, MRVII-33, and MRVII-37) with changed translated amino acid sequences have shown good agronomical traits and alterations

```
Dongan     1   MASLVLSLRIAPSTPPLGLGGGRFRGRRGAVACRAATFQQLDAVAVREEESKFKAGAAEGCNILPLKRCI 70
MRVII-9    1   MASLVLSLRIAPSTPPLGLGGGRFRGRRGAVACRAATFQQLDAVAVREEESKFKAGAAEGCNILPLKRCI 70
MRVII-33   1   MASLVLSLRIAPSTPPLGLGGGRFRGRRGAVACRAATFQQLDAVAVREEESKFKAGAAEGCNILPLKRCI 70
MRVII-37   1   MASLVLSLRIAPSTPPLGLGGGRFRGRRGAVACRAATFQQLDAVAVREEESKFKAGAAEGCNILPLKRCI 70

Dongan     71  FSDHLTPVLAYRCLVREDDREAPSFLFESVEQGSEGTNVGRYSVVGAQPAMEIVAKANHVTVMDHKMKSR 140
MRVII-9    71  FSDHLTPVLAYRCLVREDDREAPSFLFESVEQGSEGTNVGRYSVVGAQPAMEIVAKANHVTVMDHKMKSR 140
MRVII-33   71  FSDHLTPVLAYRCLVREDDREAPSFLFESVEQGSEGTNVGRYSVVGAQPAMEIVAKANHVTVMDHKMKSR 140
MRVII-37   71  FSDHLTPVLAYRCLVREDDREAPSFLFESVEQGSEGTNVGRYSVVGAQPAMEIVAKANHVTVMDHKMKSR 140

Dongan     141 REQFAPDPMKIPRSIMEQWNPQIVEGLPDAFCGGWVGFFSYDTVRYVETKKLPFSNAPEDDRNLPDIHLG 210
MRVII-9    141 REQFAPDPMKIPRSIMEQWNPQIVEGLPDAFCGGWVGFFSYDTVRYVETKKLPFSNAPEDDRNLPDIHLG 210
MRVII-33   141 REQFAPDPMKIPRSIMEQWNPQIVEGLPDAFCGGWVGFFSYDTVRYVETKKLPFSNAPEDDRNLPDIHLG 210
MRVII-37   141 REQFAPDPMKIPRSIMEQWNPQIVEGLPDAFCGGWVGFFSYDTVRYVETKKLPFSNAPEDDRNLPDIHLG 210

Dongan     211 LYNDIVVFDHVEKKTHVIHWVRVDCHESVDEAYEDGKNQLEALLSRLHSVNVPTLTAGSVKLNVGQFGSA 280
MRVII-9    211 LYNDIVVFDHVEKKTHVIHWVRVDCHESVDGAYEDGKNQLEALLSRLHSVNVPTLTAGSVKLNVGQFGSA 280
MRVII-33   211 LYNDIVVFDHVEKKTHVIHWVRVDCHESVDEAYEDGKNQLEALLSRLHSVNVPTLTAGSVKLNVGQFGSA 280
MRVII-37   211 LYNDIVVFDHVEKKTHVIHWVRVDCHESVDEAYEDGKNQLEALLSRLHSVNVPTLTAGSVKLNVGQFGSA 280

Dongan     281 LQKSSMSREDYKKAVVQAKEHILAGDIFQVVLSQRFERRTFADPFEVYRALRIVNPSPYMAYLQARGCIL 350
MRVII-9    281 LQKSSMSREDYKKAVVQAKEHILAGDIFQVVLSQRFERRTFADPFEVYRALRIVNPSPYMAYLQARGCIL 350
MRVII-33   281 LQKSSMSREDYKKAVVQAKEHILAGDIFQVVLSQRFERRTFADPFEVYRALRIVNPSPYMAYLQARGCIL 350
MRVII-37   281 LQKSSMSREDYKKAVVQAKEHILAGDIFQVVLSQRFERRTFADPFEVYRALRIVNPSPYMAYLQARGCIL 350

Dongan     351 VASSPEILTRVEKRTIVNRPLAGTIRRGKSKAEDKVLEQLLLSDGKQCAEHIMLVDLGRNDVGKVSKPGS 420
MRVII-9    351 VASSPEILTRVEKRTIVNRPLAGTIRRGKSKAEDKVLEQLLLSDGKQCAEHIMLVDLGRNDVGKVSKPGS 420
MRVII-33   351 VASSPEILTRVEKRTIVNRPLAGTIRRGKSKAEDKVLEQLLLSDGKQCAEHIMLVDLGRNDVGKVSKPGS 420
MRVII-37   351 VASSPEILTRVEKRTIVNRPLAGTIRRGKSKAEDKVLEQLLLSDGKQCAEHIMLVDLGRNDVGKVSKPGS 420

Dongan     421 VKVEKLMNVERYSHVMHISSTVTGELRDDLTCWDALRAALPVGTVSGAPKVRAMELIDQMEGKMRGPYSG 490
MRVII-9    421 VKVEKLMNVERYSHVMHISSTVTGELRDDLTCWDALRAALPVGTVSGAPKVRAMELIDQMEGKMRGPYSG 490
MRVII-33   421 VKVEKLMNVERYSHVMHISSTVTGELRDDLTCWDALRAALPVGTVSGAPKVRAMELIDQMEGKMRGPYSG 490
MRVII-37   421 VKVEKLMNVERYSHVMHISSTVTGELRDDLTCWDALRAALPVGTVSGAPKVRAMELIDQMEGKMRGPYSG 490

Dongan     491 GFGGVSFRGDMDIALALRTIVFPTGSRFDTMYSYTDKNARQEWVAHLQAGAGIVADSKPDDEHQECLNKA 560
MRVII-9    491 GFGGVSFRGDMDIALALRTIVFPTGSRFDTMYSYTDKNARQEWVAHLQAGAGIVADSKPDDEHQECLNKA 560
MRVII-33   491 GFGGVSFRGDMDIALALRTIVFPTGSRFDTMYSYTDKNARQEWVAHLQAGAGIVADSKPDDEHQECLNKG 560
MRVII-37   491 GFGGFFFRGDMDIALALRTIVFPTGSRFDTMYSYTDKNARQEWVAHLQAGAGIVADSKPDDEHQECLNKA 560

Dongan     561 AGLARAIDLAESTFVDE 577
MRVII-9    561 AGLARAIDLAESTFVDE 577
MRVII-33   561 AGLARAIDLAESTFVDE 577
MRVII-37   561 AGLARAIDLAESTFVDE 577
```

Fig. 7.10 Multiple alignment of translated OASA1 amino acid sequences. *Arrows* indicate amino acid changes in the mutant lines compared with the wild-type cv. Dongan (Chun et al. 2012)

in tryptophan biosynthesis compared with the cv. Dongan line. In the three mutant lines, the amount of tryptophan was 2.2- to 2.3-fold higher than in the wild-type, which contains 0.048-nmol tryptophan/mg protein. In particular, MRVII-33 had the highest amount of tryptophan (0.111 nmol/mg protein) among the three mutant lines (Fig. 7.11).

7.2 Genome Editing Using Engineered Nucleases

7.2.1 Types and Characteristics of Engineered Nucleases

Two of the most important factors in plant breeding programs are to diversify the genetic background to be screened and to select the individuals that have the favored alleles or traits. In traditional breeding programs, a lot of resources are usually required, including time, land, and money before the selected traits can be introduced successfully into existing commercial cultivars or elite lines. It is probably even more difficult when favored alleles are linked closely to bad alleles such as linkage dragging, which either allows no segregation between the alleles or requires the adoption of an alternative technique.

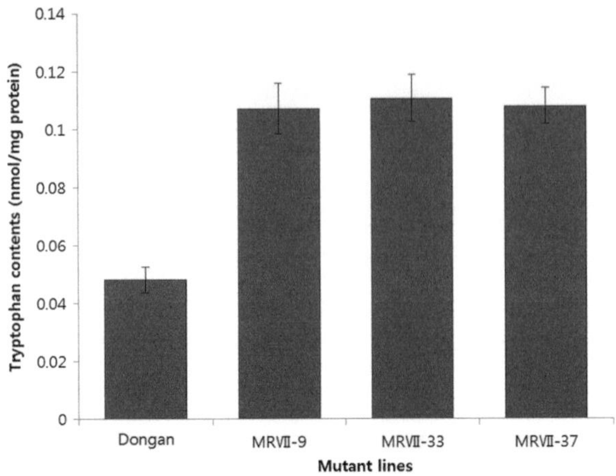

Fig. 7.11 Variation of tryptophan content between the wild-type cv. Dongan and three high-tryptophan-containing mutant lines that were genetically fixed by self-pollination (Chun et al. 2012)

Genetic engineering that encompasses transformation is an advanced breeding method that can overcome reproductive barriers and extend genetic variations. However, gene introduction via transformation occurs randomly and the modification efficiency of the traits of interest is usually very low, which hinders the wide application of the transformation tool to plant breeding. Possibly, the most important limitation to using genetic engineering in cultivar development is its limited public acceptance in many countries, including Korea.

Recently, genome editing using engineered nucleases (artificial restriction enzymes) has been applied to plant breeding. One of its greatest advantages is the targeted modification of traits, regardless of the genome locations of the genes of interest. So far, more than 30 plant genes have been targeted for gene editing and mutagenized successfully in several plant species including tobacco, maize, Arabidopsis, soybean, rice, and Brachypodium (Table 7.1). Engineered nucleases were developed to modify the genome targets by cutting into the exact positions in the genomes that contained the genes of interest. This precise and efficient technique that can induce changes in the genome positions now can be applied to a variety of cells and is likely to become a universal tool once the experimental basis is well established.

In the engineered nuclease-based genome editing method, the double-stranded DNA is first cleaved to take advantage of mutations that occur during the DNA repair process (Fig. 7.12). Cells that contain damaged DNA would be killed if breaks in the DNA double-strand are not repaired, therefore, such cells will activate recovery mechanisms such as non-homologous end joining (NHEJ). Genome editing can be classified into gene knock-out (removing the gene function), gene knock-in (adding a foreign gene), and gene replacement (i.e., gene therapy, changing corrected DNA bases in target alleles) (Fig. 7.12).

Table 7.1 Examples of plant species and target genes modified by engineered nucleases (Tzfira et al. 2012)

Species	Target	Nuclease	Selection, outcome	Reference
BY2 cells	*CHN50*	ZFN[a]	Bialaphos-resistance, HR[b]	Cai et al. (2013)
Tobacco	*CHN50*	ZFN	Bialaphos-resistance, HR	Cai et al. (2013)
Maize	*IPK1*	ZFN	Bialaphos-resistance, HR	Shukla et al. (2009)
Tobacco	Transgene	ZFN	Site-specific mutagenesis, reconstitution of GUS[c] expression	Marton et al. (2010)
Petunia	Transgene	ZFN	Site-specific mutagenesis, reconstitution of GUS expression	Marton et al. (2010)
Tobacco	*SuRA, SuRB*	ZFN	Resistance to herbicide, HR	Townsend et al. (2009)
Arabidopsis	Transgene	ZFN	Site-specific mutagenesis, PCR	Lloyd et al. (2005)
Tobacco	Transgene	ZFN	Site-specific mutagenesis, reconstitution of GUS expression	Tovkach et al. (2009)
Arabidopsis	Transgene	ZFN	Site-specific mutagenesis, reconstitution of GUS expression	Tovkach et al. (2009)
Tobacco	Transgene	ZFN	Reconstitution of kanamycin resistance and GUS expression	Wright et al. (2005)
Tobacco	*Hax3-box*	TALEN[d]	Site-specific mutagenesis, reconstitution of GUS expression	Mahfouz et al. (2011)
Arabidopsis	Transgene	ZFN	Site-specific mutagenesis, reconstitution of GUS expression	Even-Faitelson et al. (2011)
Arabidopsis	*ADH1*	TALEN	Site-specific mutagenesis, PCR	Cermak et al. (2011)
Tobacco	Transgene	ZFN	transgene removal, elimination of GUS gene	Petolino et al. (2010)
Arabidopsis	*ADH1, TT4*	ZFN	Site-specific mutagenesis, PCR	Zhang et al. (2010)

(continued)

Table 7.1 (continued)

Species	Target	Nuclease	Selection, outcome	Reference
Soybean	Transgene	ZFN	Site-specific mutagenesis, PCR	Curtin et al. (2011)
Soybean	*DCL, RDR, HEN*	ZFN	Site-specific mutagenesis, PCR	Curtin et al. (2011)
Soybean	*DCL4b*	ZFN	Site-specific mutagenesis, PCR	Curtin et al. (2011)
Arabidopsis	*ABI4*	ZFN	Site-specific mutagenesis, ABA-insensitive	Osakabe et al. (2010)
Arabidopsis	Transgene	ZFN	Site-specific mutagenesis, HR-mediated integration	de Pater et al. (2009)
Rice	Disease susceptibility (S) gene	TALEN	Agrobacterium, NHEJ[e] (gene knockout)	Li et al. (2012)
Rice	BADH2	TALEN	Transformation of protoplast, NHEJ (gene knockout)	Shan et al. (2013)
Brachypodium		TALEN	Agrobacterium, NHEJ (gene knockout)	Shan et al. (2013)

[a]*ZFN* zinc-finger nuclease
[b]*HR* homologous recombination
[c]*GUS* beta-glucuronidase
[d]*TALEN* TAL effector nuclease
[e]*NHEJ* non-homologous end joining

The engineered nuclease-created break in the double-stranded DNA is usually repaired by NHEJ, result in a functionally modified knock-out mutant. During the NHEJ repair process, some bases can be inserted or deleted, which creates frame shift mutagenesis and leads to the formation of truncated proteins where the function is lost. When a donor DNA is introduced into a cell with the engineered nuclease, a gene correction or gene replacement by a homologous recombination (HR) repair mechanism will take place. Gene correction or gene replacement is dependent on the donor DNA sequence. For instance, a gene correction will occur when the donor DNA contains the wild-type bases, while a gene replacement will occur when a mutated DNA sequence is introduced as the donor. The two distal ends of the donor DNA should have the same base pairs as the target DNA fragment for a perfect fit between them.

There are two types of engineered nucleases, zinc-finger nuclease (ZFN; Fig. 7.13a) and TAL effector nuclease (TALEN; Fig. 7.13b), that have different DNA-binding domains. Both are type IIs restriction enzymes that have a *Fok*I DNA-cleavage domain and engineered base modules linked together, which function in binding to the target gene. The *Fok*I domain has breaking activity when it exists as a dimer, therefore, ZFN and TALEN both have a pair of *Fok*I domains to ensure that successful mutagenesis takes place.

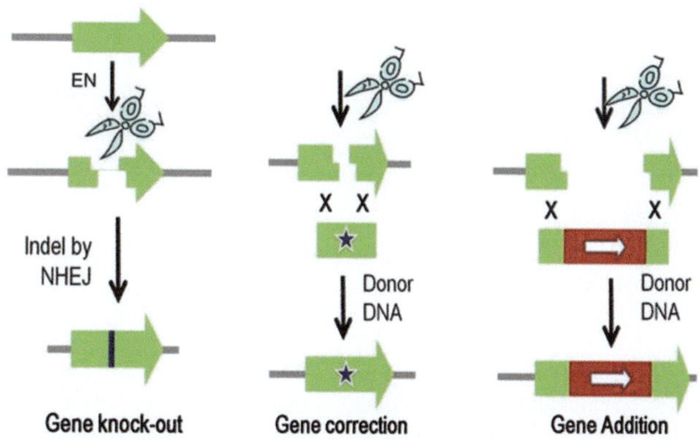

Fig. 7.12 Genome editing and DNA repair processes. *EN* engineered nuclease, *NHEJ* non-homologous end joining

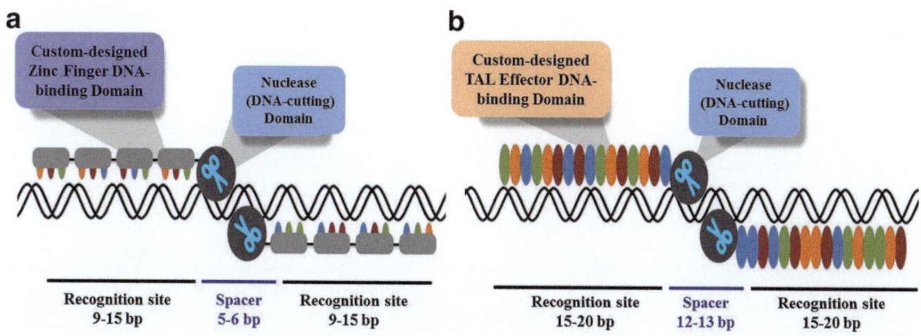

Fig. 7.13 Engineered nucleases. (**a**) zinc-finger nuclease (ZFN), (**b**) TAL effector nuclease (TALEN) (Source: Toolgen, Ltd. Seoul, Korea)

One zinc-finger module can recognize three DNA bases and one DNA-binding domain consists of three to five zinc fingers, which can bind to 9–15 bp for one pair of the engineered ZFN. Theoretically, there could be 64 types of the module because each zinc-finger module recognizes 3 bp (4^3), but in the ZFNs 30–40 zinc-finger modules are normally used.

Additionally, the diversity of gene sequence recognition is limited because the specificity and binding ability of the zinc-finger DNA-binding domain in the engineered ZFNs are limited. For this reason, the ZFNs on the target gene are designed to be 150–500 bp apart for the active nucleases to bind.

TALEN was designed after it was suggested that the TAL effector recognized host DNA during the transcription process of bacterial species found in some plants (Cermak et al. 2011). Each TAL effector module can recognize one DNA base, and 12–20 modules are joined to create the DNA-binding domain. For TALEN, there are repeat variable di-residues in the TAL effector module, which determine its specificity to the target gene. Therefore, if there are four TAL effector modules each

Table 7.2 Structural and functional characteristics of three nuclease types (Gasiunas and Siksynys 2013)

	ZFN[a]	TALEN[b]	RGEN[c]
DNA-binding module	Zinc-finger proteins	TAL effectors	crRNA or sgRNA
Cleavage module	*Fok*I	*Fok*I	Cas9
Target site expansion	18–36 bp	30–40 bp	23 bp
Targetable sequences	Guanine-rich	No limitations	End with GG (PAM[d])
Targeting frequency	High	High	High, depends on PAM
Reprogramming	Complicated: requires domain shuffling, assembly, and protein engineering	Relatively easy: requires domain shuffling, assembly, and protein engineering	Easy and fast: requires only sgRNA[e]
Off-target effects	High	Low	Variable
Cytotoxicity	Variable to high	Low	Low
Size	~1 kbp × 2	~3 kbp × 2	4.2 kbp (Cas9) þ 0.1 kbp (sgRNA)
Public resources	Zinc finger Consortium Addgene	Seoul National University (www.talenlibrary.net)	Addgene
Commercial resources	Sigma-Aldrich	Life Technologies, Cellectis	ToolGen
Polymeric state	Function as dimer	Function as dimer	Function as monomer
Miscellaneous	Sequence bias, some variants show toxicity	Large protein size	Multiplexing possibilities

[a]*ZFN* zinc-finger nuclease
[b]*TALEN* TAL effector nuclease
[c]*RGEN* third-generation engineered nuclease
[d]*PAM* protospacer adjacent motif
[e]*sgRNA* single guide RNA

of which binds specifically to one of the four nucleotides, customized DNA-binding domains can be designed with potentially high combining abilities to any target DNA sequence. Indeed, we are currently conducting experiments using a TALEN with a customized DNA-binding domain that was created by combining TAL modules that can recognize 18–20 bp in a target gene. We have found that this TALEN structure exhibits high resolution every 5–20 bp (unpublished data).

Recently, a third-generation engineered nuclease (RGEN) has been reported, which has diversified the genome editing technology that was based only on ZFN and TALEN (Table 7.2). The RGEN consists of the CRISPR-associated Cas9 protein and RNA molecules (crRNA [short CRISPR RNA] and tracrRNA [trans-activating crRNA]) where the cleavage site is determined by the about 20-bp sequence of crRNA and the protospacer adjacent motif (PAM) of Cas9. Thus, a customized RGEN can be produced by replacing the target-recognizing sequences in the crRNA.

TALEN and RGEN are more specific than ZFN because their modules can be designed for one-to-one recognition of the target bases, which allows higher engineering flexibility for precise locations (Table 7.2). TALEN can recognize a long DNA sequence, while RGEN recognizes the target sites using the stable and unique binding of RNA and DNA. However, the ZFN has the advantage of being smaller than the larger TALEN and RGEN, which is particularly important when DNA delivery is not easy or a particular vector system is necessary. The size of the nucleases is about 1-kb long, while in TALEN it is about 3-kb long and in the Cas9 protein is about 4-kb long.

The application of these engineered nucleases to plants has been limited, but is expected to expand in the future because of their efficiency and ease of use. However, the mutant selection system needs to be improved and studies into the correlation of genome modification and efficiency of the engineered nucleases are still lacking. In this chapter, some of the techniques related to editing the target traits in which the engineered nucleases have been applied to various plant species are introduced.

7.2.2 Description of Related Terms

- Engineered nuclease: Artificial restriction enzyme used to edit genome sequences within a cell. These nucleases are designed to break DNA double strands precisely. The technique uses mutagenesis, which occurs in the process of DNA repair. Currently, three types of the engineered nucleases are available: zinc-finger nuclease (ZFN), TAL effector nuclease (TALEN), and RNA-guided engineered nuclease (RGEN).
- Non-homologous end joining (NHEJ): Simple DNA repair mechanism of connecting two terminal ends after a DNA break. This process is anticipated regardless of cell cycles. Compared with homologous recombination, NHEJ frequently occurs and results in base-deletion or base-insertion mutagenesis, so it can be a mechanism of error-prone double-strand break (DSB) repair.
- Homologous recombination (HR): Process to recover from DNA break using a sister chromatid produced during the S phase of the cell cycle for genome duplication. The HR process operated only in the limited cell cycle and uses the correct duplicates so that it is an error-free DNA repair process, unlike DSB repair.
- Zinc-finger nuclease (ZFN): First-generation engineered nuclease developed during the late 1990s to mid-2000s. ZFN consists of customized DNA-binding zinc-finger domains in a modular structure. Each zinc-finger module can recognize 3 bp of a target DNA sequence and can be combined to recognize more than 9 bp. The combined zinc-finger modules are connected to a *Fok*I DNA-cleavage domain. ZFN dimmers act as DSB enzymes that can recognize a minimum of 18 bp of the target sequence.
- TAL effector nuclease (TALEN): Second-generation engineered nuclease developed using the DNA-binding TAL effector protein that was detected in one

particular plant bacteria. The TAL effector consists of a basic module of 34 amino acids. Sequence variations among the modules allow particular modules to recognize each of the four bases specifically.

- RNA-guided endonuclease (RGEN): Third-generation engineered nuclease in which each module can recognize one base of the target sequence, allowing flexibility and high specificity. Unlike the other two nucleases, RGEN is composed of one protein (Cas9) and two RNAs (crRNA+tracrRNA).
- Genome engineering: A mutagenesis approach using an engineered nuclease for gene modification at the target site.
- Gene knock-out: Technique used to modify the function of a gene by structure modification at the DNA level.
- Gene knock-in: Technique used to modify a trait by the insertion of foreign genes into a target site.
- Gene editing: Process used to change genetic information of a target gene for mutant induction.

7.2.3 Engineered Nuclease-Based Target Trait Editing Techniques

Many processes are required for targeted mutagenesis, including target gene selection, construction of engineered nucleases, confirmation of activity, construction of vectors for the engineered nucleases, delivery of the engineered nuclease into a cell, mutagenesis, plant regeneration, phenotyping, and genetic screening (Fig. 7.14).

7.2.3.1 Outcomes of Genome Editing Using Engineered Nucleases

Genome engineering using site-specific nucleases can have two purposes: gene knock-out (KO) and gene knock-in (KI), or gene correction. KO is used to eliminate gene functions, while KI or gene correction is used to insert foreign genes into a target site. When the DNA sequences of the target gene are cut and recovered by non-homologous end joining, mutations can lead to structural changes in the DNA. If an engineered nuclease is designed to cut the target gene at the protein initiation site, a frameshift mutation would have 2/3 probability of producing a truncated protein, thus producing a gene KO. In most higher plant cells, NHEJ is used to repair breaks in DNA, and KO mutants are most commonly derived from changes in target gene sequences.

For KI or gene correction, an engineered nuclease with an appropriately designed donor DNA sequence is introduced into a cell where part of the double-stranded DNA break is repaired by homologous recombination and the donor DNA is inserted exactly into the target location. Depending on the designed donor DNA, gene KO at the target location, shift change, DNA recovery, or foreign gene introduction can all occur.

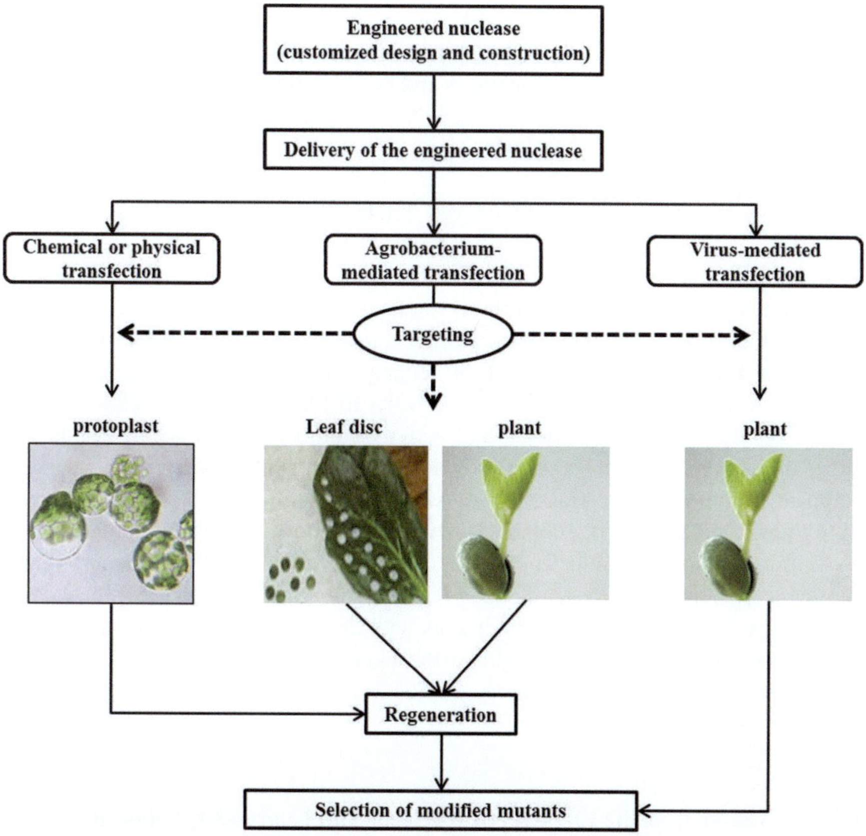

Fig. 7.14 Targeted mutagenesis using an engineered nuclease (Mahfouz and Li 2011)

7.2.3.2 Choosing the Target Gene Location

To design an engineered nuclease, it is important to determine the target sites of the gene. The target location can be determined by gene KO, and gene KI or gene correction. For gene KO, the engineered nuclease is designed to target the 5′ end of the exon site of a protein coding gene to produce a frameshift by the introduction of an indel mutation in the repair process.

For KI or gene correction, the target location is determined, for example by tagging a fluorescent protein to the target gene using the KI method. The engineered nuclease is designed generally to cut at either the target start or stop codon. Similarly for gene correction, the engineered nuclease is designed to replace a particular amino acid sequence so that the exact amino acid codon is targeted. The efficiency will be greater when the nuclease is designed to recognize the locations close to the target gene.

ToolGen, Korea provides engineered nucleases designed for a target gene. Design tools are available for users to decide the location to be targeted. For instance,

Zinc Finger Tools is publicly available and users can search DNA sequences for cleavage sites, find a suitable zinc-finger protein, prescreen zinc-finger protein DNA binding sites, and search DNA sequences and target sites for close matches (Mandell and Barbas 2006).

7.2.3.3 Construction of an Engineered Nuclease

To break the target sequence of a gene within a cell, engineered nucleases needs a programmable high binding specificity that allows them to recognize the target sequences among the billions of base pairs that make up a genome. For example, the ability of a nuclease to recognize a 16-bp sequence in the human genome is theoretically less than one time (approximately $4^{16} = 4 \times 10^9$) for 3×10^9 base pairs.

The nuclease design, synthesis, and activity confirmation is usually conducted either in a laboratory or increasingly by a commercial service. Commercial service companies include Sigma-Aldrich (http://www.sigmaaldrich.com) for ZFN; Cellectis (http://www.cellectis.com), ToolGen (www.toolgen.com), Life Technologies (http://www.lifetechnologies.com/us/en/home.html) for TALEN; and ToolGen, System Biosciences (http://www.systembio.com/) for RGEN (or CRISPR/CAS). It is advisable for users to discuss their requirements with the company in advance to acquire an engineered nuclease that is right to the experimental purpose. To design and synthesize a nuclease in the laboratory, a suitable vector system can be obtained from Addgene (www.addgene.org).

7.2.3.4 Delivery of the Engineered Nuclease and Expression Validation

Engineered nucleases can be introduced into plant cells indirectly through an Agrobacterium system or directly using a protoplast system. In the Agrobacterium system, transgenic lines carrying an expression vector of the engineered nuclease are constructed to induce transformed mutants exactly in the target gene. In the protoplast system, a vector carrying the engineered nuclease is introduced into the isolated protoplast by physical or chemical method, and then the expression of the nuclease is allowed for some time period. Finally, mutants derived from the genetically modified cells are obtained.

For ZFN and TALEN to work, dimers are required. Therefore, a method that can safely deliver two nuclease vectors should be considered, especially in the Agrobacterium system. Two methods can be used to achieve this; an expression cassette for two separate nucleases working on one target gene, or a cassette carrying two nucleases connected by the self-cleaving 2A peptide.

In Agrobacterium-mediated engineered nuclease delivery into leaf segments, the nuclease can be subcloned in the Ti-plasmid. For the protoplast system, polyethylene glycol (PEG)-mediated transformation or electroporation-mediated transformation has been mainly used (Mahfouz and Li 2011).

Transient assays have been used to detect whether the engineered nuclease operates precisely in a plant system. The bacterial two-hybrid (B2H) assay, *in vitro* system, and yeast system are known as transient assay methods (Hurt et al. 2003; Carroll et al. 2006). A surrogate assay system has been reported in plants because a lot of time and effort are required for phenotyping screening (Lloyd et al. 2005). In Arabidopsis, an assay of a reporter construct carrying the engineered nuclease was used in isolated plant protoplasts (Lloyd et al. 2005). To complement the assay methods, generally, the cleaved target locus is visualized in agarose gels after PCR amplification, and then sequenced to compare the individual bands that appear.

In the surrogate reporter system that has been reported recently (Kim et al. 2011), the plasmid carries the target DNA attached to the green fluorescence protein on the off-frame side. If the engineered nuclease is not operating properly, the downstream fluorescence protein is not expressed, but the fluorescence can be visualized under a microscope when the targeted DNA has been broken and the stop codon has shifted. A plasmid delivered into a protoplast through PEG- or electroporation-transformation generally exhibits fluorescence expression in 12–24 h, therefore, the activity of the nuclease can be verified indirectly in the surrogate reporter plasmid.

Methods used in the early stages of engineered nuclease development without a fluorescence microscope are introduced below (Hoshaw et al. 2010).

Protoplast Transformation

① In a tissue culture hood, add 100 uL MMg medium (0.4 M mannitol, 15 mM $MgCl_2$, 4 mM MES buffer) to the corner of a sterile tissue culture plate that is slightly tilted.
② Remove and add the protoplasts to the MMg medium. Mix gently by slowly pipetting up and down, using a micropipette with a 1-mL sterile tip.
③ Add 200 uL of the protoplast solution to Eppendorf tubes, making sure to keep the protoplasts evenly suspended during the transfer process.
④ Add 60 ug of DNA per tube, including positive and negative controls. Mix by inverting several times and wait for 2 min. A plasmid encoding GFP works well as a positive control where transformation can be monitored by observing GFP fluorescence after 6 h.
⑤ Add 100 uL PEG solution to each sample. Mix every 10 min for 30 min by inverting the tube several times.
⑥ Add 800 uL W5 solution and mix by inverting. Wait for 5 min.
⑦ Centrifuge for 3 min at $150 \times g$ in a microcentrifuge.
⑧ Remove the supernatant. Add 1 mL W5 solution to resuspend the protoplasts. Repeat step 7 and wash one more time.
⑨ Add 1 mL K3/S1 medium to the washed protoplasts. Transfer the solution onto cell culture plates and add 4 mL K3/S1 medium.
⑩ Seal the plants using parafilm and place at room temperature for 2 days.

Protoplast DNA Isolation

① Collect about 5 mL of the protoplasts in three Eppendorf tubes using a pipette. Remove the supernatant after centrifuging at 5,500 × *g* for 15 min.
② Follow the other steps as in the normal CTAB DNA extraction process (Keim et al. 1988).
③ Dissolve the DNA in 30-uL TE buffer. DNA can be kept at 4 °C for short-term use or −20 °C for longer periods.

Target DNA Digestion and Enrichment

① Digest the genomic DNA. Use 10 uL DNA, 5 uL of the appropriate restriction enzyme (RE) buffer, 4-uL RE, and 31 uL of ddH_2O for a total of 50 uL. Leave the digestion overnight at the temperature specific for the RE being used. Performing this digestion before the PCR step will remove wild-type DNA targets and enrich for targets with mutations.
② Amplify the digested DNA by PCR using a normal PCR mixture (4-uL DNA, 50-uL master mix, 2-uL primer 1, 2-uL primer 2, and 42 uL of ddH2O).
③ Purify the PCR product using a PCR purification kit (Qiagen, Valencia, CA).
④ Digest the purified product again using 10 uL DNA, 2.5-uL buffer, 1-uL RE, and 11.5 uL of ddH2O for 3 min at the preferred temperature for the RE. This step further removes targets with wild-type sequences.
⑤ Following digestion the DNA is separated by electrophoresis in a 2 % agarose gel. The larger RE-insensitive fragments are excised and purified from the gel using agarose gel extraction kit.
⑥ This population of fragments is cloned using a TOPO TA cloning kit (Invitrogen, Waltham, MA) according to the manufacturer's instructions.
⑦ Transform 2 uL of the ligation product into one vial of electroporation competent cells (Invitrogen, Waltham, MA). Incubate on ice for 30 min before heat shocking the cells for 30 s at 42 °C in a water bath. Add 250 uL of SOC (Super Optical Broth + 20 mM glucose) medium (Invitrogen) to the cells, shake for 1 h at 37 °C, and spread 100 uL of the transformed cells onto a kanamycin plate. Incubate the plate overnight at 37 °C.
⑧ Pick white colonies from the plate with a 200-uL pipette tip and put into a culture plate containing 1 mL selective LB medium in each well. Seal the plate and incubate overnight in a 37 °C shaker at 250 rpm.
⑨ PCR amplify the cloned fragments directly, using 1 uL of the overnight culture, 0.5-uL primer 1, 0.5-uL of primer 2, 12.5-uL master mix, and 10.5-uL water for 35 cycles.
⑩ Verify the presence of RE-insensitive fragments by digesting an aliquot of the PCR reaction, followed by electrophoresis on a 2 % agarose gel.
⑪ For those clones containing RE-insensitive inserts, the DNA is submitted for sequence analysis. Some clones may produce partially digested products and such clones should not be sequenced.

⑫ Sequences are aligned to the reference target DNA sequence using a DNA analysis software program. Mutations at the ZFN cleavage site should now be apparent.

7.2.3.5 Verification of Engineered Nuclease-Driven Genome Modification

Gene Knock-Out (KO) and Correction

Mismatch-Sensitive Nuclease Assay Gene KO and repair is a process of varying a sequence by making small indels in targeted restriction sites of restriction nucleases. The variation is hardly observable among PCR products by simple procedures. Furthermore, the modification made by restriction nuclease needs to be screened using a multi-individual screening process because of the low efficiency of this experiment. For this, the most widely used experimental method is the mismatch-sensitive nuclease assay, which uses T7 endonuclease I [NEB (New England Biolabs, Inc., Ipswich, MA), ToolGen (Seoul, Korea), Fig. 7.15] and the surveyor nuclease (Transgenomic, Inc., Omaha, NE) as described below.

① Genomic DNA preparation: The sample for a mismatch-sensitive nuclease assay is mixed DNA with more than two genotypes at the same locus. This kind of sample

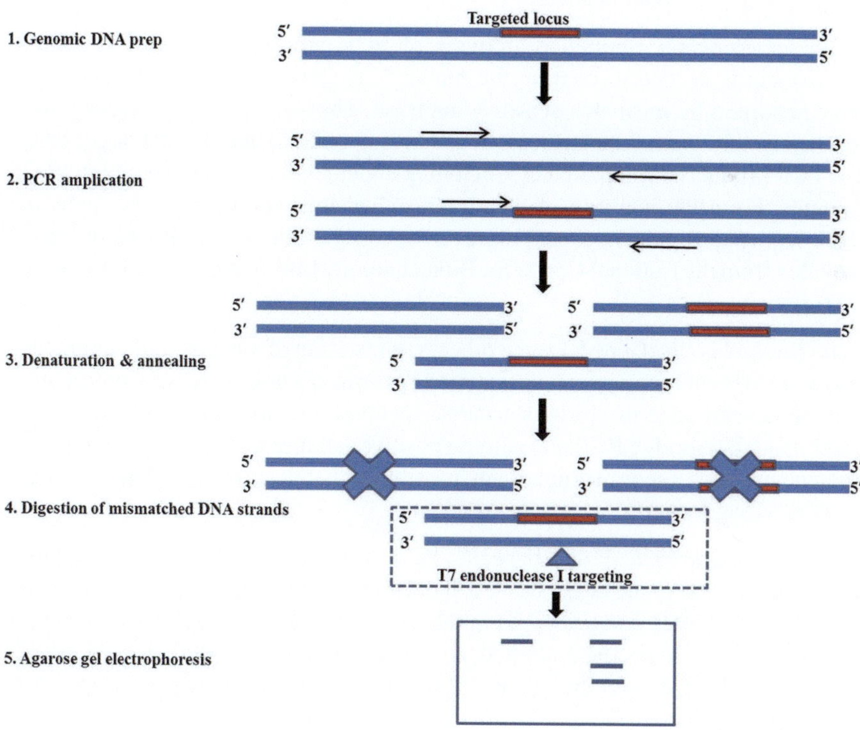

Fig. 7.15 Confirmation of mutants using T7 endonuclease I

can be, for example, a cell pool of nuclease-treated normal cells and mutated cells, or cells/individuals of homozygous/heterozygous mutants. Normally, the sensitivity of mismatch-sensitive nuclease assays is around 1 %. A variety of manual procedures or kits are available for genomic DNA preparation, but the concentration and purity of the resultant DNA should be good enough for PCR.

② Target site PCR: For successful mismatch-sensitive nuclease assays, it is important that a large enough quantity of the PCR product is available. To ensure this, nested PCRs are recommended. Generally, the size of the PCR products range from 400 to 600 bp, and the restriction sites should not be biased either towards the 5′ or 3′ end.

③ Denaturing/re-annealing (heteroduplex induction): This process is used to induce heteroduplexes between the wild-type and mutant alleles so that a mismatch-sensitive nuclease can cut the DNA sequence. The following PCR program is recommended: 95 °C/2 min to −2 °C/s to 85 °C to −0.1 °C/s to 25 °C to 16 °C until completed.

④ T7 endonuclease I reaction: Reaction mix containing 2-uL 10x NEB #2 buffer, PCR product × uL (300–1,000 ng), 0.25-uL T7E1 enzyme, up to 20 uL with ddH2O. Incubation for 20 min at 37 °C.

⑤ Agarose gel electrophoresis: The proper agarose gel concentration is determined based on the PCR product size. For PCR products of size 400–600 bp, 2–2.5 % agarose gel is recommended.

Sequence Analysis Sequence analysis of PCR amplicons captured through TA cloning is needed to confirm the mutated samples or mutant individuals that were identified by mismatch-sensitive nuclease. The same primer previously used for the mismatch-sensitive nuclease reaction is applied to amplify the target locus. The PCR products are captured in TA cloning under conditions set to produce 10–20 colonies. From the majority of the clones, other analyses are required including DNA sequence alignment using programs such as the ClustalW or Blast2 programs provided from the National Center for Biotechnology Information (NCBI, Bethesda, MD).

Gene Knock-In (KI) Gene KI is the process of inserting of a segment of donor DNA into a gap in homology arm at a site targeted by the restriction nuclease. Confirmation that the desired gene insertion is accurate involves: (1) observation of the desired event using KI-specific PCR; (2) gaining positive proof of gene KI through junction sequencing; and (3) elimination of any possibility that the donor DNA was inserted into the wrong location.

Knock-in Specific PCR F or KI-specific PCR, a primer specific to the donor DNA and a primer close to the target DNA will only amplify the segment when the resulting DNA structure is induced by KI in the modified cell (Fig. 7.16). In particular, if the sequence is KI-specific amplified, a subsequent PCR procedure is used to confirm the two locations on the 5′ and 3′ homology arms to ensure the KI operation was properly achieved.

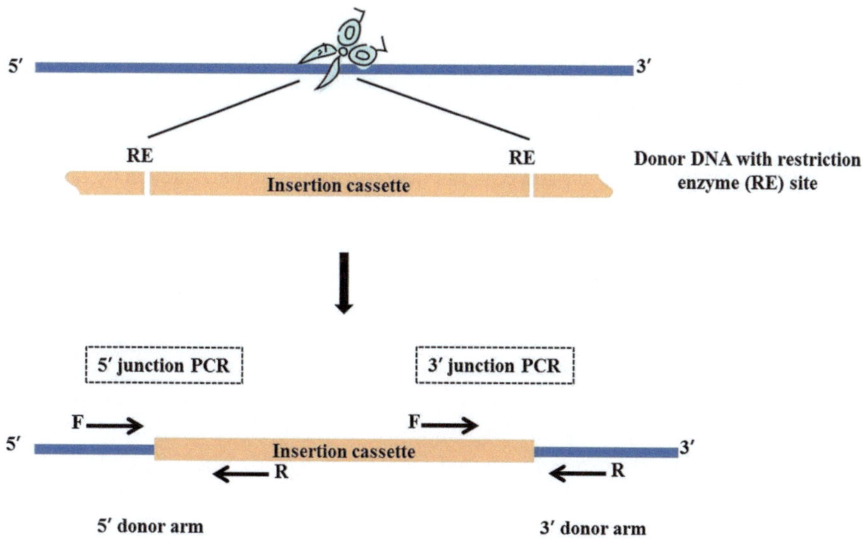

Fig. 7.16 Diagram of gene knock-in mechanism and verification by PCR

Junction Sequencing Once the KI-induced insertion of the donor DNA is confirmed, it is advisable to sequence the KI junction and confirm the configuration of the structure. Sometimes, a KI-induced mutation and a NHEJ-mediated indel are found to have been produced simultaneously.

Random Insertion If the KI operates as it is designed to, the genetic information from the donor DNA can be transferred into a space between the two homology arms. Therefore, it is necessary to perform a PCR to check if any random insertion of the donor DNA into non-target sites has occurred.

7.2.3.6 Evaluation of the Toxicity of Engineered Nucleases

If an engineered nuclease operates in other places on the genome rather than in the target gene, the viability of the cell would be compromised, leading to what is termed "cytotoxicity of the nuclease" (Cornu et al. 2008). An engineered nuclease can be cytotoxic for various reasons, including: (a) lack of DNA binding specificity, (b) mechanism difference in the homodimeric or heterodimeric activity of the *FokI-cleavage* domain, and (c) differences in the spacer length of a linker domain. To evaluate the cytotoxicity of an engineered nuclease, a cytotoxicity assay and/or a genotoxicity assay can be performed (Table 7.3). The cytotoxicity assay evaluates the impact of the engineered nuclease on the survival of cells by counting the number of cells remaining after treatment with an engineered nuclease compared with a control (i.e., the number of cells remaining after a similar treatment but without the nuclease). The cells containing the engineered nuclease can be visualized by introducing two fluorescent proteins together (e.g., GFP and DsRed-Express [REx]; a red fluorescent marker protein).

Alternative methods to reduce the toxicity of engineered nucleases have been tried (Pruett-Miller et al. 2009). For instance, an ubiquitin moiety was linked to the N terminal end of an engineered nuclease to reduce its half-life and make it less toxic (Fig. 7.17). It is not easy to isolate single plant cell, but quantification can be achieved by layering the treated cells onto a fluorescence background (e.g., GFP) using protoplasts.

7.2.3.7 Genome Editing Using Engineered Nucleases

Engineered nucleases are used to modify traits by disabling target genes or by inserting a donor DNA. The majority of the modifications can be achieved by non-homologous end joining causing frameshifts in the DNA strands and loss of gene function. ZFN, for example, caused 1–19 % mutation rates (i.e., CoDA ZFN) that were inherited by the next generation (Sander et al. 2011). Similarly in the homologous recombination, some points that need to be considered for successful mutagenesis using ZFN have been detailed by the Zinc Finger Consortium (www.zincfingers.org) and are listed below.

① If an engineered nuclease is only designed based on the cDNA, it will be disconnected after intron slicing, therefore, genomic DNA is commonly used.
② For ZFN, the *Fok*I-cleavage is active as a heterodimer, therefore, it is necessary to verify that the correct structure is present.
③ The efficiency of introducing double strand breaks depends on the concentration of ZFN that is available to bind to the target genes. Thus, the concentration of the engineered nuclease has to be carefully balanced; if the concentration is too low ZFN will not bind to the target gene, and if the concentration is too high cytotoxicity will be induced.

Table 7.3 Toxicity assays for engineered nuclease [Animal cells (HEK293T, HT-1080)]

Steps	Cytotoxicity assay	Genotoxicity assay
1	Cell culture of targeting gene (24 h)	
2	Replacement of subculture media and preparation of nuclease mixture	
3	Mixture of the cultured cell and the engineered nuclease (24 h)	
4	Replacement of fresh medium and reculture (30 h)	
5	Removing of the culture media and washing	
6	FACS buffer solution preparation	Cell centrifugation (2×)
7	Flow cytometer prep. and measurement	Addition of one and Two antibody solution and culture
8	Cytometer measurement using the remaining cells after 5 days	Flow cytometer prep. and measurement
9	Calculation of survived cells (minimum of three replications)	Toxicity evaluation based on the fluorescence intensity among the positive fluorescence responses

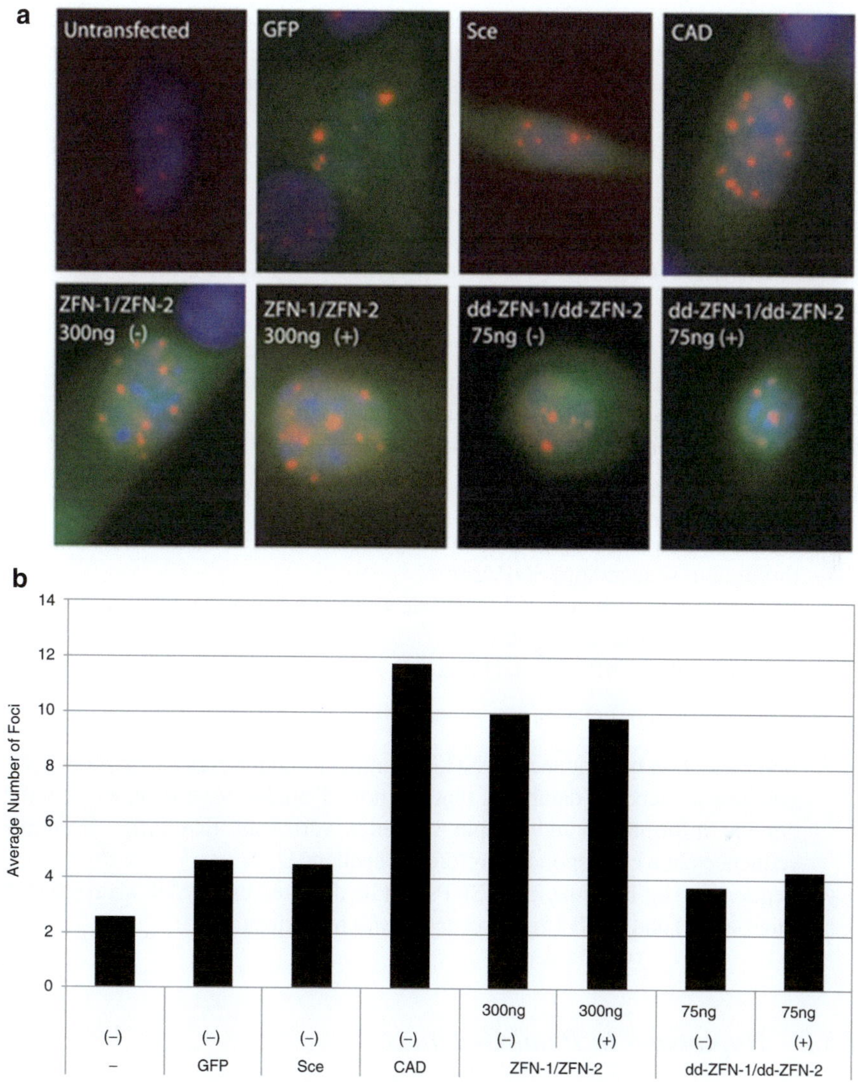

Fig. 7.17 Visualization of ZFN-induced double-strand breaks by sensitized 53BP1 foci formation. (**a**) Cells with transfected DNA (53BP1) in red staining and GFP-positive in green staining, (**b**) Average number of 53BP1 foci per transfected cell (Pruett-Miller et al. 2009)

④ If all the above conditions are met and the target genome is still not cleaved, this may indicate that the activity of the engineered nuclease is low, the specificity is low, or there is a problem with accessibility of the ZFN dimers to the binding sites on the chromatin. In such cases, the mutation rate can be improved if the nuclease is designed to cleave more than two sites on the target gene.

⑤ If the specificity is low, the engineered nuclease may function on non-targeted sites (cytotoxicity of the engineered nuclease); therefore, it is important to design the nuclease by selecting dissimilar sequences on the genome.

7.3 Haplotype Analysis

7.3.1 Single Nucleotide Polymorphisms (SNPs) in Plant Genetics

SNPs are single nucleotide mutations that can cause phenotypic polymorphisms. Experimental approaches for plant SNP detection can be divided into two types, detection of known and unknown SNPs.

1. Detection of known SNPs
   - Hybridization techniques: Microarrays, real-time PCR, and HTS (high-throughput sequencing) SNP arrays.
   - Enzyme-based techniques: Nucleotide extension, cleavage, ligation, reaction product detection, and display.
   - Comparison of SNP assay techniques.
2. Detection of unknown SNPs
   - SNP detection by sequencing: NGS techniques and platforms, and the whole-genome sequencing databases that are now available have revealed specific genetic information and simple sequence variations that differ from the sequences in a corresponding reference sequence.
   - Sequencing by hybridization: SNPs can be detected using SNP arrays (DNA microarray) and FISH (fluorescence *in situ* hybridization).

7.3.2 Haplotypes in Plant Genetics

When genes on a chromosome are tightly linked, the possibility of genetic recombination is low and the genetic combination on gametes remains the same. Haplotype defines a set of closed linked alleles on one chromosome that tend to be inherited together with no recombination among them.

In diploid individuals, two different haplotypes at a certain genetic location could be used to distinguish an allele on one of a pair of homologous chromosomes. A haplotype often represents a set of sequence variations in and around a certain gene that are statistically inherited together on homologous chromosomes.

7.3.3 SNP Haplotypes

Pure line cultivars in autogamous crops (self-fertilizing crops) are genetically stable/fixed homozygotes in all DNA pools. Thus, DNA base sequence differences are not expected to exist in homologous chromosomes. For example, cultivars A and B in self-fertilizing crops contain the DNA sequences shown below in their genomes.

A cultivar: ATTTAGGGATGCCTCAATACGCTATATGCA
B cultivar: ATTTAGCGATGACTCAATACGCTAGATGCA

These DNA sequences have three SNPs. SNP recombination rates can be estimated from meiosis of F1 hybrids. In this case, there are eight possible haplotypes: GCT, GCG, GAT, GAG, CCT, CCG, CAT, and CAG. However, recombination is less likely to occur because of the close proximity of the three SNPs; the most likely SNP combinations in the haploid gametes of the F1 plants would be GCT and CAG, which correspond to the SNPs in cultivars A and B. A SNP haplotype is a collection of SNPs that are inherited together without any recombination among the SNPs. This pattern is called a SNP haplotype block.

A thorough analysis of haplotype alleles becomes possible when natural and artificial gene mutations occur. The process of genotypic and phenotypic linkage analysis using SNP haplotype analysis is shown in Fig. 7.18.

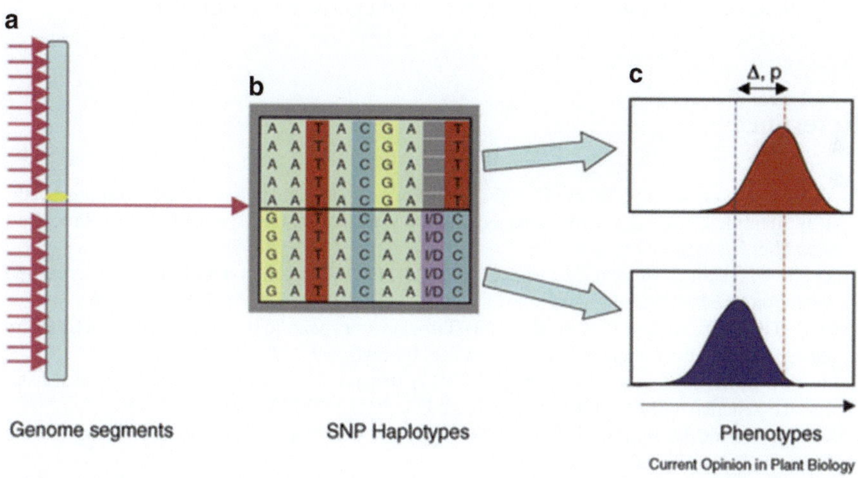

Fig. 7.18 Principles of linkage analysis. (**a**) Genotype analysis on a DNA pool where genetically variable individuals are closely linked together in a genome, (**b**) Classify genotypes into SNP haplotype group or SNP groups, (**c**) Compare statistic to chi-squared distribution of each haplotype or allelic-phenotype (Rafalski 2010)

Studies into the evolution of genes can be performed by analyzing phenotypic differences and haplotype mutations in the genes of various crop species. One such example is the haplotype analysis of the *GS3* gene in different breeds of rice (Takano-Kai et al. 2009). This study revealed that a C165A mutation in the second exon of the gene was responsible for size differentiation in the different breeds. The A allele showed a long-grain shape compared with the variety carrying the C allele and the phenotypes corresponded to the haplotypes.

The low cost of sequencing and newly developed assembly technique, along with data imputation techniques, have allowed users to create HapMaps – high resolution haplotype maps that plot the location of chromosomal haplotypes for crops of interest. In 2009, a haplotype map that confirms the haplotype of first-generation corn, which had undergone a variety of recombination changes, using 27 inbred lines, was reported (Gore et al. 2009). In 2010, Huang et al. (2010) created a high-density haplotype map using a newly assembled data-imputation technique, for 517 re-sequenced domestic rice strains, and performed genome-wide association studies (GWAS) for 14 agronomic characters of subspecies of indica rice. GWAS can also find positions that can be used to adjust targeted traits in crops, and can be easily implemented in breeding programs. Huang et al. (2012) have helped broaden the use of GWAS based on haplotype analysis to confirm 32 new loci for flowering time and grain-related traits using 950 different varieties of rice.

The study of SNPs and haplotypes based on whole genome sequencing data helps in understanding genetic diversity as well as genotype-phenotype associations, and also provides useful genetic network information. Additionally, SNPs can be used to study plant evolution, phenotypic changes, and characteristics of plant traits.

References

Alonso JM, Stepanova AN, Solano R et al (2003) Five components of the ethylene-response pathway identified in a screen for weak ethylene-insensitive mutants in *Arabidopsis*. Proc Natl Acad Sci USA 100:2992–2997

Botticella E, Sestili F, Hernandez-Lopez A et al (2011) High resolution melting analysis for the detection of EMS induced mutations in wheat *SbeIIa* genes. BMC Plant Biol 11:156

Cai CQ, Doyon Y, Ainley WM et al (2013) Targeted transgene integration in plant cells using designed zinc finger nucleases. Plant Mol Biol 69:699–709

Caldwell DG, McCallum N, Shaw P et al (2004) A structured mutant population for forward and reverse genetics in Barley (*Hordeum vulgare* L.). Plant J 40:143–150

Carroll D, Morton JJ, Beumer KJ et al (2006) Design, construction and in vitro testing of zinc finger nucleases. Nat Protoc 1:1329–1341

Cermak T, Doyle EL, Christian M et al (2011) Efficient design and assembly of custom TALEN and other TAL effector-based constructs for DNA targeting. Nucleic Acids Res 39:e82

Chun JB, Ha BK, Jang DS et al (2012) Identification of mutations in *OASA1* gene from a gamma-irradiated rice mutant population. Plant Breed 131:276–281

Colbert T, Till BJ, Tompa R et al (2001) High-throughput screening for induced point mutations. Plant Physiol 126:480–484

Comai L, Young K, Till BJ et al (2004) Efficient discovery of DNA polymorphisms in natural populations by Ecotilling. Plant J 37:778–786

Cooper JL, Till BJ, Laport RG et al (2008) TILLING to detect induced mutations in soybean. BMC Plant Biol 8:9

Cornu TI, Thibodeau-Beganny S, Guhl E et al (2008) DNA-binding specificity is a major determinant of the activity and toxicity of zinc-finger nucleases. Mol Ther 16:352–358

Curtin SJ, Zhang F, Sander JD et al (2011) Targeted mutagenesis of duplicated genes in soybean with zinc-finger nucleases breakthrough technologies. Plant Physiol 156:466–473

de Pater S, Neuteboom LW, Pinas JE et al (2009) ZFN-induced mutagenesis and gene-targeting in Arabidopsis through Agrobacterium-mediated floral dip transformation. Plant Biotechnol J 7:821–835

Dong C, Dalton-Morgan J, Vincent K et al (2009) A modified TILLING method for wheat breeding. Plant Gen 2:39–47

Even-Faitelson L, Samach A, Melamed-Bessudo C et al (2011) Localized egg-cell expression of effector proteins for targeted modification of the Arabidopsis genome. Plant J 68:929–937

Gady AL, Hermans FW, Van de Wal MH et al (2009) Implementation of two high through-put techniques in a novel application: detecting point mutations in large EMS mutated plant populations. Plant Methods 5:13

Galeano CH, Gomez M, Rodriguez LM et al (2009) CEL I nuclease digestion for SNP discovery and marker development in common bean (L.). Crop Sci 49:381–394

Gasiunas G, Siksnys V (2013) RNA-dependent DNA endonuclease Cas9 of the CRISPR system: Holy Grail of genome editing? Trends Microbiol 21:562–567

Gilchrist EJ, Haughn GW, Ying CC et al (2006) Use of Ecotilling as an efficient SNP discovery tool to survey genetic variation in wild populations of *Populus trichocarpa*. Mol Ecol 15:1367–1378

Gore MA, Chia J-M, Elshire RJ et al (2009) A first-generation haplotype map of maize. Science 326:1115–1117

Hegyi J, Schwartz RA, Hegyi V (2004) Pellagra: dermatitis, dementia, and diarrhea. Int J Dermatol 43:1–5

Hoffmann D, Jiang Q, Men A et al (2007) Nodulation deficiency caused by fast neutron mutagenesis of the model legume *Lotus japonicus*. J Plant Physiol 164:460–469

Hoshaw JP, Unger-Wallace E, Zhang F et al (2010) A transient assay for monitoring zinc finger nuclease activity at endogenous plant gene targets. Methods Mol Biol 649:299–313

Huang X, Wei X, Sang T et al (2010) Genome-wide association studies of 14 agronomic traits in rice landraces. Nat Genet 42:961–967

Huang X, Zhao Y, Wei X et al (2012) Genome-wide association study of flowering time and grain yield traits in a worldwide collection of rice germplasm. Nat Genet 44:32–39

Hurt JA, Thibodeau SA, Hirsh AS et al (2003) Highly specific zinc finger proteins obtained by directed domain shuffling and cell-based selection. Proc Natl Acad Sci USA 100:12271–12276

Kadaru SB, Yadav AS, Fjellstrom RG et al (2006) Alternative Ecotilling protocol for rapid, cost-effective single-nucleotide polymorphism discovery and genotyping in rice (*Oryza sativa* L.). Plant Mol Biol Rep 24:3–22

Keim P, Olson TC, Shoemaker RC (1988) A rapid protocol for isolating soybean DNA. Soybean Genet Newslett 15:150–152

Kim DS, Lee IS, Jang CS et al (2004) Development of AFLP-derived STS markers for the selection of 5-methyltryptophan-resistant rice mutants. Plant Cell Rep 23:71–80

Kim HJ, Um EJ, Cho SR et al (2011) Surrogate reporters for enrichment of cells with nuclease-induced mutations. Nat Methods 11:941–943

Kisaka H, Kisaka M, Kameya T (1996) Characterization of interfamilial somatic hybrids between 5-methyltryptophan resistant rice (*Oryza sativa* L.) and 5MT-sensitive carrot *(Daucuscarota* L.); expression of resistance to 5MT by the somatic hybrids. Breed Sci 46:221–226

Krattinger S, Wicker T, Keller B (2009) Map-based cloning of genes in Triticeae (wheat and barley). In: Muehlbauer GJ, Feuillet C (eds) Genetics and genomics of the Triticeae. Springer, New York, pp 337–357

Kurowska M, Daszkowska-Golec A, Gruszka D et al (2011) TILLING – a shortcut in functional genomics. J Appl Genet 52:371–390

Li X, Song Y, Century K et al (2001) A fast neutron deletion mutagenesis-based reverse genetics system for plants. Plant J 27:235–242

Li X, Lassner M, Zhang Y (2002) Deleteagene: a fast neutron deletion mutagenesis-based gene knockout system for plants. Comp Funct Genomics 3:158–160

Li T, Liu B, Spalding MH et al (2012) High-efficiency TALEN-based gene editing produces disease-resistant rice. Nat Biotechnol 30:390–392

Lloyd A, Plaisier CL, Carroll D et al (2005) Targeted mutagenesis using zinc-finger nucleases in Arabidopsis. Proc Natl Acad Sci USA 102:2232–2237

Lochlainn SÓ, Amoah S, Graham NS et al (2011) High Resolution Melt (HRM) analysis is an efficient tool to genotype EMS mutants in complex crop genomes. Plant Methods 7:1–9

Lucki I (1998) The spectrum of behaviors influenced by serotonin. Biol Psychiatry 44:151–162

Mahfouz MM, Li L (2011) TALE nucleases and next generation GM crops. GM Crops 2:99–103

Mahfouz MM, Li L, Shamimuzzaman M et al (2011) De novo-engineered transcription activator-like effector (TALE) hybrid nuclease with novel DNA binding specificity creates double-strand breaks. Proc Natl Acad Sci USA 108:2623–2628

Mandell JG, Barbas CF (2006) Zinc Finger Tools: custom DNA-binding domains for transcription factors and nucleases. Nucleic Acids Res 34:W516–W523

Marton I, Zuker A, Shklarman E et al (2010) Non-transgenic genome modification in plant cells. Plant Physiol 154:1079–1087

McCallum CM, Comai L, Greene EA et al (2000) Targeted screening for induced mutations. Nat Biotechnol 18:455–457

Men AE, Laniya TS, Searle IR et al (2002) Fast neutron mutagenesis of soybean (*Glycine soja* L.) produces a supernodulating mutant containing a large deletion in linkage group H. Genome Lett 1:147–155

Nieto C, Piron F, Dalmais M et al (2007) EcoTILLING for the identification of allelic variants of melon *eIF4E*, a factor that controls virus susceptibility. BMC Plant Biol 7:34

Oldroyd GE, Long SR (2003) Identification and characterization of *nodulation-signaling pathway 2*, a gene of *Medicago truncatula* involved in nod factor signaling. Plant Physiol 131:1027–1032

Oleykowski CA, Mullins CRB, Godwin AK et al (1998) Mutation detection using a novel plant endonuclease. Nucleic Acids Res 26:4597–4602

Osakabe K, Osakabe Y, Toki S (2010) Site-directed mutagenesis in Arabidopsis using custom-designed zinc finger nucleases. Proc Natl Acad Sci USA 107:12034–12039

Perry JA, Wang TL, Welham TJ et al (2003) A TILLING reverse genetics tool and a web-accessible collection of mutants of the legume *Lotus japonicus*. Plant Physiol 131:866–871

Petolino JF, Worden A, Curlee K et al (2010) Zinc finger nuclease-mediated transgene deletion. Plant Mol Biol 73:617–628

Pruett-Miller SM, Reading DW, Porter SN et al (2009) Attenuation of zinc finger nuclease toxicity by small-molecule regulation of protein levels. PLoS Genet 5:e1000376

Rafalski JA (2010) Association genetics in crop improvement. Curr Opin Plant Biol 13:174–180

Rogers C, Wen J, Chen R et al (2009) Deletion-based reverse genetics in *Medicago truncatula*. Plant Physiol 151:1077–1086

Saika H, Oikawa A, Matsuda F et al (2011) Application of gene targeting to designed mutation breeding of high-tryptophan rice. Plant Physiol 156:1269–1277

Sander JD, Dahlborg EJ, Goodwin MJ et al (2011) Selection-free zinc-finger-nuclease engineering by context-dependent assembly (CoDA). Nat Methods 8:67–69

Sato Y, Shirasawa K, Takahashi Y et al (2006) Mutant selection from progeny of gamma-ray-irradiated rice by DNA heteroduplex cleavage using Brassica petiole extract. Breed Sci 56:179–183

Searle IR, Men AE, Laniya TS et al (2003) Long-distance signaling in nodulation directed by a *CLAVATA1*-like receptor kinase. Science 299:109–112

Shan Q, Wang Y, Chen K et al (2013) Rapid and efficient gene modification in rice and brachypodium using TALENs. Mol Plant 6:1365–1368. doi:10.1093/mp/sss162

Shukla VK, Doyon Y, Miller JC et al (2009) Precise genome modification in the crop species *Zea mays* using zinc-finger nucleases. Nature 459:437–441

Slade AJ, Fuerstenberg SI, Loeffler D et al (2004) A reverse genetic, nontransgenic approach to wheat crop improvement by TILLING. Nat Biotechnol 23:75–81

Takano-Kai N, Jiang H, Kubo T et al (2009) Evolutionary history of GS3, a gene conferring grain length in rice. Genetics 182:1323–1334

Till BJ, Reynolds SH, Greene EA et al (2003) Large-scale discovery of induced point mutations with high-throughput TILLING. Genome Res 13:524–530

Till BJ, Burtner C, Comai L et al (2004a) Mismatch cleavage by single-strand specific nuclease. Nucleic Acids Res 32:2632–2641

Till BJ, Reynolds SH, Weil C et al (2004b) Discovery of induced point mutations in maize genes by TILLING. BMC Plant Biol 4:12

Till BJ, Zerr T, Comai L et al (2006) A protocol for TILLING and Ecotilling in plants and animals. Nat Protoc 1:2465–2477

Till BJ, Cooper J, Tai TH et al (2007) Discovery of chemically induced mutations in rice by TILLING. BMC Plant Biol 7:19

Tovkach A, Zeevi V, Tzfira T (2009) A toolbox and procedural notes for characterizing novel zinc finger nucleases for genome editing in plant cells. Plant J 57:747–757

Townsend JA, Wright DA, Winfrey RJ et al (2009) High-frequency modification of plant genes using engineered zinc-finger nucleases. Nature 459:442–445

Tsai H, Howell T, Nitcher R et al (2011) Discovery of rare mutations in populations: TILLING by sequencing. Plant Physiol 156:1257–1268

Tzfira T, Weinthal D, Marton I et al (2012) Genome modifications in plant cells by custom-made restriction enzymes. Plant Biotechnol J 10:373–389

Urnov FD, Rebar EJ, Holmes MC et al (2010) Genome editing with engineered zinc finger nucleases. Nat Rev Genet 11:636–646

Wang J, Sun J, Liu D et al (2008) Analysis of *Pina* and *Pinb* alleles in the micro-core collections of Chinese wheat germplasm by Ecotilling and identification of a novel *Pinb* allele. J Cereal Sci 48:836–842

Wang TL, Uauy C, Robson F et al (2012) TILLING in extremis. Plant Biotechnol J 10:761–772

Wright DA, Townsend JA, Winfrey RJ Jr et al (2005) High-frequency homologous recombination in plants mediated by zinc-finger nucleases. Plant J 44:693–705

Zhang L, Fetch T, Nirmala J et al (2006) *Rpr1*, a gene required for *Rpg1*-dependent resistance to stem rust in barley. Theor Appl Genet 113:847–855

Zhang J, Lu Y, Yuan Y et al (2009) Map-based cloning and characterization of a gene controlling hairiness and seed coat color traits in *Brassica rapa*. Plant Mol Biol 69:553–563

Zhang F, Maeder ML, Unger-Wallace E et al (2010) High frequency targeted mutagenesis in *Arabidopsis thaliana* using zinc finger nucleases. Proc Natl Acad Sci USA 107:12028–12033

Zinc finger consortium. www.zincfingers.org

Chapter 8
Isolation and Functional Studies of Genes

Mi-Ok Woo, Kesavan Markkandan, Nam-Chon Paek, Soon-Chun Jeong, Sang-Bong Choi, and Hak Soo Seo

Abstract Determining and analyzing the functional roles of genes is a prerequisite for genetic engineering. A wide variety of strategies are employed to isolate genes and to characterize their functions. This chapter describes hybridization techniques, such as Southern, Northern, and Western blotting; tools for mutant analysis, including mutant selection and phenotypic analysis; and the analytical methods of activation tagging and overexpression of target genes. Besides *in planta* biological approaches, a variety of biochemical methods have been described in previous research, such as heterologous protein expression, protein–protein (and other ligands) interactions, domain analysis, and protein structure. Meanwhile, the cloning of a gene controlling the targeted trait using linkage maps is a fundamental tool in gene functional studies and allows *in vitro* manipulation of the gene. A helpful guide with step-by-step procedures for map-based cloning, from the selection of a target gene to the complementation test, and a case study in rice are also described.

8.1 Isolation and Functional Characterization of Genes of Interest

Isolating and characterizing genes of interest are the very first steps in genetic engineering crop plants. Owing to the rapid progress in genome research, many novel genes have been reported over the last decade. However, yet more genes still await

---

Author contributed equally with all other contributors.

M.-O. Woo • K. Markkandan • N.-C. Paek • H.S. Seo (✉)
Department of Plant Science, Seoul National University, Seoul, Republic of Korea
e-mail: miok1004@snu.ac.kr; kesavan@snu.ac.kr; ncpaek@snu.ac.kr; seohs@snu.ac.kr

S.-C. Jeong
A Bio-Evaluation Center, Korea Research Institute of Bioscience and Biotechnology, Cheongwon, Republic of Korea
e-mail: scjeong@kribb.re.kr

S.-B. Choi
Division of Bioscience and Bioinformatics, Myongji University, Yongin, Republic of Korea
e-mail: choisb@mju.ac.kr

H.-J. Koh et al. (eds.), *Current Technologies in Plant Molecular Breeding*,
DOI 10.1007/978-94-017-9996-6_8

functional characterization. Even in *Arabidopsis*, we know the biological functions of no more than 10 % of approximately 30,000 genes. Comparative genomics extends the genetic information deduced from model plants to other closely related species by assuming that such information can be used to infer the functions of homologous genes.

Prior to the analysis of gene function, genes of interest must be selected and isolated; however, finding useful genes is often described as finding a needle in a haystack. Once certain genes are functionally characterized and found to be agronomically useful, they can be transferred to other plants. Owing to improved molecular techniques, isolating and characterizing genes of interest is now much easier than before.

8.1.1 Purpose of Gene Isolation

Traditional breeding requires many years to produce novel cultivars with targeted traits and is limited to characteristics from the same species or others within a genus. In contrast, genetic engineering enables the transfer of genes between different species and even kingdoms. While classical breeding involves thousands of genes, genetic engineering focuses on one or a few selected genes. Therefore, analyzing the functional roles of one or a few genes is a prerequisite for genetic engineering. Gene functions are characterized and confirmed through a variety of techniques. As in other eukaryotic cells, plant cells transcribe RNAs, which are processed before exiting the nucleus to produce poly(A) mRNA. Early in the molecular biology era, molecular biologists frequently constructed cDNA libraries to isolate the target mRNA using nucleic acid probes and antibodies. This process involves hybridization techniques such as Southern, northern, and western blotting. The advance of technology has meant that it is possible to isolate genes of interest using polymerase chain reaction (PCR) and DNA synthesis within a single day. Since the mid-1990s, the development of microarrays, transcriptomes, and genome data have greatly reduced the effort needed to find agronomically valuable genes.

8.1.2 Identification of Genes with Useful Functions

A wide variety of strategies are employed to identify gene functions, and multiple tools are often combined to investigate target genes. One of the prevalent and most powerful tools is mutant analysis, which involves mutant selection and phenotypic analyses. However, plant knock-out mutants are not easy to generate by homologous recombination, despite ongoing efforts. Thus, instead of producing new target mutants, plant researchers screen preexisting mutant pools. Two strategies for screening mutants are forward genetics and reverse genetics. The former identifies the genotype using phenotypic observation, which often involves screening random mutant pools, and the latter is an approach to discover a gene's function by analyzing its phenotypic effects. Mutant pools can be generated either by the insertion of foreign DNA, such as transfer DNA (T-DNA) and transposons, into the genome, or by point mutations via chemical agents and other mutagens. Point mutations can be identified by map-based cloning or TILLING, while insertion mutations are analyzed by sequencing the flanking region of the inserts. Representative T-DNA and transposon-tagging populations in *Arabidopsis* and rice are being widely used for functional genomics studies. Double or triple mutants generated from individual single-mutant lines can be valuable in determining epistatic relationships among genes. Alternative approaches for mutant analysis encompass antisense RNA, RNAi, artificial miRNA, and VIGS strategies, which have been effectively used for lethal genes and genes without source mutant lines.

When loss-of-function mutations are lacking, the functional roles of genes can be identified by activation tagging or overexpression. However, some plant species are recalcitrant to transformation and are subjected to transient expression similar to VIGS. When overexpressed, target genes are transcribed either under constitutive or tissue-specific promoters. Intracellular targeting methods have been developed for particular target proteins that localize the expressed proteins to subcellular organelles, such as the chloroplast or nucleus. For this application, marker proteins such as GUS and GFP are often coexpressed to monitor the tissues in which proteins are expressed or to localize the target organelles.

Besides *in planta* biological approaches, a variety of biochemical methods have been developed, including heterologous protein expression, protein–protein (and other ligands) interaction, domain analysis, and protein structure. Plant proteins can be expressed in heterologous *Escherichia coli* and viral systems as well as in eukaryotic systems for proteins whose activities are modulated by modifications. Yeast has been used widely for screening and assays via protein–DNA (one hybrid), protein–protein (two hybrid), and protein–RNA (three hybrid) methods. These studies are usually combined with *in vitro* analyses, such as the gel mobility shift assay (electrophoretic mobility assay; EMSA), pull-down assay, or immunoprecipitation (IP). These various strategies using plant systems are also useful, and transient assays, such as protoplast transient assays and VIGS, can be also performed *in vivo*.

Unlike animal cells, plant cells have walls that interfere with single cell-based experiments; thus, cell biological studies such as immune histochemistry are not practical. GFP assays have contributed to identifying gene functions. For example, bimolecular fluorescence complementation (BiFC), bioluminescence resonance energy transfer (BRET), and fluorescence resonance energy transfer (FRET) are routinely used in the laboratory. In addition to GFP, a variety of fluorescence markers have been employed to determine the functional sites of target genes in the last 15 years. Technology is developing rapidly in this field now that confocal microscopy is readily available.

8.2 Experimental Techniques for the Analysis of Gene Functions

8.2.1 Isolation and Quantification of Plant DNA

8.2.1.1 Purpose

To analyze gene functions, plant DNA must be extracted and purified in as native form as possible. Section 8.2.1 describes a method for the isolation and purification of genomic DNA from plant tissues.

8.2.1.2 Methods

Reagents The following reagents are required: CTAB extraction buffer, phenol/chloroform/isoamyl alcohol (PCI, 25:24:1, v/v/v), liquid nitrogen, ethanol (95 %, 75 %), 3 M sodium acetate (NaOAc; pH 5.2), Tris-EDTA buffer (TE, pH 7.4), and mortar and pestle.

Method

1. Grind frozen young leaf tissue (<200 mg) into a fine powder with a mortar and pestle, then add 500-μL CTAB extraction buffer, transfer to 1.5-mL microfuge tube, and incubate for 5 min at room temperature with gentle agitation (30 rpm).
2. Spin at 20,000 × *g* for 1 min.
3. Transfer the supernatant to a new microfuge tube, add an equal volume of PCI, mix at 30 rpm for 5 min, and then spin at 20,000 × *g* for 1 min. Repeat this step.
4. Add 0.1 volume of 3 M NaOAc and 2 volumes of ice-cold ethanol to the supernatant and spin the mixture at 20,000 × *g* for 10 min.
5. Decant the supernatant, add 0.5 mL of 70 % ethanol, and spin briefly. Repeat this step.
6. Decant the ethanol, dry the pellet, dissolve in 50 μL TE, and store at −20 °C until use.

8.2.1.3 Quantification of DNA

The isolated DNA must be quantified. This can be done by ultraviolet spectroscopy, NanoDrop (NanoDrop, Wilmington, DE, USA), and fluorimetry. The basic principles of these are similar, and only the spectrophotometric method is presented here.

1. Dilute 5 μL of the DNA sample to a volume of 1 mL with distilled water and transfer to a quartz cuvette.

2. Measure the absorbance of the solution at wavelengths of 260 nm and 280 nm. If necessary, scan the sample from 230 to 320 nm.
3. Calculate DNA concentration of double-stranded DNA based on the relationship: 1 OD at 260 nm = 50 μg/mL. This estimation is affected by contaminants like RNA and very low molecular weight DNA in the solution. The ratio of OD for A280/A260 represents the quality of DNA.
4. Prepare 10 ng/μL working stocks of samples in 100 μL aliquots.

8.2.1.4 Notes

- Pre-incubate the motor/pestle and spatula in liquid nitrogen.
- Avoid vigorous mixing of extracts containing ground powder, which can fragment the DNA.
- Broad-spectrum pronase or Proteinase K can be added to the extract per the manufacturer's instructions immediately after cell lysis.
- Ensure that proteins and PCI have been completely removed by extracting the DNA solution with chloroform/isoamyl alcohol (24:1) after PCI extraction.

8.2.2 Southern Blot Hybridization

8.2.2.1 Purpose

The Southern blot technique was developed by Edwin M. Southern in 1975 to detect the presence of specific DNA fragments in a sample. This method is based on hybridization between a single-stranded DNA (or RNA) probe and denatured single-stranded target DNA. For hybridization, probes should be complementary to the target sequence and unable to bind non-complementary sequences. The target DNA may be a single gene, multiple genes, or a larger part of the genome. By combining Southern blotting and cloning, genes of interest can be isolated for further studies. Moreover, other applications, such as restriction fragment length polymorphism (RFLP) and DNA fingerprinting, can be used.

8.2.2.2 Method

1. Digest DNA (genomic or other) with restriction enzyme(s) and separate by gel (usually agarose) electrophoresis. Long DNA can be fragmented using HCl then denatured into single strands by incubation with NaOH.
2. Transfer DNA to a membrane (or nitrocellulose) using a capillary or electrical apparatus.
3. Incubate the blot with a single-stranded DNA probe, which will base pair with its complementary target to form double-stranded DNA. The probe is either

radioactive for autoradiography or linked to alkaline phosphatase or horseradish peroxidase for chemiluminescence detection.
4. After hybridization and the washing away of unlabeled probes with buffer containing Sodium Dodecyl Sulfate (SDS), expose the blot directly to X-ray film or a phosphor imager. For enzyme-linked probes, incubate the blot with a colorless substrate that the attached enzyme converts to a colored product that can be directly seen or that gives off light that is detected by X-ray film or an imager.
5. If DNA fragments are destined for cloning, excise the corresponding bands from the agarose gel and elute the DNA fragments using dialysis, gel extraction, or squeezing.

8.2.2.3 Notes

- The agarose solution may boil over during stirring and heating. Be careful of burning.
- Air bubbles can be removed from the blot using a glass test tube or glass rod.

8.2.2.4 Troubleshooting

- If restriction digestion does not work properly, add spermidine to the reaction buffer at a final concentration of 0.004 mM and incubate overnight.
- When high molecular weight DNA (15–20 kb) is not transferred, depurinate the gel in 0.2 N HCl solution until loading dye becomes yellow (ca. 10 min). To remove the HCl before denaturation, wash the gel with distilled water several times.
- If strong background signals are observed:
  - Increase the pre-hybridization incubation time.
  - Keep the membranes in the dark during hybridization.
  - Avoid the precipitation of SDS by keeping the solution warm.
  - Use uncharged nylon membranes or nitrocellulose membranes.

8.2.3 Cloning and General Transformation with Plasmid Vectors

8.2.3.1 Purpose

The essence of recombinant DNA technology is to prepare large numbers of identical DNA molecules. For this purpose, a single recombinant DNA molecule that is usually composed of a vector plus an inserted DNA fragment is introduced into a host cell for amplification. Plasmids are commonly used as vectors in cloning. Most plasmid vectors contain a replication origin, a drug-resistance gene, and cloning sites, and can replicate in bacterial cells on selection medium. Recently, a variety of commercial vectors have been developed and are available for particular cloning purposes.

8.2.3.2 Methods

Competent Cells and Transformation Transformation is the process of introducing foreign DNA (e.g., plasmids or BAC) into a bacterium. Bacterial cells that can accept foreign DNA are called competent. Some bacteria are naturally competent (e.g., *Bacillussubtilis*), whereas others, such as *E. coli*, are not. Non-competent cells can be made competent and then transformed via one of two main methods: chemical transformation and electroporation. In 1970, Morton Mandel and Akiko Higa discovered a way to make *E. coli* more competent. Their calcium chloride method is widely used today for high-efficiency transformation. Through this and other efforts, transformation efficiency, which was initially 10^5–10^6 cells/μg supercoiled DNA, has increased to 10^6–10^9.

Electroporation generally gives higher transformation efficiencies (up to 6×10^{10}–1×10^7 cells/μg supercoiled DNA) and is less laborious than chemical transformation. However, it is more expensive, requiring cuvettes to transfer the charge to the cell suspension and specialized apparatus to deliver the charge. Salt can disturb electroporation, so samples can be lost if excess salt is carried over into the cuvette. In these days, many companies provide various competent cell kits designed for specific cloning purposes.

When preparing competent cells for chemical transformation, TSB or FSB buffer have been used; however, CCMB80 buffer has recently become more convenient (US patent, Molecular cloning 4th ed. Vol 1, p. 167). To prepare competent cells with high transformation efficiency, picking a single colony and growing it at low temperature (16–20 °C) are helpful.

Cloning into Plasmid Vectors Most plasmids for cloning contain multiple cloning sites and so can accommodate DNA fragments unidirectionally. For instance, if a plasmid was linearized by digestion with *Eco*RI and *Bam*HI, the target inserts cut at the same enzyme sites can be directly ligated into the plasmid. However, if the situation does not allow, enzyme sites can be added to both ends of the target inserts by PCR. If inserts have blunt-ended termini, the vector-to-insertion ratio is important (ideally 1:3 molar ratio) for efficient ligation. To prevent self-ligation, the vector needs to be de-phosphorylated using alkaline phosphatase, which directly competes with the inserts and lowers the efficiency of the cloning reaction.

Cloning of PCR Products To produce insert DNA with the necessary restriction sites for directional cloning, PCR can be used in cloning. In general, PCR products amplified by non-proofreading Taq polymerase have an additional unpaired adenosine base at the 3′ end and can be cloned into a T-vector with an unpaired thymidyl base at the 3′ end. Although T-vectors can be prepared in a laboratory, commercial forms are commonly used. Inserts containing enzyme sites at either end also can be directly cloned into T-vectors and then subcloned into target plasmid after restriction enzyme digestion. Alternatively, PCR products can be digested with enzyme(s) before ligation reaction and then cloned into backbone plasmids. In this case, extra sequence (usually 3–6 nt) at the 5′ ends of primers are needed to assist the restriction enzyme digestion.

TOPO/TA Cloning The TOPO/TA cloning method combines the advantages of TA cloning, which uses sticky overhanging adenine ends, with the ligation activity of topoisomerase I. This process allows direct ligation of PCR products in a few minutes. Currently, topoisomerase derived from the *Vaccinia* virus is being broadly used for cleavage and ligation of DNA. Commercial TOPO cloning kits are available.

Gateway Recombinant Cloning System Gateway cloning system serves a fast and efficient direction to clone genes, and basically it conducts the site-specific recombination reactions enabling the bacteriophage λ to integrate and excise itself in and out of a bacterial chromosome. Gateway protocol includes the BP and LR clonase reactions. The BP reaction is utilized by the BP Clonase II enzyme mix that consists of the phage integrase and the integration host factor. The BP clonase mix substitutes a DNA fragment of interest (e.g., PCR products) flanked by two *att*B sequences into a donor vector (pDONR) carrying two *att*P sequences. After recombination of the matching *att*B and *att*P sequences, the DNA fragment is replaced into the donor backbone, resulting in an entry clone (pENTR), and is flanked by two *att*L site sequences. Entry clones are used for the LR reaction that is utilized by the LR Clonase II enzyme mix. The LR clonase mix inserts the DNA fragment of interest flanked by two *att*L sequences into various destination vectors (pDEST) carrying two *att*R sequences. Commercial Gateway system can use with protocol, list of destination vectors, clonase, and etc (http://www.invitrogen.com).

8.2.4 RNA Extraction and Northern Blot Analysis

8.2.4.1 Purpose

This protocol describes a method for RNA extraction, handling, and northern blot hybridization. In general, the RNA in a cell is 80–85 % rRNA and 15–20 % small RNAs, including tRNA. An additional 1–5 % of total RNA comprises mRNAs that are hundreds to several kilobase long with poly (A) tails, which enable purification by oligo(dT) affinity columns. When handling RNAs, RNase must be eliminated to prevent degradation. As RNase contains an intra-chain disulfide bond, it is not denatured by boiling or chelators. The best way to avoid RNase contamination is to isolate and purify the RNA sample quickly.

8.2.4.2 Extraction of Total RNA from Plant Tissues

Preparation Prior to the experiment, autoclave the apparatus and clean the bench and centrifuge using RNase ZAP (R2020; Sigma Aldrich, St. Louis, MO, USA) or a similar solution. It is important to change gloves frequently while handling RNA samples.

Extraction Methods

- Methods using monophasic lysis reagents include the single step protocol developed by Chomczynski and Sacchi (1987) and more recently developed reagents such as QIAzol, TriPure, and TRIzol (15596-026; Life Technologies, Carlsbad, CA, USA).
- Spin column-based methods are similar to DNA extraction, and many types of silicated membrane columns are available, including the RNeasy Plus kit (Qiagen, Venlo, Netherlands).

8.2.4.3 Northern Blot

Northern blot analysis provides a lot of information about broad range of RNA information such as identity, size, and abundance, allowing a deeper understanding of gene expression levels. This method was named for its similarity to the Southern blot technique. The first step in a northern blot is to separate the RNA, which ensures that the strands are unfolded. The RNA molecules are then fractionated by their sizes on a gel, and transferred from the gel onto a blotting membrane. Next, RNA on the membrane is hybridized with a complementary DNA or RNA probe that is labeled by radioactive isotope or a fluorescent dye.

Probe Synthesis Probes labeled with ^{32}P-dNTP (e.g., ^{32}P-dCTP) give the best quality and highest detection limits, but those labeled with non-radioactive nucleotides such as Digoxygenin or Biotin (cat # 60-01-01; Kirkegaard & Perry Laboratories) are preferred by many researchers for safety reasons.

Stringency When target genes are highly homologous to the probes, membranes can be hybridized and washed under high stringency conditions (lower concentrations of salt or/and SDS, higher washing temperatures). However, if homology is weak (~65 % identity), the stringency should be reduced. Radioactivity of probes can be up to 2×10^6 cpm/mL.

8.2.4.4 Troubleshooting

- If a high background signal is observed, the following can be considered:
  - Increase the prehybridization incubation time.

  - Avoid the precipitation of SDS by keeping the solution warm.
  - Use an uncharged nylon membrane or a nitrocellulose membrane.
  - Do not use oxidized or impure formamide.
- When the exposed band has white dots, air bubbles were not completely removed during transfer. Bubbles can be removed using glass rods.

8.2.5 Real-Time PCR

8.2.5.1 Purpose

Real-time PCR is a technique used to monitor the progress of a PCR. It enables PCR-based quantification of a relatively small amount of DNA or RNA, and is therefore called quantitative PCR or qPCR. Real-time PCR is based on the detection of fluorescence produced by a reporter molecule that increases as the reaction proceeds. Reporter molecules include dyes that bind to double-stranded DNA (e.g., SYBR®Green; cat # 4367659, Life Technologies) or sequence specific probes (e.g., Molecular Beacons or TaqMan® Probes).

8.2.5.2 Methods

Principle of Quantitative Analysis Using Real-Time PCR Real-time PCR monitors and quantifies PCR products during the exponentially amplifying state. The procedure uses the same general principle as PCR, but the amplified product is monitored by fluorescence as the reaction progresses in "real time". In standard PCR, the product is only checked at the end. Two common ways to monitor the products of quantitative PCR are to use (1) non-specific fluorescent dyes that bind with any double-stranded DNA (e.g., SYBR Green), and (2) sequence-specific DNA probes composing of oligonucleotides that are conjugated with a fluorescent reporter which allows detection only after hybridization of the probe with its complementary sequence to quantify messenger RNA (mRNA) and non-coding RNA in cells or tissues.

Real-Time PCR Instruments and Detection Real-time PCR (RT-PCR) instrument consists a thermal cycler, an optical system to excite fluorophores and capture emitted fluorescence. There is only one difference between real-time PCR and usual PCR machines, that is the fluorescence detection system. Now, a wide variety of real-time PCR machines are available, of which some are designed for rapid reactions and others to accommodate 96-well plates.

In general, two types of detection techniques are used in qPCR: intercalation of double-stranded DNA-binding dyes and the use of probes labeled with fluorescent dyes. Intercalation does not require gene-specific probes and is relatively inexpensive, but the detection specificity is not high because of occasional non-specific amplification. In contrast, probe-based techniques are more expensive but provide higher specificity based on the sequence identity of the primers.

Detection Based on Fluorogenic dsDNA Binding This technique uses small molecules that bind to double-stranded DNA and can be divided into two classes, intercalators and minor-groove binders. SYBR® Green dyes (Molecular Probes, Life Technologies) are representative intercalators. SYBR Green dyes are inexpensive, easy to use, and have high signals. Their disadvantage is non-specific binding to any double-stranded DNA, including primer-dimers and other dsDNA contaminants, resulting in overestimation of the target products. Therefore, primers should be well designed so that a single band corresponding to target DNA is produced.

Detection Using a Probe Labeled with Fluorescent Dyes This technique takes advantage of the 5′ nucleolytic activity of Taq DNA polymerase and requires an oligonucleotide probe and primer set. An oligonucleotide probe is consisted containing a fluorescent reporter dye on the 5′ end and a quencher dye on the 3′ end. Although the probe is intact, the excited fluorescent reporter dye transfers energy to the nearby quencher dye molecule by fluorescence resonance energy transfer (FRET), thereby resulting in non-fluorescence. The probe binds downstream from one of the primer sites and is cleaved by the 5′ nuclease activity of TaqDNA polymerase as this primer is extended to 5′ end. This divides the fluorescent and quenching dyes, and thus no FRET arises. Fluorescence intensity rises in each cycle, proportional to the extent of probe cleavage.

TaqMan probes (Life Technologies) have been widely adopted for real-time PCR and for detecting sequence variation. They are classified as TaqMan® probes (e.g., 5′ FAM and VIC; 3′ TAMRA or similar dye as the quencher) and TaqMan® MGB probes that binds minor grooves. The disadvantage of TaqMan assays is that different probes must be synthesized for different sequences.

More recently, the cycling probe technique (CPT) was developed for typing single nucleotide polymorphisms (SNPs) using real-time PCR. CPT is based on the TaqMan protocol but employs a DNA–RNA chimera probe with fluorescent and quencher dyes at 5′ and 3′ ends, respectively. After probe hybridization, matched probes with template are cleaved by RNase H, but mismatched ones are not, enabling the detection of SNPs in the target DNAs.

Primer and Probe Design Primer design is an important first step in qPCR assays. If primers anneal poorly or to more than one sequence, the quality and reliability of results will be significantly reduced. Prior to real-time PCR, the primers must be tested to ensure that they amplify a band of the right size and that no non-specific bands are produced. Many primer and probe design software programs are available.

When designing primers:

- The T_m should be 58–60 °C.
- The last five bases at the 3′ end should have no more than two G′s or C′s to reduce non-specific priming.

- The annealing temperatures of the primers should be as similar as possible.
- Primer pairs should be designed for target regions with a minimal number of potential primer-dimers and hairpin structures.
- The ideal amplicon size is 50–150 bp and should not exceed 400 bp.
- Primers that span exon–exon junctions are preferred for qPCR.

When designing probes:

- The T_m should be 68–70 °C (10 °C higher than primers).
- The 5′ end should contain no G′s. (Overall, G+C should be 30–80 %).
- The strand that matches the probe with more C than G bases should be selected.
- Primers should be 15–30 bases in length. TaqMan MGB probes should be as short as possible but longer than 13 nt.

8.2.5.3 Notes

- While handling RNA, wear gloves and a face mask.
- When preparing the reaction mixture, make enough for one extra reaction to ensure a sufficient amount of master mix.
- When PCR is continued using RT-PCR, the RT-PCR solution should not exceed 10 % of the PCR mix.

8.2.6 Microarray and RNA Sequencing

8.2.6.1 Purpose

RNA-Seq and microarray are two popular methods for genome-wide transcriptome profiling. Although microarray is more commonly used than RNA-Seq, it has several limitations (e.g., background hybridization of transcripts present in low abundance and probes that only work for homologous genes). Today, tremendous amounts of data have been accumulated for model plants. Representative databases are GRAMENE (http://www.gramene.org) and Genevestigator (https://www.genevestigator.com/gv/). RNA-Seq is performed via the direct sequencing of transcripts using next generation sequencing (NGS) and is replacing microarrays for whole-genome transcriptome profiling. It has considerable advantages over microarray, thus enabling the detection of novel transcripts, allele-specific expression, and splice junctions. However, RNA-Seq shows variability depending on the computational algorithm.

8.2.6.2 Method

Both microarray and RNA-Seq require a specific apparatus and analytical tools. Researchers should consider which is most appropriate for their research purposes.

Microarray Major steps if the microarray protocol include (1) preparation of the chip (e.g., oligomers, cDNA), (2) hybridization of the chips with fluorescently-labeled cDNA probes derived from sample RNA, (3) scanning the chips and imaging the spots, and (4) normalization of fluorescence signals.

RNA-Seq The RNA-Seq method needs the conversion of RNA into cDNA, which is flanked by adaptors and then directly sequenced by NGS. Depending on the preparation protocols of RNA, all types of RNA can be analyzed including noncoding RNA. The procedure of mapping the resulting short sequencing reads onto the reference genome can quantify the expression levels of genes relative to the condition of interest or absolute levels. Although microarray has been used in whole transcriptome analysis, currently RNA-Seq is emerging as a preferred strategy as it does not need the reference source of the transcriptome.

8.2.6.3 Notes

In all steps of both microarray and RNA-Seq, RNase-free conditions must be maintained.

8.2.7 Expression of Foreign Genes in Escherichia coli

8.2.7.1 Purpose

The most frequently used host for production of enzymes and other proteins is *Escherichia coli* using recombinant DNA technology. It is preferred because of its relative simplicity, inexpensive and fast high-density cultivation. However, expression and purification of proteins in bacterial cells are not always easy to success, and often produce insoluble and nonfunctional proteins. Moreover, unlike eukaryotic cells, *E. coli* cannot express proteins modified by glycosylation or phosphorylation. Many factors affect the successful production of proteins during cloning, expression, and mass production. Ways to increase the amount of soluble protein include the use of a strong promoter and the control of transcription and translation efficiency. A general outline is shown Fig. 8.1.

8.2.7.2 Methods

Selection of Target Genes and Expression Vectors Strong promoters are preferable but should be repressed under uninduced conditions. An example is pET vectors (cat # 69749-3; Novagen, Tokyo, Japan), which transcribe high amounts of mRNA and protein (Fig. 8.2).

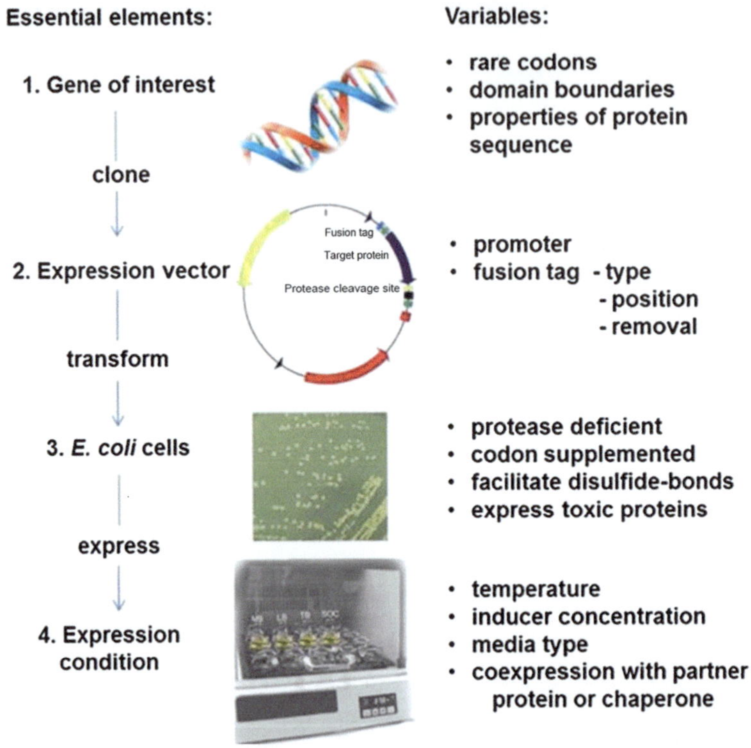

Fig. 8.1 Expression of recombinant proteins in *Escherichia coli* (Modified from Hannig and Makrides 1998)

Cloning of Target Genes Once target genes are available, they can be cloned into multi-cloning site (MCS) while preserving the reading frame. To match the restriction enzyme site between target genes and MCS, the enzyme sites are frequently added at the ends of the inserts by PCR. After cloning, the reading frame and sequence identity should be verified by sequencing.

Transformation into Host Cells There are several *E. coli* strains for expressing proteins. In general, strains with lower protease activity (e.g., BL21 -ompT$^-$/Ion$^-$, cat # 70232-3) are preferred. For genes that use nonbacterial genetic codes, codon-optimized strains (e.g., BL21 codon plus, Rosetta) can be considered for primary expression. For proteins containing higher concentrations of arginine, isoleucine, glycine, leucine, and proline, strains that express those tRNAs at higher levels should be considered.

Induction For pET vectors, *Lac* repressor is de-repressed in the presence of allolactose, resulting in transcription of target genes. IPTG has a similar effect but is not hydrolyzed and is widely used for induction. *Escherichia coli* can be cultured at 37 °C until the OD_{600} reaches 0.4–0.5 before induction, transferred to a shaker at <25 °C with the addition of 0.1–1 mM IPTG, and cultured for 4–5 h (Fig. 8.3).

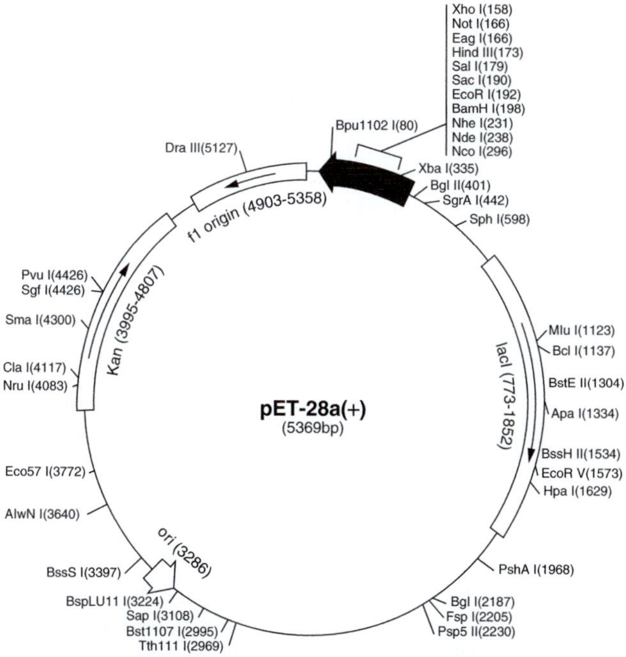

Fig. 8.2 pET28a map with C-terminal His6 and Multi Cloning Site (MCS) (Adopted from Novagen)

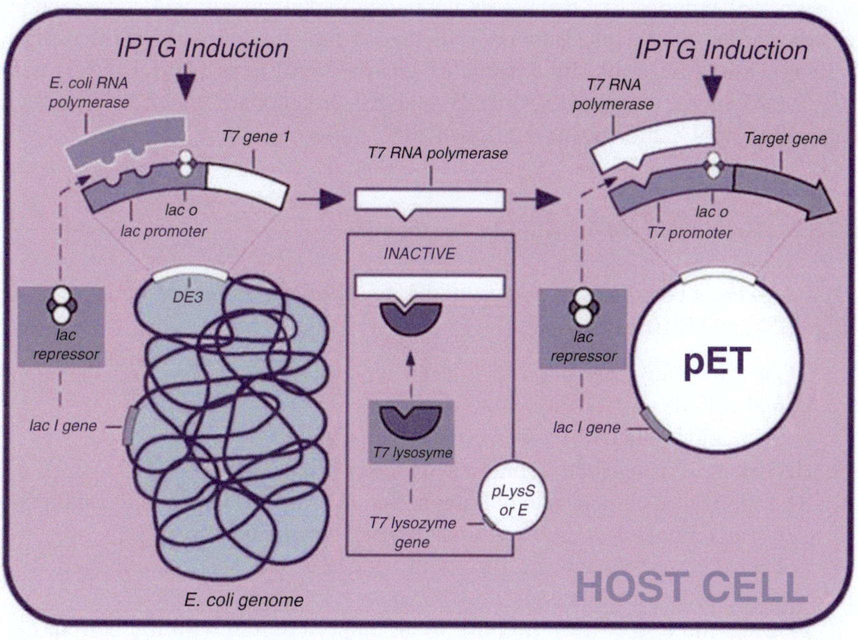

Fig. 8.3 Control elements of the pET expression system (Adopted and modified from Novagen)

8.2.7.3 Troubleshooting

- If the recombinant protein is not produced or goes to an inclusion body, culture the cells below 20–22 °C overnight, or grow for a shorter period with a low IPTG concentration.
- To increase solubility, co-express the target proteins with chaperones (e.g., DnaK-DnaJ, GroES-GroEL) or with folding catalysts (e.g., DsbA, DsbB, DsbC, DsbD).
- When the solubility of target proteins is low, change or use another tagging fusion protein, such as thioredoxin, glutathione S-transferase (GST), and maltose-binding protein (MBP).

8.2.8 Protein Purification with Affinity Chromatography

A wide variety of protein purification methods are being used, including size-exclusion chromatography (gel filtration), ion-exchange chromatography based on charge, and affinity chromatography based on ligand specificity. Affinity chromatography is one of the most popular techniques and has been used to purify recombinant proteins expressed in *E. coli*. For affinity purification, expressing fusion proteins with tags such as His6, GST, and MBP, is common. For instance, poly-histidine tag can be positioned on either the N- or C-terminus. Expressed His-tagged proteins can be purified and detected relatively easily because the tag binds to several types of immobilized metal ions, including nickel, cobalt, and copper, under specific buffer conditions. Unbound His-tagged proteins are washed off using a binding/wash buffer typically consists of Tris-buffered saline (TBS, pH 7.2) with 10–25 mM imidazole. Finally, only His-tagged proteins are eluted using a high concentration of elution agent (e.g., >100 mM imidazole).

8.2.8.1 Brief Protocol for Affinity Purification

The following method can be used with the most popular tagging systems, such as His6, GST, and MBP tags.

1. Clone the target genes into MCS of the expression vector with a tag that is in-frame with the target fusion protein.
2. Express and induce the target recombinant proteins.
3. Harvest cells using centrifugation and collect cell pellets at 4 °C.
4. Resuspend the cell pellets in binding buffer and sonicate under cold conditions.
5. Centrifuge lysate at 10,000 × *g* at 4 °C to pellet cellular debris.
6. Collect the supernatant and mix with resin that has been equilibrated with binding buffer for >30 min.
7. Transfer the lysate–resin mixture to an empty column with the bottom cap still attached.

8. Remove the bottom cap and collect the flow through for SDS–PAGE analysis.
9. Wash twice with wash buffer.
10. Elute the recombinant protein with elution buffer.

8.2.8.2 Troubleshooting

When the expressed protein is absent after elution:

- Ensure that protein was expressed properly by analyzing the crude supernatant debris (step 4) on an SDS-PAGE gel.
- Confirm whether the expressed protein is soluble using the supernatant from step 5.
- Analyze the resin that is collected at step 10 using SDS-PAGE to determine whether the expressed proteins were eluted.

When the expressed proteins form inclusion bodies:

- Grow the cells at lower temperature for a longer time (double the induction time for every 7 °C).
- Reduce the IPTG concentration.
- Resuspend the pellet at step 5 in guanidine-HCl or urea solution, induce protein refolding by dialysis in binding buffer, and then proceed with step 5.

When multiple contaminated bands are observed:

- Repeat the washing at step 9.
- Reduce the concentration of elution agent (e.g., imidazole) in the elution buffer at step 10.

8.2.9 Antibody Production and Purification

Separation of pure proteins is the first step to produce an antibody. In addition, an antibody must be capable of binding to the target protein and not to proteins that lack the tag. If the antibody is pure and has high target specificity, it will not interact with other proteins and can be used to check expression level and protein stability.

8.2.9.1 Types of Antibody

There are two types of antibody, polyclonal and monoclonal. Polyclonal antibodies bind to different epitopes of an antigen and are normally used for western blotting but are not suitable for large volumes of samples because antigen specificity is low. In contrast, monoclonal antibodies can be used for uniform and high specificity binding to antigens and also bind rapidly. The production of monoclonal antibodies takes a long time, however, and are only done in mouse and rat.

For polyclonal antibodies, the selection of animal is important, but this protocol describes the rabbit, which is commonly used.

1. Blood serum should be collected before immunization and refrigerated.
2. The antigen is diluted to 1 mg/mL.
3. Make antigen and Freund's adjuvant emulsion.
4. Sterilize the rabbit's fur with 70 % ethanol before and after injecting the antigen.
5. Booster immunizations are given at 2–4 week intervals.
6. Two weeks after the second boost, 50 mL of serum is collected and tested for antibody binding capacity.
7. Two weeks after the third boost, the remaining serum is collected.

For monoclonal antibodies, mouse or rat is used. The antigen is injected into myeloma cells or lymphocytes, and the monoclonal antibody is harvested from spleen cells (Gerdes et al. 1983).

1. Make Freund's complete adjuvant and antigen emulsion complex and inject 100–200 μL into the abdomen.
2. Booster immunizations are given at 2–4 week intervals.
3. After 2 weeks, add three additional cells that fused antigen 3–4 days prior to intravenous injection.
4. The rodent is anesthetized, then a cervical dislocation is made, and the spleen is carefully excised.
5. The hybrid cells made by culturing the spleen cells fused with myeloma cells are extracted.

8.2.9.2 Troubleshooting

- If there is more background than the desired band, the antibody must be purified. This can be achieved by affinity purification with the antigen and antibody bound to CnBr resin.
- If the antibody does not detect the target protein, the antibody can be produced using only the N-terminal or C-terminal region of an antigen. Antibodies can be produced even with the C-terminal end, where the chance of detection is high.

8.2.10 Western Blot

Western blot is an analytical technique widely used to detect the amount specific protein in a sample, especially the antibody response towards the specific antigen. The total protein is separated by polyacrylamide gel electrophoresis (denatured condition) and subsequently transferred onto a membrane (nitrocellulose or PVDF). The membrane is treated with a specific antibody to detect the amount of the target protein (Burnette 1981).

8.2.10.1 Protocol

1. The plant sample is finely ground with a pestle and mortar in liquid nitrogen.
2. Total protein is extracted using extraction buffer and then quantified using Bradford assay.
3. SDS sample buffer is added to the protein and incubated at 100 °C for 3–5 min.
4. Perform SDS-PAGE
5. Transfer the protein in the gel to a PVDF membrane.

   A. Immerse the PVDF membrane in methanol.
   B. Then, immerse in transfer buffer.
   C. In the transfer cassette, cover the 3 M paper on the sponge.
   D. Keep the gel and membrane over 3 M paper. Make sure that there are no air bubbles between the gel and membrane.
   E. Keep the 3 M paper over the membrane.
   F. Pour the appropriate transfer buffer and run the system at 250 mA for 1 h.

6. Once the transfer is completed, move the membrane into TBST containing 1–5 % BSA or skim milk for blocking, and shake for 1 h.
7. Add the appropriate concentration of antibody (primary) and shake for 1 h.
8. Wash the membrane with TBST three times at 10 min intervals, then discard the TBST.
9. Repeat the blocking for 1 h with TBST containing 1–5 % BSA or skim milk. To reduce time, go to step 10.
10. Add an appropriate secondary antibody to conjugate with the primary antibody. Incubate for 1 h.
11. Wash the membrane with TBST three times at 10 min intervals.
12. For detection:

    A. Make the ECL solution with luminol and enhancer that catalyzes hydrogen peroxide.
    B. Soak the membrane in ECL solution.
    C. Keep the membrane in the western film cassette, wrap it, and take it to the dark room to expose to photosensitive X-ray film.
    D. Develop the X-ray film.

8.2.10.2 Troubleshooting

If there is too much background signal:

- Increase the number of washings.
- Use blocking solution with 5 % skim milk.
- Reduce the transfer time.

If there is weak signal, add more antibody.

8.2.11 In Vivo *and* In Vitro *Pull Down*

This method is important to detect the protein–protein interaction of fusion proteins. For example, if protein X (bait) binds with protein Y (prey), protein X is modified with another tagged protein to form a fusion protein, which is added to the beads (resin) that can bind proteins. Thus, protein X–protein Y can bind to the beads and this protein–protein interaction is determined. Detection of this bait–prey fusion protein with tagged protein beads by precipitation is called a pull-down assay (Sambrook et al. 1989). An *in vitro* pull-down assay can be used to confirm the direct interaction of two proteins. Beads are bound with 3–5 μg of fusion protein and should be washed to remove non-specific binding. An *in vivo* pull-down assay can be done for more amount proteins. More than 1 mg total protein should be used because of non-interacting proteins present in the sample.

8.2.11.1 Protocol

1. Conjugate the purified bait and prey proteins in binding buffer.
2. Add appropriate beads that can bind the tagged bait protein.
3. Precipitate the beads.
4. Wash the beads to remove non-specific proteins.
5. Elute the binding proteins by incubating at 100 °C for 3–5 min or with elution buffer.

8.2.11.2 Troubleshooting

Determining whether the substances present in the buffer affect the native structure and hydrophobic interaction of these proteins is critical. The binding buffer uses pH and detergent and salt concentrations to maintain the protein's native structure. Salt can be in the form of 100–200 mM $NaCl_2$ or $MgCl_2$. Whether the detergent is ionic or non-ionic should be determined before use. If a non-ionic detergent is used for the binding buffer, it should have a low critical micelle concentration.

8.2.12 *Immunoprecipitation*

Immunoprecipitation (IP) is used to identify protein–protein interactions by precipitating the protein that binds with a specific target protein and has high affinity to a specific antibody. IP involves the interaction of specific individual proteins, whereas complex immunoprecipitation (Co-IP) involves large protein complexes. When the protein interacts with DNA, the process is called chromatin immunoprecipitation (ChIP), and similarly RNA immunoprecipitation (RIP) involves

RNA–protein interactions. Co-IP is used in plants to detect the proteins or complexes binding with specific target proteins (Bonifacino and Dell'Angelica 2001).

8.2.12.1 Protocol

1. Extract total protein from the plant samples.
2. Pre-clearing: Add only beads to remove the proteins that bind to the beads.
3. For protein binding, add 0.5–1-mg bait protein, an antibody that binds to it, and the relevant beads.
4. For antibody binding, add an antibody with affinity to the target proteins to form an antibody–antigen–bead complex.
5. Wash the bead complex to remove any proteins that are non-specifically bound.
6. Use a low-pH elution buffer to elute the antibody–protein complex.

8.2.12.2 Troubleshooting

The main concern here is contamination of the eluent with proteins other than the desired protein. This can be mitigated at any stage prior to the final elution. Addition of other antibodies in the pre-clearing step can remove non-specific unwanted proteins. To increase specificity, lower amounts of an antibody are always recommended. The washing buffer should be fresh, and the number of washings can be increased. To reduce non-specific antibody binding, increase the concentration of detergent, reduce the binding time, or increase the temperature.

8.2.13 Two-Dimensional Gel Electrophoresis

Two-dimensional gel electrophoresis (2DE) separates proteins based on two characteristics, namely isoelectric point (pI or IEP) and molecular weight (MW). Total proteins are separated horizontally by isoelectric focusing (IEF) and vertically by SDS-PAGE (Smejkal and Lazarev 2010). Both sample preparation and selection of the tissue or organ to be extracted require skill (Isaacson et al. 2006), because many factors can affect the protein separation during IEF and SDS-PAGE and must be removed. Some seed extracts have high lipid and carbohydrate contents that affect the pI value of the proteins, resulting in spot migration (Fig. 8.4).

8.2.13.1 Protocol

1. During protein extraction, as many impurities as possible should be removed to obtain a pure protein.
2. Then, 50 ng of protein is dissolved in 300 μL of IEF buffer.

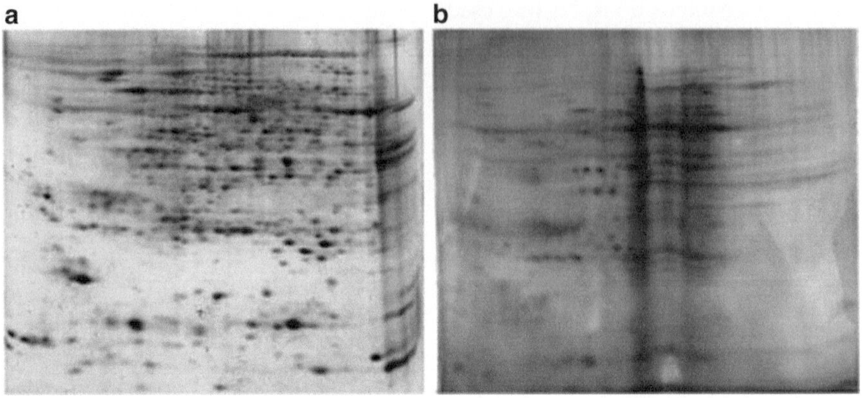

Fig. 8.4 Proteins separated by two-dimensional electrophoresis. (**a**) Sample of purified proteins, (**b**) Crude protein

3. The first separation distinguishes the proteins based on pI value using an IEF strip (gel).
4. Once IEF completed, the strip should be equilibrated in equilibration buffer before SDS-PAGE.
5. The second separation distinguishes proteins by molecular weight via SDS-PAGE.
6. Separated proteins are silver stained.

8.2.13.2 Troubleshooting

Protein spots in the gel should be circular. If they are oblong, the problem could be impurities or poor sample preparation. Samples containing large amounts of carbohydrates, lipids, and/or nucleic acids will not separate well, which can be rectified by phenol extraction.

The entire experimental procedure requires a careful attention. Even a small air bubble can affect protein migration during IEF.

Other proteins in a sample may obscure a specific protein in 2DE. These proteins are called masking proteins. Examples include, glycin in precursor, which is especially rich in seeds, and rubisco in leaves. However, these masking proteins can sometimes be eliminated. Rubisco can be removed from the sample by treating with $CaCl_2$ or protamine sulfate (Krishnan and Natarajan 2009; Kim et al. 2013). Sometimes masking proteins are not completely removed, depending on the sample and working environment. In this case, stronger or longer treatment, or even IT, are recommended before the 2DE experiment.

8.2.14 Detection of Post-translationally Modified Proteins

8.2.14.1 Objective

Most proteins undergo modification after translation to create post-translationally modified proteins (PTM). About 120 modification processes have been reported, including proteolytic cleavage of the signal peptide; attachment of functional groups like phosphate, methyl, acetyl, adenyl, or amide; or binding of sugars, biotin, ubiquitin, or SUMO to the target protein by the formation of disulfide bonds (Wold 1981; Chapman-Smith and Cronan Jr. 1999; Hay 2005; Kerscher et al. 2006).

Secreted proteins are synthesized as precursors with N-terminal methionine and activated by proteolytic cleaving of the signal peptide after reaching the intracellular organelle (Heijne 1990). Phosphorylation regulates protein function and activity by adding or removing a phosphate from an amino acid residue, changing the conformation and affecting interactions with other proteins and receptors (Boyer et al. 1977). Glycosylation attaches glycans to nitrogen, phosphate, or hydroxyl oxygen in an amino acid. Glycosylation functions in protein folding or to improve protein stability (Hart 1997). Ubiquitin is conjugated to defective target proteins that are designated for degradation via 26S proteasome. Ubiquitin conjugation is involved in cellular processes, including the cell cycle, division, and DNA repair (Finley and Chau 1991). Disulfide bonds are covalent bonds between two proteins, or a combination to be performed from the oxidation of two cysteine residues in sulfhydryl groups; these are important in forming the secondary and tertiary structures in a protein (Gruber et al. 2006).

Several methods have been developed to detect or identify PTMs, including dyeing a phosphorylated protein or a glycoprotein with SYPRO RUBY after 2DE (Berggren et al. 2002; Orsatti et al. 2009), detecting modified proteins by western blot using specific antibodies, or Edman degradation, a method of sequencing amino acids in a peptide by labeling the amino-terminal residue and analyzing by HPLC–MS/MS. Here, simple staining methods like PRO-Q Diamond and SYPRO RUBY (cat # P-33300, S12000; Life Technologies) are discussed.

8.2.14.2 Protocol

Phosphoprotein Staining by Pro-Q Diamond Solution

1. Transfer the 2DE-separated gel into a clean container.
2. Using an orbital shaker, fix the gel in 100 mL of fixing solution (50 % methanol, 10 % acetic acid) twice, at 30 min intervals.
3. Wash the gel with 100 mL distilled water three times at 10 min intervals.
4. Stain the gel with 60 mL of Pro-Q Diamond stain solution for 60–90 min.
5. Transfer the gel into a new container and destain with Pro-Q Diamond destain solution or 20 % acetonitrile and 50 mM sodium acetate (pH 4) solution three times at 30 min intervals.

6. Transfer the gel into another container and wash twice with distilled water at 5 min intervals.
7. Equilibrate by transferring the gel into another container and washing with distilled water three or more times at 5 min intervals.
8. View the gel at 532–560 nm using visible-light laser-based or xenon arc lamp-based gel-scanning equipment.

Glycoprotein Staining by SYPRO Ruby Solution

1. Transfer the 2DE-separated gel into a clean container.
2. Using an orbital shaker, fix the gel in 100 mL of fixing solution (50 % methanol, 7 % acetic acid) twice at 30 min intervals.
3. Stain the gel overnight with 60 mL of SYPRO Ruby gel stain solution.
4. Transfer the gel into a new container and wash with 100 mL of washing solution (10 % methanol, 7 % acetic acid) for 30 min.
5. Transfer the gel into another container and wash with distilled water for 30 min.
6. Equilibrate by transferring the gel into another container and washing with distilled water three or more times at 5 min intervals.
7. View the gel at 490 nm using a 300-nm UV or blue-light transilluminator or laser scanner.

8.2.14.3 Troubleshooting

- If dark or uneven staining occurs, the destaining solution was not effective. Wash the gel for 15–30 min.
- Linear or circular patterns in the stained gel are caused by contamination of the solution or the distilled water, contamination of the hands of the person handling the gel, or incomplete staining during orbital shaking.
- If no staining occurs, stain may have been removed by SDS or the methanol and acetic acid were not completely removed during washing.
- When all proteins are stained, the staining solution has decomposed or the gel scanning was not optimized.

8.2.15 Yeast One-Hybrid and Yeast Three-Hybrid Systems

8.2.15.1 Objective

This technique involves the expression of yeast GAL4-transcription activation domain (GAL4 AD prey protein) with DNA-binding protein (bait) fusion to identify interaction between protein and DNA (cis-acting element) or protein and RNA. The bait protein, DNA, or RNA is cloned upstream of a reporter gene in a plasmid and transformed into yeast. The prey library fused to the activation domain will be co-transformed into the reporter strain in yeast. If bait and prey proteins interact, the reporter gene should activated by activation domain to be selected on nutrient medium. This technique is one of the best methods to identify interaction partners (Chapman-Smith and Cronan 1999; Sengupta et al. 1999).

8.2.15.2 Protocol

1. The DNA or RNA is cloned into a target–reporter construct then linearized with a restriction enzyme.
2. Prepare YPDA (yeast peptone dextrose adenine) medium and SD (synthetic defined) medium.
3. Grow the yeast in YPDA medium and transform the linearized target–reporter construct.
4. To one yeast cell harboring BD and AD vectors, two vectors will be co-transformed into yeast or mating yeast cells that containing each vectors.
5. Select the yeast colonies by growing on SD medium lacking amino acids.
6. If an interaction takes place between bait and prey proteins, the reporter gene can express inside the yeast genome. Usually histidine, adenine, uracil, and X-α-gal are used as reporter systems and yeast cells will grow only in media lacking these amino acids.

8.2.15.3 Troubleshooting

- If there is too much background signal or false-positive colonies grow on the medium, solutions include: increasing the medium selection strength; using an appropriate selection marker (reporter gene) based on the yeast strain (for example, selection can be made using medium lacking histidine, leucine, tryptophan and adenine or X-α-gal); or adding an appropriate amount of 3-aminotriazole, which inhibits the synthesis of histidine, into the medium.
- If there is poor transformation or low mating efficiency, test whether the bait protein is toxic to the yeast.
- If no interaction proteins are identified some proteins may not be stably expressed in yeast, the target protein may inhibit the GAL4-interacting domain, or the target protein did not move into the nucleus or fold properly in yeast. To rectify this, the LexA reporter system can be used rather than GAL4 system.
- If the reporter gene expresses in the AD fused library, it may directly interact with its own promoter. In this case, use a different yeast strain.

8.2.16 *Subcellular Localization*

8.2.16.1 Objective

Eukaryotic cells comprise the cell wall, cytoplasm, nucleus, mitochondria, Golgi apparatuses, endoplasmic reticulum, vacuoles, cytoskeleton, nucleoplasm, nuclear matrix, and ribosomes. Functional roles of proteins vary by location within the cell. Bacteria also have subcellular localizations, albeit fewer than eukaryotic cells. Subcellular localization techniques are widely used to study where particular genes are expressed in the cell when the products are fused to YFP, GFP, or CFP tags.

8.2.16.2 Materials

Cell (Fig. 8.5)

Buffers

Cell culture medium includes PBS, phosphate buffer, Transfection reagents, and plasmids. Equipment comprises a fluorescent microscope and CO_2 incubator. The electroporation buffer contains 0.4 M sucrose (13.7 %), 2.4 g/L HEPES, 6 g/L KCl, and 600 mg/L $CaCl_2{\bullet}2H_2O$, pH 7.2 (with KOH).

8.2.16.3 Protocol

1. Prepare cells for transfection.
2. Do the cell transfection via electroporation.
   - Wash about 1 h.
   - Use 0.25 mg of the plasmids and do the transfection for 1–1.5 h.
3. Incubate the cells until ready for fluorescence or western blot.
   - Incubate for 12–36 h after transfection.
   - Incubate at 30 °C with 5 % CO_2 to promote maturation of the fluorophore and to increase the signal.
4. Wash with PBS to remove dead cells or debris and add fresh medium.
5. Check the signal.

8.2.16.4 Troubleshooting

- Confirm that the ORF is in frame when constructing expression vectors.
- Use cells at a higher density.

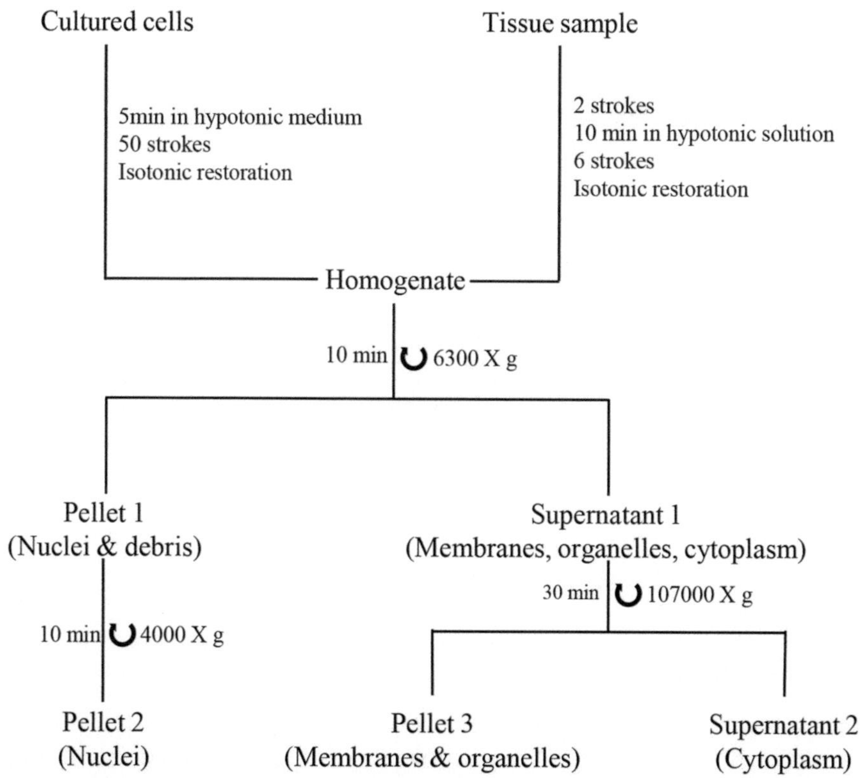

Fig. 8.5 Schematic diagram shows cell organelle separation

- Use more plasmid.
- Sucrose content is important to isolate high quality protoplasts.
- Culture time should be altered according to the gene.

8.2.17 Transient Assay

This method is used to investigate gene function by fusing a reporter gene to the target gene and then transfected the construct into cells. This method reveals the enzyme activity of the reporter gene and thereby identifies the expression or activation of protein, RNA, and DNA. The most commonly used reporter enzymes are 3-glucuronidase (GUS), chloramphenicol acetyltransferase (CAT), luciferase (LUC), and neomycin phosphotransferase (NPTII).

8.2.17.1 Materials

Protoplast Transformation HEPES-buffered saline (HBS) consists of 10 mM HEPES, pH 7.2, 150 mM KCl, and 4 mM $CaCl_2$ supplemented with sufficient mannitol to stabilize the protoplast.

NPTII Assays The callus medium is Murashige and Skoog (MS) medium containing 1 mg/L NAA (naphthalene acetic acid), 2,4-D (2,4-dichlorophenoxyacetic acid), 0.1 mg/l BAP (6-benzylaminopurine), and 0.8 % agar with different antibiotic concentrations.

GUS Assays The lysis buffer is 50 mM sodium phosphate buffer, pH 7.0, 10 mM EDTA, 0.1 % (v/v) Triton X-100, and 10 mM 3-mercaptoethanol. The GUS histochemical stain is 1.2 mM 5-bromo-4-chloro-3-indolyl glucuronide (X-Gluc) in DMSO, 100 mM potassium phosphate buffer, pH 7.0, 10 mM EDTA, 0.5 mM potassium ferricyanide, 0.5 mM potassium ferro cyanide, and 0.1 % Triton X-100.

CAT Assays The CAT reaction mix contains 50 μL 1 M Tris-HCl, pH 7.8, 10 μL 60 mCi/mmol, ^{14}C-labeled chloramphenicol, and 20 μL 3.5 mg/mL acetyl coenzyme A and must be freshly made. The CAT lysis buffer is 50 mM sodium phosphate buffer, pH 7.0, 10 mM EDTA, 0.2 % Triton X-100, and 10 mM p-mercaptoethanol.

LUC Assay The LUC cell lysis reagent contains 0.1 M phosphate buffer, pH 7.8, 1 % Triton X-100, 2 mM EDTA, and 1 mM DTT. The LUC assay buffer is 30 mM Tricine, pH 7.8, 3 mM ATP, 15 mM $MgSO_4$, and 10 mM DTT.

8.2.17.2 Protocol

Protoplast Isolation BE, Col, Ler and C24 genotypes are good for protoplast isolation. Plants are grown under long-day conditions so that they will flower quickly and so 3–4 week-old plants can be used.

1. Finely excise leaves (0.5–1-mm thickness) without otherwise injuring them.
2. Place the leaves in the enzyme solution and keep under vacuum infiltration for 5–30 min to allow digestion to proceed.
3. Add $CaCl_2$ (50 mM) to maintain the protoplasts before filtering. Store at room temperature.

Transfection

1. All steps are performed at 23 °C.
2. Use 10-μl DNA (10–20-μg DNA if less than 5 kb in length).

3. Use 100-μL protoplasts (use steps 4–8) or callus, seedlings, or whole plants (go to step 9) and 110 μL PEG/$CaCl_2$ mix.
4. Incubate at 23 °C for 5–30 min.
5. Add 0.44 mL W5 solution and mix gently.
6. Centrifuge at 100 × *g* and remove PEG.
7. Resuspend the protoplasts gently in 100 μL water and add 1 mL WI or W5.
8. Incubate the protoplasts.
9. Electroporate the protoplasts (See 8.2.17.4).
10. Transfection (2–6 h for RNA analysis, 2–16 h for protein labeling or enzyme activity analysis).

8.2.17.3 Solutions

1. The enzyme solution contains 1–1.5 % cellulase R10 (RS is too strong), 0.2–0.4 % macerozyme R10, 0.4 M mannitol, 20 mMKCl, 20 mM MES, pH 5.7. Heat the enzyme solution at 55 °C for 10 min, cool to room temperature, then add: 10 mM$CaCl_2$, 5 mM B-mercaptoethanol, and 0.1 % BSA.
2. The PEG solution (40 %, v/v) contains 4 g PEG4000, 3 mL H_2O, 2.5 mL 0.8 M mannitol, and 1 mL 1 M $Ca(NO_3)_2$ or $CaCl_2$.
3. The washing and incubation solution (WI) contains 0.5 M mannitol, 4 mM MES, pH 5.7, and 20 mM KCl.
4. The W5 solution contains 154 mM NaCl, 125 mM $CaCl_2$, 5 mM KCl, and 2 mM MES (pH 5.7).
5. The MMg solution contains 0.4 M mannitol, 15 mM $MgCl_2$, and 4 mM MES (pH 5.7).

8.2.17.4 Electroporation

1. Use 40-μg plasmid DNA.
2. Use 4–6 × 10^4 protoplasts/300 μL of 0.3 M mannitol + 4 mM MES, pH 5.7 + 150 mM KCl.
3. Shock with 300–400 V/cm, 5 msec, 200 μF, 1–2 pulses.

8.2.17.5 Notes

- When isolating protoplasts, the plant leaves should not be injured.
- Adding the enzyme solution may stress the leaves, so do the experiment as quickly as possible.
- Use leaves before the plant flowers.
- Always check the incubator conditions, temperature, humidity, intensity of light, water, nutrients, plant, enzymes, genes, and leaves.
- Ensure that the plasmid DNA is of high quality.

8.2.18 *Reporter Gene System*

A reporter gene is used to identify protein localization and quantitatively study the gene expression in the cell. For example, beta-galactosidase, GFP (confirmed by fluorescence), or His-tag (confirmed by an antigen–antibody reaction), can be used as reporter genes to determine the expression level of a particular gene. The GUS gene can be used to identify gene expression in a cell but cannot be used to study expression over a time period in living cells (Table 8.1).

8.2.18.1 GFP (Green Fluorescent Protein)

The green fluorescent protein (GFP) isolated from *Aequorea Victoria* and *Renilla reinformis* emits bright green fluorescence when exposed to light (Tsien 1998). GFP is a very stable molecule composed of 238 amino acid residues (27 kDa) and is acidic, compact and globular in structure. It is very stable up to 65 °C in the presence neutral buffer between pH 5.5–12. The advantage of GFP is that its expression can be seen inside cell compartments (e.g., chloroplasts, mitochondria, vacuole, membranes, cell membrane, cytoplasm), but large amounts must be used when gene expression is weak (Fig. 8.6).

8.2.18.2 GUS (*beta*-Glucuronidase)

Beta-glucuronidase (GUS) is one of the most effective molecules for studying gene regulation in plant molecular biology. Gene or promoter expression can be identified using a GUS reporter gene at the tissue-specific level (e.g., stem, leaf, vein, root, flower, guard cell) (Fig. 8.7).

Working Solution The working solution contains 100 mM X-gluc, 90 % acetone (−20 °C), and rinse solution (50 mM sodium phosphate buffer with 0.05 mM $K_3F_e(CN)_6$, and 0.05 mM $K_4F_e(CN)_6$).

Table 8.1 Reporter genes used in plant systems

Reporter gene	Reporter construct	Mechanism
β-galactosidase (*lacZ*)	Target gene promoter + *lacZ*	Catalyzes the hydrolysis of X-gal and produces deep blue color
Green florescent protein (GFP)	Target gene promoter + *gfp*	Green florescence can be detected under UV light
Luciferase (*luc*)	Target gene promoter + *luc*	Firefly luciferase catalyzes the bioluminescent oxidation of luciferin and produces light
β-glucuronidase (GUS)	Target gene promoter + *gus*	Blue coloration is produced when provided with the appropriate substrate

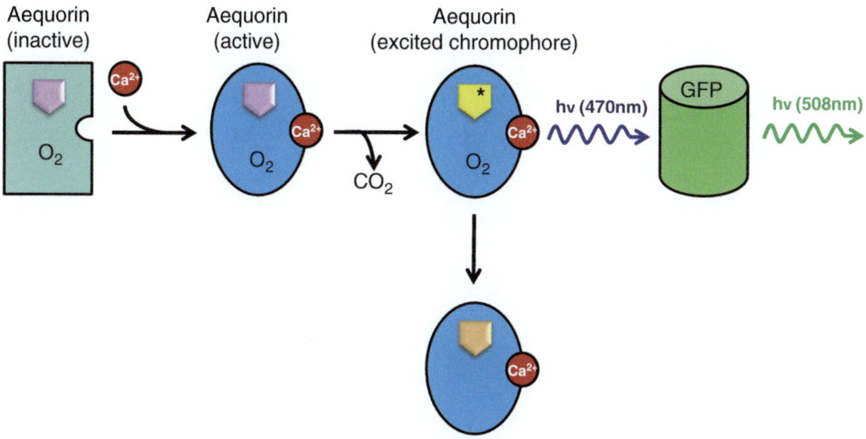

Fig. 8.6 Green fluorescent protein signaling mechanism (Haute 2003)

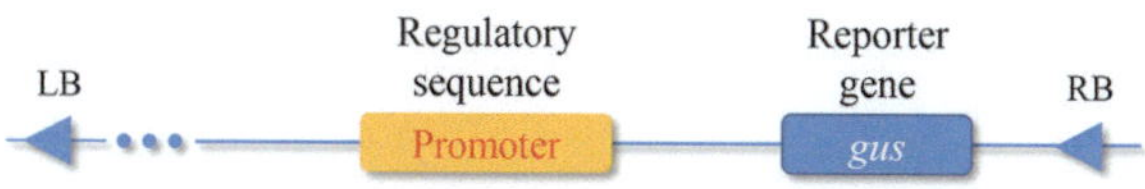

Fig. 8.7 β-Glucuronidase (GUS) reporter gene

GUS Staining Protocol

1. Fix the tissue in cold 90 % acetone (−20 °C) and incubate on ice for 20 min.
2. Wash the tissue with rinse solution 2–3 times to remove the acetone.
3. Pour off the rinse solution, add 100 mM X-gluc, and incubate the sample at 37 °C for 2–3 days.
4. After 2–3 days, wash the sample with 70 % ethanol to remove the color.

8.2.18.3 Luciferase

Enzymes play an important role in bioluminescence, which is widely distributed in nature. Luciferase is known from protozoa, bacteria, and the light-emitting organs of mushrooms. Luminescence is emitted through the oxidation of luciferin by an oxygen molecule. Luciferase is used to detect trace amount of ATP, because it requires molecular oxygen and ATP (Fig. 8.8).

8.2.18.4 Troubleshooting

- When doing GUS staining, always perform a positive control to test the efficiency of the GUS solution.

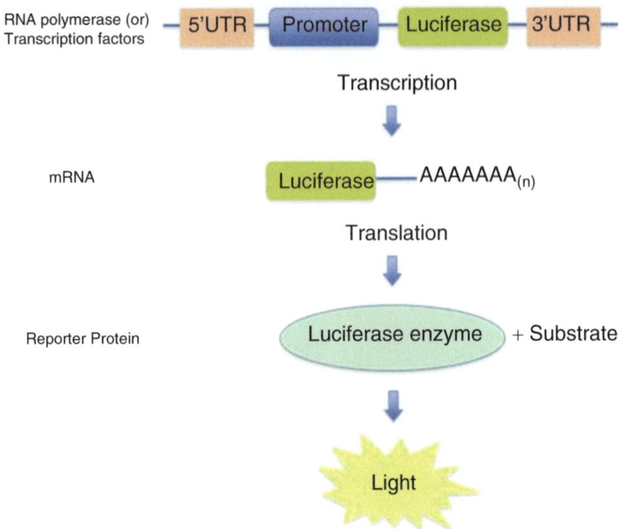

Fig. 8.8 Luciferase reporter assay

- Ensure that the GUS solution is made properly.
- In transgenic plants, T-DNA is entered into a random position so do the GUS staining with randomly sampled tissues.
- Ensure that the promoter, gene, and reporter gene are cloned in frame.
- Do the experiment without background autofluorescence.
- Use care when setting the fluorescence detection and sensitivity adjustment of the microscope.

8.2.19 Live Cell Imaging

8.2.19.1 Introduction

Live cell imaging allows the study of functional and physiological characteristics of live plant cells, including structure and dynamics under normal and stress conditions. This technique is ideal for small flowering plants, such as *Arabidopsis thaliana*, which can be examined with different live microscopy techniques. *Arabidopsis* is small, so light can penetrate inside cells. Its nucleus has a heterochromatin domain in the chromocenter and five pairs of chromosomes. Thus, it can be viewed easily under a microscope. In addition, the availability of powerful genetic tool can facilitate the investigation of the molecular mechanisms of various cell features. Furthermore, plants depend on light, temperature, osmotic pressure, humidity, gravity, and nutrients for gene expression and have physiological behavior called a circadian rhythm. The green leaves of plants are usually thick and may exhibit strong auto fluorescence. Therefore, studying live cell imaging in plants is needed to optimize culture conditions first.

8.2.19.2 Requirements

In preparing an optical microscopy system for live-cell images, detector sensitivity, the required speed of image acquisition, and specimen viability are the primary considerations. The relatively high light intensities and long exposure times are typically considerations in capturing images of fixed cells and tissues. In virtually all cases, live-cell microscopy represents a compromise between achieving the best possible image quality and preserving the health of the cells. Rather than unnecessarily oversampling time points and exposing the cells to excessive levels of illumination, the spatial and temporal resolutions set by the experiment should be limited to match the goals of the investigation.

8.2.19.3 Experimental Considerations for Live Cell Imaging

Temperature Many biological developments are temperature sensitive, so the temperature should be regulated by using a heater.

Oxygen Most living organisms need oxygen, whereas closed chambers can be depleted in oxygen; the chamber media should be changed.

pH Plant metabolism can vary over a period of time, resulting in pH changes. To rectify this problem, monitor the chamber pH, and use HEPES (10–25 mM) buffer media.

Humidity Changes in humidity may occur owing to heat, affecting pH and salinity. Use of a closed chamber configuration (perfusion chamber) and/or humidifier can be helpful.

Fluorescence Signal Strength Probes with low concentration or weak fluorescence may produce weak image signals. To enhance the signal, increase the pixel dwell time, open the confocal pinhole aperture, and always use the average frame and adjust the objective back aperture illumination.

8.2.19.4 Troubleshooting

Check that the instrument is working properly. DNA quantity and quality are important; in onion, they are known to affect the final data.

8.2.20 *Scanning Electron Microscopy (SEM)*

8.2.20.1 Objective

Scanning electron microscopy (SEM) was developed to observe the microstructure of a sample surface. Additionally, SEM can be used to evaluate the constituent elements of the sample and the denseness of the tissue surface. SEM can be used for energy dispersive spectroscopy (EDS), which provides elemental and chemical analysis of a sample in a short period of time. SEM is widely used to observe organisms such as eukaryotes (plants and animals), biological molecules (proteins, nucleic acids, polysaccharides, antibodies), and microorganisms (viruses, bacteria) (Goldstein and Harvey 1975; Hawes 1991).

8.2.20.2 Protocol

1. Fix samples by immersion in 2 % paraformaldehyde, 2 % glutaraldehyde, and 0.05 M sodium cacodylate buffer (pH 7.2) at 4 °C for 2–4 h.
2. Wash samples three times at 10 min intervals with 0.05 M sodium cacodylate (pH 7.2) solution at 4 °C.
3. Immerse in 0.05 M sodium cacodylate buffer (pH 7.2) solution containing 1 % osmium tetroxide at 4 °C for 2 h.
4. Wash twice with distilled water at room temperature.
5. Immerse dehydrated samples for 10 min at room temperature with 70 %, 80 %, 90 %, and 30 %, 50 %, 100 %, 100 %, 100 % ethanol series.
6. Wash the sample twice with 100 % hexamethyldisilazane or tetramethylsilane, and then dry for 15 min.
7. Wash twice with 100 % isoamyl acetate for 15 min, and then air dry.
8. Dry the samples completely in a critical point dryer.
9. Immobilize the sample by mounting on a metal stub.
10. Surface coat the sample with gold or tungsten particles.
11. Observe under SEM.

8.2.20.3 Troubleshooting

- If the surface of the sample is not clearly visible, check that the sample was dried completely. The sample must undergo a series of drying procedures before being subjected to critical point drying. This is especially important for seeds, which are solid and contain more water than other plant parts.
- If image resolution is not good, check the focus and calculate the saturation of the peak filament. Image noise occurs when the power is kept low during image magnification. Adjust the working distance by increasing the power of the short beam.

- If the sample has edge effects, there may be a shade of binary signal strength. In this case, the power of the beam is kept low to reduce the edge effect.
- Heat from the high-energy beam can damage the sample. If this occurs, beam power should be lowered or the working distance increased.

8.2.21 Transmission Electron Microscopy (TEM)

Transmission electron microscopy (TEM) operates on the same basic principles as light microscopy but uses electrons instead of light. An image is represented from the interaction of the electrons transmitted through the specimen; the image is magnified and focused onto an imaging device. The sample is cut into thin slices to fit in the observation tube. Living cells cannot be observed under TEM.

8.2.21.1 Protocol

1. For the primary fixation, slice a small piece of tissue (about 1 mm thickness) and immerse in fixative solution for 2–4 h at 4 °C. The sample should be as small as possible.
2. Wash three times in PBST solution at 30 min intervals or once overnight at 4 °C.
3. For the secondary fixation, treat the sample until it turns dark black in 1 % OsO_4/PBST solution for 2–4 h at 4 °C.
4. Wash three times in PBST solution at 30 min intervals or once overnight at 4 °C.
5. Dehydrate in 25 % EtOH (20 min), 50 % EtOH (20 min), 75 % EtOH (20 min to overnight at 4 °C), 90 % EtOH (20 min), 95 % EtOH (30 min), and 100 % EtOH (1 h, 2–3 times).
6. Infiltrate with 25 %, 50 %, 75 %, 100 %, 100 %, and 100 % Spurr's resin for 1 h each at room temperature.
7. For embedding curing, put the sample and 100 % Spurr's resin on an embedding mold. Place the mold in a closed oven for 48 h at 60 °C or 8 h at 70 °C.
8. Cut the sample into 100-nm thicknesses with an ultra-microtome and transfer onto a 150-mesh grid. Dry for 15 min, and then stain with 8 % aqueous uranyl acetate and lead citrate for 10 min.
9. Analyze the sample under a transmission electron microscope.

8.2.22 Immunocytochemistry

Immunocytochemistry is a highly productive method in biomedical research in which an antibody binds specifically to an antigen (protein) within cells (Baron 1984). A fluorescent dye or horseradish peroxidase fluorescent antibody can produce a color reaction by combining with ferritin or colloidal gold, which localized the protein and makes it visible under a light microscope. The commonly used antibodies are human or rabbit polyclonal antibodies or a mouse monoclonal antibody. Although, direct detection of antigens is also available using a single antibody, combinations of secondary antibodies produce better results.

8.2.22.1 Immunocytochemistry versus Immunohistochemistry

Immunocytochemistry differs from immunohistochemistry in the kind of samples used. Immunocytochemistry is performed on intact cells with their surrounding extracellular matrix removed. In contrast, immunohistochemical samples are sections of tissue in which each cell is surrounded by tissue architecture and adjacent cells (Javois 1999; Amsterdam et al. 2012).

8.2.22.2 Protocol

There are direct and indirect methods used to identify the distribution of proteins in the sample. In the direct method, only a primary antibody is used to localize the antigen of interest, whereas in the second method, a primary antibody is coupled with a secondary antibody that recognizes the antigen substance.

8.2.22.3 Sample Preparation

The antibody coupled with antigen must produce a signal for accurate visualization. Additionally, the natural shape of the cell should be maintained throughout the experiment. For this purpose, paraformaldehyde is generally used for cell immobilization. Based on the cell type, a chemical linker or cytocentrifugation is used to fix the cells on a slide. Immunoglobulin passes poorly through the cell membrane to the nucleus; to overcome this obstacle, surfactants such as Tween-20 or Triton X-100 are added to increase cell permeability.

8.2.22.4 Immunocytochemistry Method

To remove background signal caused by antibody reactions, a non-specific antigen is blocked using a blocking buffer containing fat-free milk, bovine serum albumin (BSA), or gelatin material. The primary antibody is added to the antigen, and then the secondary antibody coupled with linker (detection substance) is added. The sample is observed under a microscope.

8.2.22.5 Troubleshooting

The biggest problem in immunocytochemistry is background signal. Care should be taken throughout the procedure, as it affects the color. Another problem is the binding of antigen–antibody that results in weak fluorescence. Weak binding affinity of antibodies to the antigen also cause problems and increase background signal. High quality standard antibodies should be used. However, increasing the concentration of the blocking substance also reduces the background signal. Changing the type of the color-developing material will reduce autofluorescence.

8.2.23 Fluorescence Resonance Energy Transfer (FRET) Microscopy

8.2.23.1 Objective

Fluorescence resonance energy transfer (FRET) is a technique used to investigate gene expression levels as compared with energy transfer of the two chromophores, the part of a molecule responsible for its color. The investigation is based on nonradiative dipole–dipole coupling of a donor and an acceptor chromophore. Measurements of FRET analysis can be compared with the efficiency of different wavelengths. FRET is mainly used in biological and biochemical fields by analyzing the wavelengths emanating from the chromophore.

8.2.23.2 Materials

Materials include: Fibronectin, Akind FRET probes, pECFP-C1 obtained from Clontech (discontinued, cat. no. 6076-1), fetal bovine serum (FBS), penicillin/streptomycin, Lipofectamine 2000, Fluoro-Gel, and paraformaldehyde.

8.2.23.3 Protocol

1. Grow the cells.
2. Add 2 mL of 0.25 % (wt/vol) trypsin-EDTA to the cells and incubate at 37 °C.
3. Add 4–6 mL of growth medium and mix well.
4. Mix 1.5 μg of cDNA with Lipofectamine 2000 at the ratio of 2.5:1.
5. Dilute the cDNA in 250 μL of Opti-MEM (250 μL of Opti-MEM contains 3.75 μL Lipofectamine 2000 and 7.5 μL cDNA).
6. Incubate at 23 °C for 5 min to separate the cDNA from Lipofectamine.
7. Wash the coverslip in PBS.
8. After fixing the cells, incubate the mixture at 23 °C for 15 min.
9. Observe the image under a microscope

8.2.23.4 Notes

- Treat the medium with insoluble enzyme to remove it.
- The experiment should be conducted rapidly.
- Steps at room temperature should be minimized.
- Keep the sample in PBS until it is moved to dark conditions.
- Samples should always be maintained in the dark.

8.2.24 Bimolecular Fluorescence Complementation (BiFC)

Bimolecular fluorescence complementation (BiFC) analysis allows direct visualization of protein–protein interactions in living cells. Two proteins are cloned to the each fluorescent protein of the N-terminal fragment and C-terminal fragment. If two proteins interact with one another, these fragments will become adjacent to or coupled with each other and form a fluorescent protein complex. This technique can be used in animals, plants, and yeast, and the resulting protein interaction is measured as fluorescence signal using simple equipment at any position within the cell. The fluorescence signal can be observed with even weak protein interactions. However, combination of candidate proteins is necessary for the control experiments, as they may receive a false positive signal due to non-specific binding (Fig. 8.9).

8.2.24.1 Protocol (Using Onion Epidermal Cells)

1. Yellow fluorescent protein (YFP) truncated at residue 155 (YN155 vector) and the fragments of YFP truncated at residue 173 (YC173 vector) are cloned by gene A and gene B, respectively.
2. Plasmids A-YN155 and B-YC173 are mixed 1:1 (25 μg each), then $CaCl_2$ and Spermidine are added and mixed well with the gold particles.
3. The gold-coated plasmids are transformed into onion epidermal cells using a gene gun, followed by incubation in the dark for 16 h. Signal is then observed under a confocal laser fluorescence microscope.

8.2.24.2 Notes

- The fluorescent protein complex formation does not occur instantly and separation (rapidly changing) of interaction proteins is not suitable for real-time analysis.
- The fluorescence signal may not be observed, even if the two proteins interact.
- Combination with each other proteins should be studied to form a fluorescent protein–protein complex.
- The fluorescent protein fragments linked to each other should not affect the function of proteins and their intracellular localization.

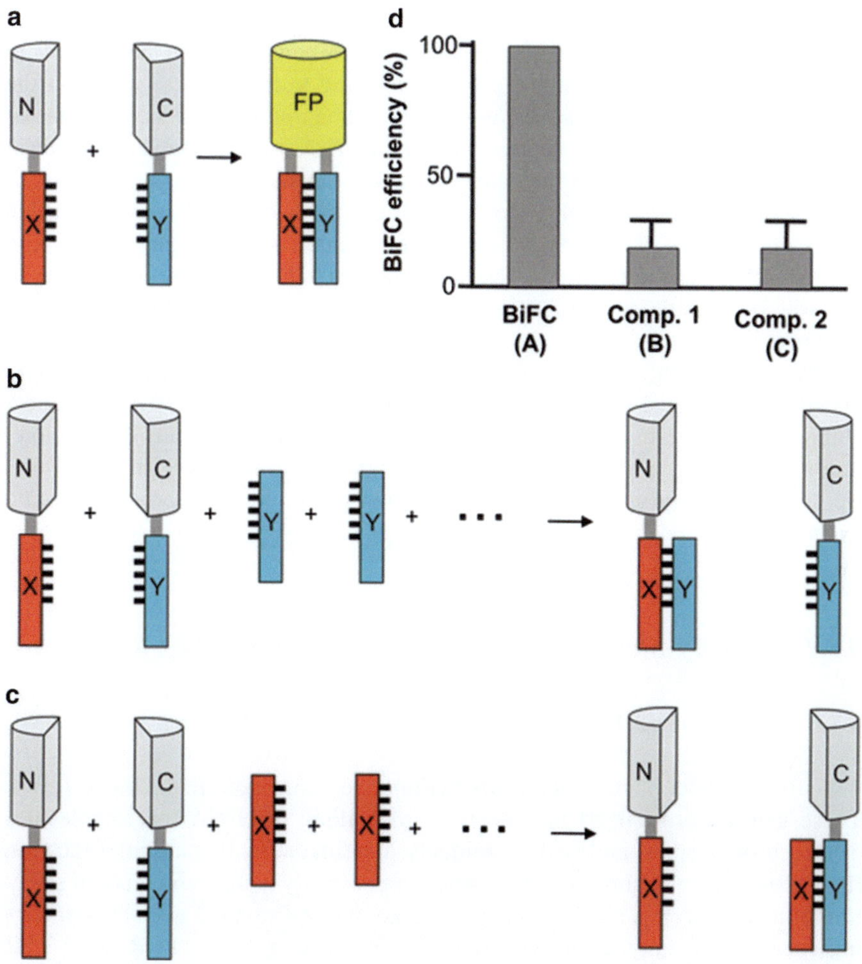

Fig. 8.9 Principle of the bimolecular fluorescence complementation (BiFC) assay (Kodama and Hu 2012). (**a**) Schematic diagram of the BiFC assay for visualizing the interaction between proteins X and Y, (**b**) Competition assay using non-fused protein Y as the competitor. When the competitor Y protein is present, the interaction between the protein X and protein Y fusions is subjected to competition by the Y protein, leading to a reduction in BiFC efficiency as shown in (D), (**c**) A similar competition experiment can be designed using non-fused protein X as the competitor, (**d**) Hypothetical result of a BiFC competition assay showing the percentage of BiFC efficiency from the experiments in A, B, and C

8.3 Map-Based Cloning of Genes

Gene function is verified by characterizing isolated genes *in vitro* or by a complementation test, which introduces the isolated gene into an organism. These procedures are dogma, similar to Koch's postulate for identifying pathogens. In other words, using an *in vitro* isolated gene to verify gene function is a commonly accepted framework in current biological research. Representative examples that verify function *in vitro* are enzyme-encoding gene studies in which the amino-acid sequence of the purified enzyme can be compared to the DNA sequence of the enzyme-encoding gene. However, as genes encoding transcription factors or membrane proteins are seldom functional *in vitro*, their verification frequently requires a complementation test, in which the isolated gene is transformed into a variant of the host or model organism that shows alteration or loss-of-function of the targeted trait. Consequently, cloning that allows *in vitro* manipulation of a gene is a key tool for research into gene function. Map-based cloning is the most difficult cloning to perform, but it is the most robust and straightforward method and often provides big rewards.

Map-based cloning refers to cloning a gene using information from both genetic and physical maps. Map-based cloning is also referred to as positional cloning, but this term may be too abstract. Cloning of a gene controlling a target trait using maps is a long, complicated, and often labor-intensive process. Here, we describe a helpful guide with step-by-step procedures and provide specific examples.

8.3.1 Step-by-Step Guide to Map-Based Cloning (Fig. 8.10)

8.3.1.1 Selection of the Targeted Gene for Map-Based Cloning

The simple model plant *Arabidopsis thaliana* contains more than 20,000 genes. These genes are arranged in a precise order along the DNA macromolecules called chromosomes, and each is uniquely identifiable by its particular location (locus) on a specific chromosome. Genes can be easily isolated by random selection of a clone from a cDNA library or isolation of a disrupted gene by associating a changed phenotype with a tag sequence obtained by inverse PCR or tail-PCR of a genetic line selected from a T-DNA transformant population. However, finding an association between an isolated cDNA clone and a phenotype is difficult using the cDNA approach. T-DNA tagging provides an easier means to determine the association between a gene and a phenotype, but the results are difficult to apply in breeding because of public concerns about the potential risks of genetically modified organisms. Moreover, research on agronomically-important genes begins with determining which gene is regulating an observed quantitative (rarely, qualitative) trait. The first step in detecting a particular chromosomal location of a candidate gene is the construction of a genetic map in the gene's vicinity. Thus, a map-based cloning approach is usually the most appropriate for crop research.

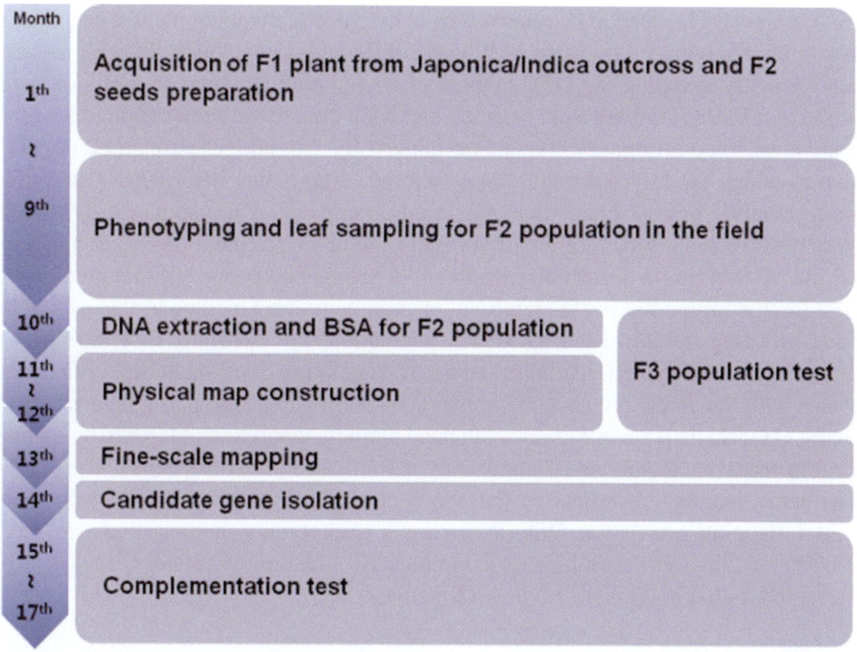

Fig. 8.10 Flow chart for map-based cloning in rice

8.3.1.2 Genetic Map Construction

A genetic map is constructed by estimating the recombination frequency between the targeted genetic trait (measured by phenotype, e.g., morphology, photoperiod, seed-coat color) and molecular markers. Artificial mapping populations, such as F2 populations, recombinant inbred lines, and backcrossed populations, are used to construct a genetic map for crop plants. Approximately 100 individuals per population are usually sufficient to map a qualitative trait that follows a Mendelian inheritance pattern. The parents of the population need to be carefully selected. Distantly related lines have high polymorphism levels but are not desirable because highly variable non-targeted traits from either parent may mask the phenotype of the targeted trait. Ideally, the parents would show high polymorphism levels only in the chromosomal region containing the targeted trait. For example, if one is studying plant height using a population derived from a cross between wild and cultivated crop lines, viny (prostrate) growth inherent in the wild line would make plant height measurements impractical.

The genetic mapping population and phenotype scoring need to be simplified to map-base clone each locus. We consider cloning of a major QTL as an example.

First, minor QTLs need to be removed from the genetic mapping population, which is commonly achieved by using an advanced backcross population. In other words, after roughly mapping the QTL in BC1 or F2 populations, lines that do not have additive minor QTL alleles are selected and backcrossed with the recipient line (or another selected no-minor-QTL and no-major-QTL line) to generate an F2 population in which the targeted major QTL follows a Mendelian inheritance pattern. A more reliable, though more time consuming, approach is to use an introgression line, which has a segment of donor chromosome in the recipient parent, as a parental line. Given that a quantitative trait can sometimes be converted to a qualitative trait for phenotype scoring, a simplified phenotype scoring method is also desirable for generating the advanced backcross population. For example, Li et al. (2006) observed during map-based cloning of a gene regulating grain shattering in rice that plants with the heterozygous and homozygous dominant genotypes at their targeted major QTL shed all mature grains with hand tapping, whereas plants with the homozygous recessive genotype did not or only partially shed mature grains under vigorous hand shaking, regardless of the genotypes at two other minor QTLs in their initial mapping population. With this reliable phenotyping method, they screened 12,000 seedlings for recombinants to circumscribe the targeted major QTL to a 1.7-kb region with a previously unknown function.

8.3.1.3 Conversion of the Genetic Map to a Physical Map

One centiMorgan (cM) in a genetic map corresponds to approximately 200 kb in Arabidopsis or 250 kb in rice, but can differ across genomic regions and across species. As a 200-kb DNA sequence can contain up to 20-40 genes, a target gene cannot be pinpointed even after mapping its approximate location on a chromosome. However, the resulting tightly-linked markers can be used for marker-assisted selection.

The next step in map-based cloning is converting the genetic map to a physical map. Until recently, constructing a physical map was just a dream for crop researchers, because reference genome sequences were not available. However, the reference genome sequences of most major crop plant species have now been reported using next-generation sequencing platforms. Thus, old methods for constructing a physical map, such as BAC library screening, are not described here. With a reference genome sequence, a physical map can be acquired by BLAST-searching sequences of molecular markers to the reference genome sequence.

8.3.1.4 Fine-Scale Physical Map Construction

Given a target gene located in a 200-kb region delimited by markers at either side of the gene, the physical region can be narrowed down as follows. First, seeds of individuals heterozygous for the marker loci on both sides of the targeted gene region are collected from the genetic mapping population. The seeds are planted and their DNA extracted. Plants homozygous for one marker and heterozygous for the other are selected and grown for phenotypic evaluation. A phenotype score is usually obtained from the selected individuals. If this is not possible, 10–15 of their progeny are evaluated. At the same time, simple sequence repeat (SSR) and SNP markers are progressively generated by searching SSR motifs in the reference genome sequence or by resequencing. This keystep can cause confusion; although progeny from different generations are used, map-based cloning uses an identical genetic mapping population to construct both the genetic and physical maps. Therefore, if the eventual objective of the study is gene cloning, seeds of the initial population must be retained. In other words, some researchers misunderstand that map-based cloning requires thousands of individuals in an initial genetic mapping population. However, in some cases an initial mapping population of 100–200 individuals can be sufficient to complete all map-based cloning procedures for many crop species, although larger populations will help narrow down the target region.

8.3.1.5 Candidate Gene Selection

As physical mapping progresses, the region delimited by molecular markers can be narrowed down to a short area containing a few genes or even one. The size of the delimited zone depends on where the targeted gene is located on the chromosome and the number of informative recombinants that were identified. A telomeric region with a high recombination rate can be narrowed down to an extremely short region containing only one gene, whereas near the centromere, which has a low recombination rate, narrowing a region to 1000-kb and 10 or more genes may be difficult.

Once the physical map cannot be narrowed down further, the delimited region is sequenced, and gene annotation is performed. In some cases, gene annotation will reveal that one of the genes in the delimited region is already functionally known in another organism and may code for a trait that is similar to our targeted trait, indicating that the known-gene homolog is a strong candidate gene.

8.3.1.6 Complementation Test

The final confirmation that a cloned gene is the targeted gene occurs via an *in vitro* functional study or complementation test by transformation into a mutant, regardless of whether the function of the candidate gene is known. Of course, the function of a gene requiring map-based cloning will rarely be verified by an *in vitro* experiment. The recessive parent from genetic mapping or a mutant is most desirable for the complementation test. If the host species cannot be readily transformed, other model species, such as *Escherichia coli* or *A. thaliana* with a mutant gene homologous to the candidate gene can be used as an alternative.

8.3.1.7 Further Study of the Cloned Gene

Rescuing a phenotype using a complementation test and a subsequent sequence analysis were usually sufficient during the 1990s to publish in a prestigious journal, such as *Science* or *Nature*. However, additional studies, such as gene expression profiling, which may be influenced by the cloned gene, are now required to publish a cloned gene. As gene functions vary, methods to further characterize function are diverse, so the method used should be carefully considered.

8.3.2 Case Study of Map-Based Gene Cloning

The practical process for map-based gene cloning is described here using mutant rice and based on a study by Woo et al. (2008).

8.3.2.1 Acquisition of F1 Plants Derived from *japonica/indica* Outcrosses

Prior to acquiring the F1 plant, the mutant should be characterized by accurate comparison of its phenotype with that of the wild type. In addition, precise information about the pedigree of the mutant is required. Next, as a map is constructed using marker polymorphism, the candidate region cannot be narrowed down if the F2 population contains no polymorphic markers. Therefore, to create a mapping population for rice, the researcher should know whether each parent originated as *japonica* or *indica*, and then the mutant parent should be crossed with approximately three parents showing high marker polymorphisms relative to the mutant. For example, in the case of a *japonica*-origin mutant, an *indica*-origin wild type should be used as the cross parent in reciprocal outcrossing (Fig. 8.11).

As shown in Fig. 8.11, to obtain the F1 hybrid, a mutant parent is crossed with each of three selected parents, and each F2 and F3 population is maintained through serial self-pollination. One F2 population of three combinations is preferred for in map-based cloning. The other F2 populations may be used for further work in narrowing down of the mapping region by providing more recombinants.

After obtaining and planting F1 hybrid seeds, whether the mutated gene is recessive or dominant can be determined by identifying the F1 phenotype. Most mutated genes are recessive, in which case the F1 plant appears as a normal wild type. For the F2 mapping population, if possible, 1500–2000 F2 seeds should be prepared.

In this case study, a male-sterile mutant, Hwacheong *ms-h*, was induced via chemical mutagenesis using N-methyl-N-nitrosourea from a Korean *japonica* cultivar, Hwacheongbyeo (Koh and Heu 1995). The F1 hybrids from a cross between the Hwacheong *ms-h* mutant (temperate *japonica*) and Milyang 23 (*tongil* type rice, derived from an *indica* × *japonica* cross and similar to *indica* in its genetic make-up)

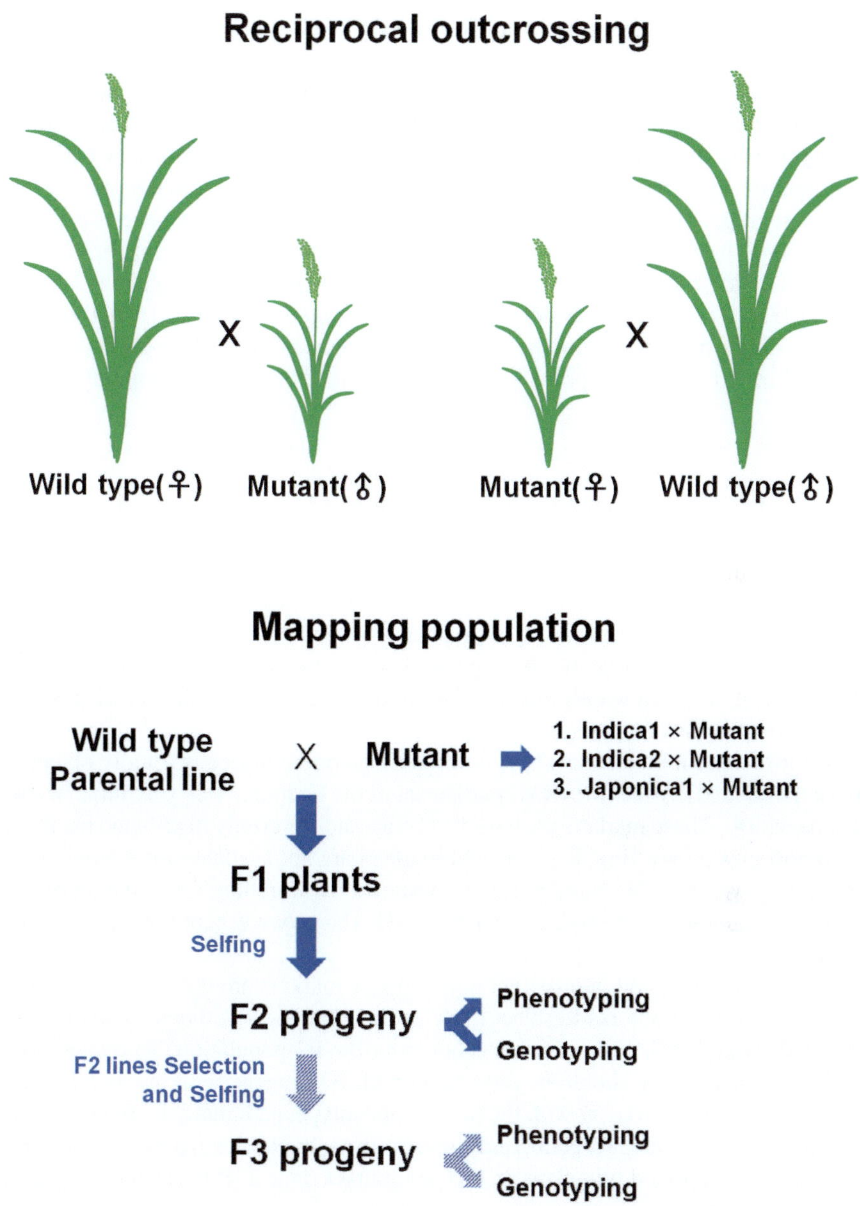

Fig. 8.11 Reciprocal outcrossing and the procedure for developing a mapping population

appeared to have normal fertility, indicating a recessive mutation. To map the *ms-h* gene, an F2 population was derived from the F1 hybrid between the *ms-h* mutant and Milyang 23, and 1,051 F2 plants were evaluated for phenotypic segregation of male fertility and sterility by examining spikelet fertility (Woo et al. 2008).

8.3.2.2 Genetic Segregation Analysis in the F2 Population

To analyze the genetic segregation ratio of the mutated gene, all F2 plants should be evaluated for phenotypic segregation with whole numbering and recording. Analysis is commonly performed with achi-square test according to Mendelian segregation hypotheses. If F1 hybrids are normal (wild type phenotype) and F_2 plants segregate in a ratio of 3:1 (normal: mutant), the mutated gene is a single recessive gene. The segregation ratio in the F2 mapping population for *ms-h* fit 3:1, indicating that the trait was controlled by a single recessive gene (Koh and Heu 1995). For incomplete dominance, the segregation ratio is 1:2:1 (normal: intermediate: mutant phenotype). Additional analysis in the F3 population can be performed for more precise results if those from the F2 population are unclear.

8.3.2.3 Bulked Segregant Analysis and Physical Map Construction

Bulked segregant analysis (BSA) is used to identify molecular markers associated with a trait in an organism. This measures the variation present in pools of segregants that have been sorted according to phenotype and uses the correlation between these measurements and the pool phenotype to assign a likely map location. Therefore, it has been widely adopted for rapid identification of molecular markers in specific regions of a genome.

Information on molecular markers for gene mapping in rice is usually obtained from the Rice Microsatellite (RM) marker set of the Gramene web site (http://www.gramene.org). These markers are based on SSRs and are evenly distributed on all 12 chromosomes of rice (Fig. 8.12). If RM markers are not available for the region of interest, new molecular markers can be designed by searching for polymorphisms between *japonica* and *indica* using NCBI (http://www.ncbi.nlm.nih.gov) or Gramene.

Previously identified inter-parent polymorphic markers are then surveyed in the two parents and two bulks to select markers linked to the phenotype. The markers selected through BSA are used for genotyping the F2 population. To reduce time and costs, genotyping should be done first on all F2 progeny with the mutant phenotype, then on F2 progeny with the normal phenotype, and finally on the F3 population. Figure 8.13 shows genotyping results from PCR with markers for the F2 population, using wild type (lane 1; A type), mutant (lane 2; B type), and F1 plants (lane 3; H type) as controls.

The table for physical map construction should be made by recording these genotyping results as A, B, and H type, combined with each phenotype (Fig. 8.14).

To map the *ms-h* gene, DNA was extracted from the fresh young leaves of rice. Five fertile and five male-sterile individuals were selected from the F2 population and made into separate DNA bulks using equal amounts of DNA from each of the

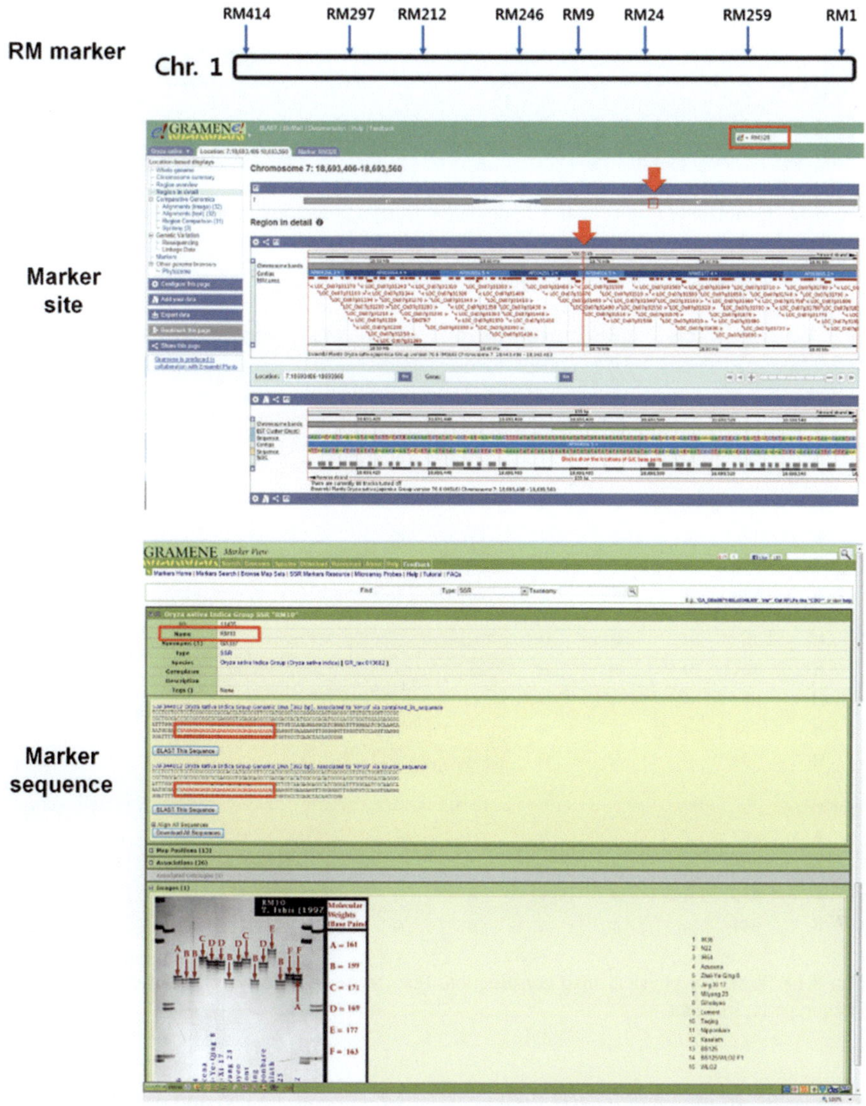

Fig. 8.12 Rice Microsatellite (RM) marker search and identification on the Gramene web site

five plants (Michelmore et al. 1991). A total of 58 RFLP markers distributed evenly on the 12 chromosomes were pre-surveyed between the two parents. Polymorphic markers were then surveyed in the two parents and two bulks to identify those linked to male fertility. After confirming that the gene was located on chromosome 9, other 24 RFLP markers for that chromosome were surveyed for polymorphism in the parents of the mapping population. Of these, 11 markers were mapped around the

Fig. 8.13 Genotyping results of an F2 population using a polymorphic marker. Lane 1; wild type (A type), Lane 2; mutant (B type), and Lane 3; F1 plant (H type)

		SSR	STS	STS	STS	STS	STS	STS	SSR	STS	SSR
Phenotype	F2	RM6835	AP3956	AP3956	AP3956	AP4259	AP4259	AP4259	RM8257	AP4259	RM320
			40kb	49.4 k	72.5 kb	9.5 kb	15.8 kb	42.4 kb	55 kb	60.1 kb	
A	343	H	A	A	A	A	A	A	A	A	A
A	539	H	A	A	A	A	A	A	A	A	A
B	61	H	B	B	B	B	B	B	B	B	B
B	137	B	B	B	B	B	B	B	B	H	H
B	336	B	B	B	B	B	B	B	B	H	H
B	580	B	B	B	B	B	B	B	B	B	H
H	97	H	H	H	H	H	H	H	H	A	A
H	150	H	H	H	H	H	H	H	H	H	B
H	282	A	A	A	H	H	H	H	H	H	H
H	298	A	H	H	H	H	H	H	H	H	B
H	588	H	H	H	H	H	H	H	H	H	A
Phenotype	F3	RM6835	AP3956	AP3956	AP3956	AP4259	AP4259	AP4259	RM8257	AP4259	RM320
A	382	H	H	H	H	H	A	A	A	A	A
A	534	A	A	A	A	A	A	A	H	H	H
A	653	A	A	A	A	A	A	A	A	H	H
A	654	A	A	A	A	A	A	A	H	H	H

Fig. 8.14 Table for physical map construction. Phenotype A, wild type; Phenotype B, mutant; Phenotype H, F1 plant

ms-h locus. *RG451* and *RZ404* flanked the *ms-h* gene at 2.5 cM and 3.3 cM, respectively (Fig. 8.15; Koh et al. 1999).

To identify additional markers closely linked to the *ms-h* gene, 15 sequence-tagged site and 12 cleaved amplified polymorphic sequence (CAPS) markers within the interval containing the *ms-h* gene were developed based on rice genome sequences. As a result of the map-based cloning experiment, the *ms-h* gene was narrowed down to within a 60-kb region (Fig. 8.15; Woo et al. 2008).

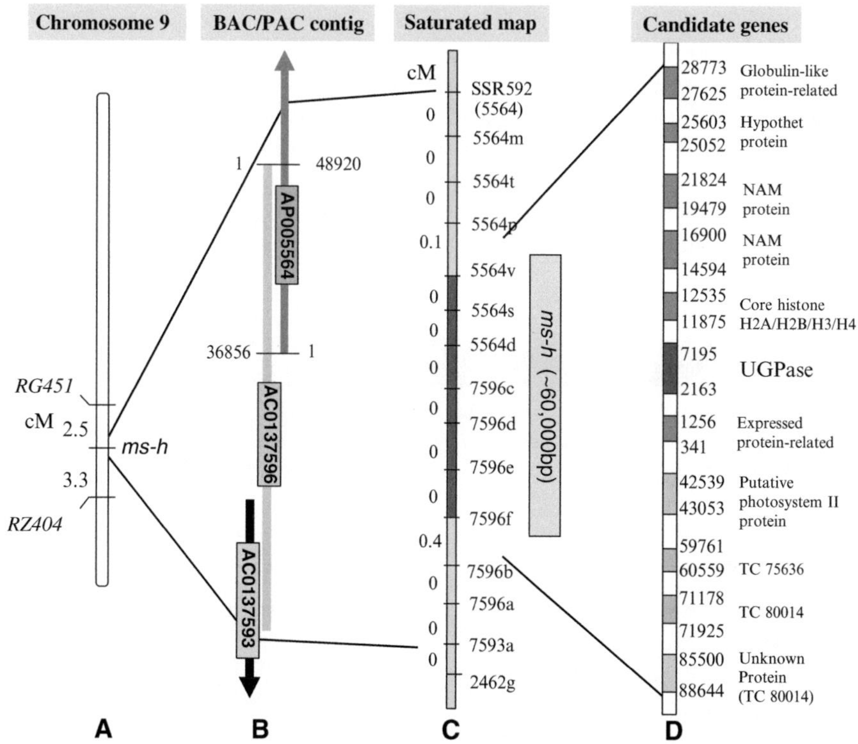

Fig. 8.15 Saturated map of the region containing the *ms-h* locus and candidate genes (Woo et al. 2008)

8.3.2.4 Candidate Gene Cloning and Mutation Point Identification

Candidate genes in the target interval based on fine-scale mapping are detected by *in silico* genome annotation (http://rgp.dna.affrc.go.jp, http://www.gramene.org). To identify the best candidate among these genes, all must be sequenced in both the wild type and mutant and then compared with their homologs. Once the mutation (e.g., insertion, deletion, or substitution) is detected and the best gene selected, it should be validated using phenotype-characterized F2 progeny DNA. Common methods include a comparison of PCR product size for insertion/deletion mutations and derived CAPS (dCAPS) marker analysis for point mutations.

In the *ms-h* gene mapping, 11 candidate genes were identified within the 60-kb target interval. As a result of comparative sequencing of all 11 gene candidates in the *ms-h* mutant and in the wild type (Hwacheong), a single point mutation from guanine to adenine leading to amino acid change in the *UGPase1* gene was identified. To further explore the association between this SNP in *UGPase1* and the male-sterile phenotype of the *ms-h* mutant, a dCAPS marker to detect the functional base substitution was designed and used to validate the gene in the F2 population.

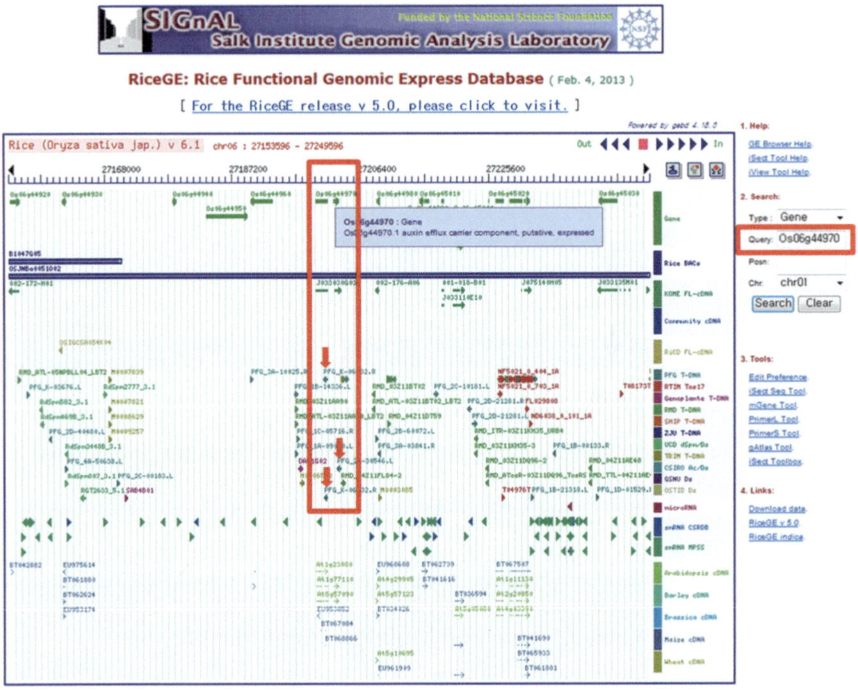

Fig. 8.16 T-DNA mutant line search and selection on the SALK web site

Fig. 8.17 (continued) morphology of a vector-transformed plant (*left*) and UGPase1-RNAi plant (*right*) at the heading stage, **(f)** I_2–KI staining of pollen grains from a vector-transformed plant at the heading stage, **(g)** I_2–KI staining of pollen grains from a UGPase1-RNAi plant at the heading stage, **(h)** Phenotype of Hwacheong *ms-h* mutants after ripening: plants containing empty vector (*left*) and complemented by the introduction of pUGP1COM (*right*), **(i)** Triple enlargement of part of photo **(h)**, **(j)** Panicles of a vector-transformed plant and the complemented plant at anthesis (*left*) and after ripening (*right*), **(k)** Flower and anther morphology of an empty vector-transformed plant (*left*) and the complemented plant (*right*) at the heading stage, **(l)** I_2–KI staining of pollen grains from a empty vector-transformed plant at heading, **(m)** I_2–KI staining of pollen grains from the complemented plant at heading. The scale bar corresponds to 100 μm (Woo et al. 2008)

Fig. 8.17 Transgene constructs and phenotypes of transgenic plants. (**a**) Schematic diagrams of the pUGP1RNAi and pUGP1COM constructs used for the complementation test, (**b**) Hwacheong plants after ripening containing the empty vector (*left*) and transformed by pUGP1RNAi (*right*), (**c**) Double enlargement of a part of photo (**b**), (**d**) Panicles of a vector-transformed plant and a UGPase1-RNAi plant at anthesis (*left*) and after ripening (right), (**e**) Flower and anther

8.3.2.5 Complementation Test

Finally, a complementation test should be conducted to ensure that a given cloned gene is the target. This test should confirm that overexpression of the cloned target gene in transformed mutant plants restores the wild phenotype. Moreover, exploiting double-stranded RNA-mediated interference to silence the target gene in wild-type plants and produce the mutant phenotype will be more convincing.

In the event of difficulty in transforming plants, mainly owing to problems with callus induction and regeneration, T-DNA mutant stock corresponding to the cloned gene from the SALK web site (http://signal.salk.edu) can be ordered and used for confirmation (Fig. 8.16).

Likewise, in a two-way confirmation test for the cloned *ms-h* gene, suppression of UGPase by introducing a *UGPase1*-RNAi construct in wild-type plants nearly eliminated seed set, similar to the *ms-h* mutant phenotype, whereas overexpression of *UGPase1* in *ms-h* mutant plants restored male fertility (Fig. 8.17; Woo et al. 2008).

Glossary

Antigen Any substance that causes the immune system to produce antibodies against it.

Callus Undifferentiated or unorganized mass of cells.

CentiMorgan (cM) The unit for measuring genetic linkage. It is defined as the distance between chromosome positions (loci or markers) for which the expected average number of intervening chromosomal crossovers in a single generation is 0.01. It is often used to infer distance along a chromosome.

Chromatography Method of separating compounds in a mixture based on movement speed.

Cis-acting element Region of non-coding DNA that binds a nucleotide sequence to regulate the transcription of nearby genes. These main regulate gene transcription.

Cleaved Amplified Polymorphic Sequences (CAPS) An amplified DNA fragment is digested by restriction enzymes. Sequence differences in the DNA fragments of different individuals can be revealed by electrophoresis.

Derived Cleaved Amplified Polymorphic Sequences (dCAPS) A modification of CAPS (or alternatively, PCR-RFLP) for detecting single nucleotide polymorphisms (SNPs). In a dCAPS assay, mismatches in PCR primers are used to create restriction endonuclease (RE)-sensitive polymorphisms based on the target mutation. dCAPS is useful for genotyping known mutations and for genetic mapping of isolated DNAs.

Detergent Surfactants, compounds with a hydrophobic (water repellent) and a hydrophilic (water loving) part.

Electroporation Method by which DNA is introduced into cells using a pulse of high voltage current.

F1 hybrid The first filial generation of offspring of distinctly different parental types.

Freund adjuvant emulsion A medium developed by J. Freund and used to cause a strong immune response in experimental animals.

Gene gun Method developed to introduce biological material into living cells. DNA is coated with 1 μm tungsten or gold microparticles and accelerated through the cell wall or cell membrane. This technique is useful to study gene regulation.

Genetic marker A gene or DNA sequence with a known location on a chromosome that can be used to identify individuals or species. It is an observable variation that may arise from mutation or alterations in the genomic loci. A genetic marker may be a short DNA sequence, such as the region near a single nucleotide polymorphism, or a long one, like a minisatellite.

Glycosylation Attachment of a glucose molecule to proteins that alter the protein's function by changing its structure. The complex is called a glycoprotein.

Heterozygous Having two different alleles at a given locus.

Homozygous Having identical alleles at a given locus.

Hydrophobic interaction Interaction between a polar solution such as water and a non-polar substance such as alcohol.

In vitro Latin term meaning an experiment performed in a controlled environment, such as a test tube or Petri dish, rather than in living cells.

Inclusion body Insoluble protein commonly formed by excess expression of a foreign protein in a host cell. Must be activated by folding (refolding) but loss of active protein occurs.

Isoelectric point (pI) The pH of a solution or dispersion at which the net charge on the molecules or colloidal particles (polyprotic acid) is zero.

Ligand A small molecule or ion that binds to a protein to form a complex.

Linkage The tendency of certain genes to be inherited together.

Multiple cloning sites (MCS) A genetically engineered collection of several different restriction enzyme cleavage sites that allow convenient insertion of DNA fragments into a plasmid.

Nucleoplasm A general term for the plasma within the nuclear membrane.

Pedigree A line of ancestors.

Peroxisome An organelle found in the liver and kidneys of vertebrate cells; in plant leaves and seeds; and in protozoa, yeast and fungi.

Post-translational modification A generic term for any modification of a protein that occurs after translation and release from ribosomes. Adding a carbohydrate or acid to the protein deforms it.

Promoter Particular sites on the DNA template that bind RNA polymerase to initiate transcription. A promoter has a regular common base sequence (consensus sequence). In prokaryotes, promoters present 10 base pairs before the pribnow box and 35 base pairs before a representative base sequence from the transcription initiation site TTGACA.

Reporter gene A gene used to label a target gene so that its expression level and intracellular location can be easily measured. LacZ, GFP, GUS, and luciferase are commonly used marker genes that can be seen easily.

Ribosome A complex of proteins and RNA that carries out translation.

Sequence Tagged Site (STS) A short (200 to 500 base pair) DNA sequence that has a single occurrence in the genome and whose location and base sequence are known.

Simple Sequence Repeat (SSR) The repeating sequences of 2–5 base pairs of DNA. This is also known as Microsatellites or Short Tandem Repeats (STRs).

Trans-acting element Proteins that normally bind to DNA sequences in different regions of the genome. The main function is to regulate transcription.

Vacuoles A membrane-bound organelle that is present in all plants. It has no common shape or size, and its structure varies according to the needs of the cell.

Vector A DNA molecule that carries foreign DNA into a host cell, replicates inside a bacterial (or yeast) cell, and produces many copies of itself. Also called a cloning vehicle.

References

Amsterdam A, Raanan C, Schreiber L et al (2012) Use of multiple biomarkers for the localization and characterization of colon cancer stem cells by indirect immunocytochemistry. Int J Oncol 41(1):285

Baron J (1984) Immunocytochemistry. Practical applications in pathology and biology: Edited by Julia M. Polak and Susan Van Noorden, J. Wright & Sons, Ltd, London. In. Anal Biochem. 140(1):303

Berggren KN, Schulenberg B, Lopez MF et al (2002) An improved formulation of SYPRO Ruby protein gel stain: comparison with the original formulation and with a ruthenium II tris (bathophenanthroline disulfonate) formulation. Proteomics 2(5):486–498

Bonifacino JS, Dell'Angelica EC (2001) Immunoprecipitation. Curr Protoc Immunol 8(3):1–28

Boyer PD, Chance B, Ernster L et al (1977) Oxidative phosphorylation and photophosphorylation. Annu Rev Biochem 46(1):955–966

Burnette WN (1981) Western blotting – electrophoretic transfer of proteins from sodium dodecyl sulfate-polyacrylamide gels to unmodified nitrocellulose and radiographic detection with antibody and radioiodinated protein-A. Anal Biochem 112(2):195–203

Chapman-Smith A, Cronan JE (1999) The enzymatic biotinylation of proteins: a post-translational modification of exceptional specificity. Trends Biochem Sci 24(9):359–363

Chomczynski P, Sacci N (1987) Single-step method of RNA isolation by acid guanidinium thiocyanate phenol chloroform extraction. Anal Biochem 162:156–159

Finley D, Chau V (1991) Ubiquitination. Annu Rev Cell Biol 7(1):25–69

Gerdes J, Schwab U, Lemke H et al (1983) Production of a mouse monoclonal-antibody reactive with a human nuclear antigen associated with cell-proliferation. Int J Cancer 31(1):13–20

Goldstein JI, Harvey Y (1975) Practical scanning electron microscopy. Plenum Press, New York, pp 107–108

Gruber CW, Cemazar M, Heras B et al (2006) Protein disulfide isomerase: the structure of oxidative folding. Trends Biochem Sci 31(8):455

Hannig G, Makrides SC (1998) Strategies for optimizing heterologous protein expression in *Escherichia coli*. Trends Biotechnol 16(2):54–60
Hart GW (1997) Dynamic O-linked glycosylation of nuclear and cytoskeletal proteins. Annu Rev Biochem 66(1):315–335
Haute GV (2003) Green fluorescent protein. In: Purification and structural analysis of macromolecules, pp 1–9
Hawes C (1991) Electron microscopy of plant cells. Academic, London
Hay RT (2005) SUMO: a history of modification. Mol Cell 18(1):1–12
Heijne G (1990) Signal peptides. eLS. doi:10.1038/npg.els.0005050
Isaacson T, Damasceno CMB, Saravanan RS et al (2006) Sample extraction techniques for enhanced proteomic analysis of plant tissues. Nat Protoc 1(2):769–774
Javois LC (1999) Immunocytochemical: methods and protocols. Springer, New York
Kerscher O, Felberbaum R, Hochstrasser M (2006) Modification of proteins by ubiquitin and ubiquitin-like proteins. Annu Rev Cell Dev Biol 22:159–180
Kim YJ, Lee HM, Wang Y et al (2013) Depletion of abundant plant RuBisCO protein using the protamine sulfate precipitation method. Proteomics 13(14):2176–2179
Kodama Y, Hu CD (2012) Bimolecular fluorescence complementation (BiFC): a 5-year update and future perspectives. Biotechniques 53(5):285–298
Koh HJ, Heu MH (1995) Agronomic characteristics of a mutant for genic male sterility-chalky endosperm and its utilization on F1 hybrid breeding system in rice. Korean J Crop Sci 40(6):684–696
Koh HJ, Son YH, Heu MH et al (1999) Molecular mapping of a new genic male-sterility gene causing chalky endosperm in rice (Oryza sativa L.). Euphytica 106:57–62
Krishnan HB, Natarajan SS (2009) A rapid method for depletion of Rubisco from soybean (Glycine max) leaf for proteomic analysis of lower abundance proteins. Phytochemistry 70(17):1958–1964
Li C, Zhou A, Sang T (2006) Rice domestication by reducing shattering. Science 311:1936–1939
Michelmore RW, Paran I, Kesseli KV (1991) Identification of markers linked to disease resistance genes by bulked segregant analysis: a rapid method to detect markers in specific genomic regions by using segregating populations. Proc Natl Acad Sci U S A 88:9828–9832
Orsatti L, Forte E, Tomei L et al (2009) 2-D Difference in gel electrophoresis combined with Pro-Q Diamond staining: a successful approach for the identification of kinase/phosphatase targets. Electrophoresis 30(14):2469–2476
Sambrook J, Fritsch EF, Maniatis T (1989) Molecular cloning, a laboratory manual. Cold Spring Harbor Laboratory, Cold Spring Harbor, New York
Sengupta DJ, Wickens M, Fields S (1999) Identification of RNAs that bind to a specific protein using the yeast three-hybrid system. RNA 5(4):596–601
Smejkal GB, Lazarev A (2010) Separation methods in proteomics. CRC Press, Boca Raton
Tsien RY (1998) The green fluorescent protein. Annu Rev Biochem 67:509–544
Wold F (1981) *In vivo* chemical modification of proteins (post-translational modification). Annu Rev Biochem 50(1):783–814
Woo MO, Ham TH, Ji HS et al (2008) Inactivation of the UGPase1 gene causes genic male sterility and endosperm chalkiness in rice (Oryza sativa L.). Plant J 54:190–204

Chapter 9
Plant Transformation Methods and Applications

Young Hee Joung, Pil-Son Choi, Suk-Yoon Kwon, and Chee Hark Harn

Abstract Conventional plant breeding uses crossing, mutagenesis, and somatic hybridization for genome modification to improve crop traits by introducing new beneficial alleles from crossable species. However, because of crossing barriers and linkage drag, conventional plant breeding methods are time-consuming and require several generations of breeding and selection. To feed the several billion people living on this planet, the main aim of breeders is to increase agricultural production. Hence, new technologies need to be developed to accelerate breeding through improving genotyping and phenotyping methods. Molecular breeding technologies and their applications are discussed in some of the previous chapters. In this chapter, genetic modification technologies are introduced, including the protocols that have been used for the genetic transformation for ten major crops. Two other genetic modification methods are also introduced: (1) cisgenesis by which only beneficial alleles from crossable species are transferred into a recipient plant to enhance the use of existing gene alleles; and (2) reverse breeding to increase the available genetic diversity in breeding germplasm by blocking chromosome recombination during cell division.

Author contributed equally with all other contributors.

Y.H. Joung
School of Biological Sciences & Technology, Chonnam National University, Gwangju, Republic of Korea
e-mail: yhjoung@jnu.ac.kr

P.-S. Choi
Oriental Pharmaceutical Development, Nambu University, Gwangju, Republic of Korea
e-mail: cps6546@hanmail.net

S.-Y. Kwon
Korea Research Institute of Bioscience and Biotechnology, Daejeon, Republic of Korea
e-mail: sykwon@kribb.re.kr

C.H. Harn (✉)
Nongwoo Bio Co., Yeoju, Republic of Korea
e-mail: chharn@nongwoobio.co.kr

H.-J. Koh et al. (eds.), *Current Technologies in Plant Molecular Breeding*,
DOI 10.1007/978-94-017-9996-6_9

9.1 Principles of Gene Transfer

9.1.1 Transformation Using Agrobacterium

9.1.1.1 Background

Over the last 30 years, DNA transfer has played a dominant role in the production of transgenic plants and in gene function analysis studies. Most transgenic plants were obtained using *Agrobacterium*-mediated approaches. *Agrobacterium tumefaciens* is a plant pathogenic soil bacterium that infects wounded cells and causes a plant disease called crown galls in many dicotyledons and some monocotyledons (Zhu et al. 2003). Chilton et al. (1977) devised a plant transformation method using T-DNA of the *A. tumefaciens* Ti-plasmid that is stably integrated into the genome of plant cells where it modifies the genomic characteristics. Plant transformation methods using *Agrobacterium* are more effective and economical than other methods. Therefore, it is currently the most commonly used method because it provides a simple way of inserting a foreign gene into a plant genome. The recent development of various binary vector systems, selection markers, and advances in plant tissue-culture techniques not only have improved the efficiency of the plant transformation method using *Agrobacterium*, but also have enlarged the range of *Agrobacterium*-applicable hosts. Indeed, the transformation efficiency of plants has been far enhanced by the increasing infectiousness of *Agrobacterium*. To understand the mutual mechanisms between *Agrobacterium* and their plant hosts, it is important to identify the transformation-related regulatory factors involved and to determine their functions. The current understanding of these mechanisms is described in this chapter (Nandakumar et al. 2011; Zhao et al. 2014).

9.1.1.2 T-DNA Insertion by *Agrobacterium*

Attachment of *Agrobacterium* to Plant Cells When plant tissues are wounded, they exude organic acids, amino acids, saccharides, and other small molecules that can invoke chemotaxis in *Agrobacterium* and boost the secretion of acetylated acidic polysaccharides. Subsequently, *Agrobacterium* cells adhere onto the surface of plant cells. The attachment process in which cellulose fibers are synthesized and secreted is regulated by *attR* and *cel* genes in the *Agrobacterium* genome, resulting in solid adhesion of the bacteria to the surface of plant cells. The attachment process of *Agrobacterium* is known to be related indirectly with *chvA, chvB,* and *pscA* genes in the bacterial genome, as well as with arabinogalactan proteins, cellulose synthase-like proteins, and cell wall proteins in host plants (Zhu et al. 2003; Zhao et al. 2014).

Activation of Virulence (*vir*) Genes To recognize plant cells, *Agrobacterium* uses a two-way signaling system, which consists of the VirA protein and the VirG:VirA protein pair that directly perceive phenolic compounds like acetosyringone secreted from wounded plant cells. These compounds induce autophosphorylation of a VirA domain. The phosphate group of the VirA is then transferred to VirG, which binds to the enhancement elements of *vir* genes in the Ti-plasmid and regulates their transcription (Zhu et al. 2003; Zhao et al. 2014).

T-DNA Processing and T-Strand Formation VirD2 is a nuclear localization signal binding protein that is covalently bound to the 5′ end of the T-strand. VirD2 recognizes the left and right borders of T-DNA sequences, makes a nick between the third and fourth bases of antisense T-DNA border sequences, and forms covalent bonds at the 5′ end of single-stranded DNA to form the T-strand. The VirD1 protein can change the structure of T-DNA, which alleviates the tension and helps to stimulate T-strand formation by VirD2 (Filichkin and Gelvin 1993; Zhao et al. 2014).

Transport of T-strand and Vir Proteins and T-complex Formation VirD2 protein transfers the T-strand into plant cells through a VirB channel, which consists of VirB and VirD4 proteins. The VirB channel is a filamentous pilus, which connects the *Agrobacterium* and host plant cell that functions as a transporter complex through cell membranes. VirE2, which is transferred into the cytoplasm of the infected plant cell, combines with the single stranded T-DNA, to form a T-strand/protein polymer called the T-complex. The T-complex protects the T-strand from the deoxyribonucleases that exist in the plant cytoplasm and is an ideal structure to transport the large T-strand to the nucleus of a plant cell (Zhao et al. 2014).

T-complex Transfer to a Nucleus of a Plant Cell The T-complex is larger than the nuclear pores in the nuclear membrane of plant cells and is transferred to the plant nucleus by active transport. VirE2, which surrounds the T-complex, and plant-derived importin-α proteins, which specifically recognize nuclear localization sequences (NLS) in the VirD2 protein play important roles in the active transport. In *Arabidopsis thaliana*, VirD2 has been shown to conjugate specifically with NLS of AtKAPα, a member of the karyopherin-α family, and is then transferred to the plant nucleus (Zhao et al. 2014). VirE2 is essential for T-DNA transport into the plant cell nucleus. VirE2 does not combine with AtKAPα, but with the plant-derived proteins, VIP1 (VirE2 interacting protein) and VIP2, then its transfer to the nucleus is mediated by karyopherin-α. The over-expression of the *VIP1* gene particularly increases the import of T-DNA to the nucleus, and as a result the transformation efficiency is correspondingly enhanced (Tzfira and Citovsky 2000; Zhao et al. 2014).

Insertion of T-DNA into Plant Genomes VirE2 is not involved in insertion of T-DNA, but it is needed to protect the T-DNA from plant deoxyribonucleases. VirE2 secures the integrity of the T-DNA during its transportation from the cytoplasm to the nucleus, and regulates its integration into the plant chromosome. Before the T-DNA is inserted into the plant chromosome, the VirE2 surrounding the T-strand has to be removed. VirF is a defining factor of host-specificity in *Agrobacterium*. It functions as an F-box protein and shows target protein specificity in the proteolysis-related Skp1p–cullin–F-box protein (SCF) complex. When transferred into plant

Table 9.1 Functions of virulence (*Vir*) genes in T-strand insertion into plant genomes

Gene	Functions
VirA, VirG	Perception of phenolic compounds from plant wounds/induction of virulence (*Vir*) gene expression
VirD2	Endonuclease cutting T-DNA border to initiate T-strand synthesis
VirD1	DNA topoisomerase processing T-DNA
VirD2	Attached to 5′ of T-strand/ formation of T-DNA complex/ transport of T-DNA complex through nuclear pores
VirC	Combined to overdrive region to induce effective T-strand synthesis
VirE2	Single-strand DNA binding protein protecting T-strand from nuclease
VirE1	Plays the role of chaperone to stabilize VirE2 in *Agrobacterium*
VirB, VirD4	Components of T-DNA transference

cells, VirF was found to be involved in the proteolysis of VirE2 and VIP1 in the nucleus. VirD2 is known to be related to the precise insertion of T-DNA into plant genomes. Because T-DNA insertion into plant genomes is an illegitimate recombination, the host DNA repair and recombination-related genes are expected to influence the insertion of the T-DNA. In recent studies, *rat5* (resistance to *Agrobacterium* transformation) mutants were isolated and analyzed, and proteins like histone H2A, which are related to chromosome structure regulation, were found to affect directly the insertion of T-DNA into chromosomes (Mysore et al. 2000; Zhao et al. 2014) (Table 9.1).

9.1.2 Direct Gene Transfer

9.1.2.1 Particle Bombardment

Particle bombardment systems are used to introduce gold powder or tungsten-coated DNA directly into plant genomes by acceleration. This technique was applied to gramineous crops, and large numbers of transformants were developed. Particularly, insect-resistant corns in which a *Bacillus thuringiensis* (Bt) toxin gene was introduced were commercialized and the technique was used on a commercial scale (Gahakwa et al. 2000). Before a foreign gene was introduced into plant cells, the DNA was coated with tungsten or gold, and then particle bombardment was performed. As a direct gene-introduction technique, shooting the foreign gene into target plant tissues at high speed was a very important step. Regardless of the insertion of a foreign gene into a plant genome, this method has wide applicability because it can stabilize transformant production for studies on, for example, transient expression, pathogenicity, and defense mechanisms of viruses. This technique is different from the *Agrobacterium*-mediated method because it is free from biological limitations, is applicable to various plant materials, and can introduce multiple genes simultaneously. Success or failure for this technique depends on the optimization of transformation systems such as their form, preparation, acceleration of particles, selection of target tissues, and balance between number and size of particles, damage degree of plant tissue, and quantity of DNA transferred into target cells. If the quantity of DNA is low, the efficiency of transformation is also low. However, if the quantity of DNA is too high, the number of introduced genes will be so high that recombination phenomenon is evoked, resulting in gene silencing of the introduced gene. The bombardment procedure is described below.

Micro-carrier Coating Gold or tungsten particles are normally used as micro-carriers. The particles are pre-processed by sterilizing with ethanol and then washing with sterilized water. Plasmid DNA is coated with micro-carrier in 2.5 M $CaCl_2$ and 0.1 M spermidine solution, and cleared with alcohol. Micro-carrier coated with DNA is loaded onto a membrane and the alcohol is evaporated.

Target Tissues of Plant Materials Mature or immature embryos, embryogenic calli, and leaf discs are mostly used as target tissues. The plant tissue is prepared and preprocessed by placing it in the center of media with high osmotic density for about 4 h. (The time is varied according to the plant tissue being used.) High concentrations of maltose or mannitol are mainly used to obtain high osmotic densities and to reduce the damage to plant tissue caused by particle impact.

Conditions of Bombardment Prepared plant tissues are placed in a vacuum chamber (chamber pressure 27 mmHg) and the distance between the micro-carrier and the stopping plate is adjusted to about 13 cm. Gold- or tungsten-coated DNA is dispensed onto a micro-carrier membrane and transferred into a bombardment machine under vacuum. At 1,100 psi, micro-carriers are shot at the plant target tissue.

Selection of Transformed Plants and Regeneration The day after bombardment, the plant target tissues are cultivated on organogenesis or embryogenesis media according to the tissue type. To obtain young plantlets, shoots that survive on selection media are then transferred to root-induction media. Plantlets are passed through acclimation procedures to become mature plants. Leaves or tissues of the mature plants are subjected to molecular analysis to check whether they have been transformed.

9.1.2.2 Polyethylene Glycol (PEG)-Mediated Transformation

Foreign DNA can be introduced to protoplasts (cells that lack plant cell walls) by adding a di-cation (Ca^{2+}) and PEG, which destabilize plant protoplasts and allow the foreign DNA to enter the protoplasts. The foreign DNA then migrates to the plant nucleus and is inserted into the genome. Because the manipulation and cultivation of plant protoplasts is difficult, PEG-mediated transformation is not an efficient transformation method. In addition, the foreign DNA in the target tissue is easily broken down during the transformation process, and recombination events always occur. Because of these problems, this method is used only in some special cases.

9.1.2.3 Electroporation-Mediated Transformation

Electric shocks can be used to introduce foreign genes into plant cells or protoplasts. Plant tissues are cultured in a buffer mixture containing foreign DNA, and electric shocks are applied. As a result, DNA is introduced through pores in the protoplasts and can then be inserted into the plant genome. Protoplasts were usually used in the early studies, but cells, callus tissues, immature embryos, and immature inflorescence have also been used. This technique is mainly applied to gramineous crops such as rice, wheat, and corn, and has a strong point that tissues from recalcitrant crops can be cultured and subjected to electric shocks. Transformation efficiency varies depending on the condition of the plant tissues, field strength, and treatment of the target tissues. Linear DNA rather than supercoiled DNA, and buffer mixture containing spermidine, which induce condensation of DNA, can increase transformation efficiency.

9.1.2.4 Transformation Using Silicon Carbide Fibers

Transformation using silicon carbide fibers is a simple technique that does not need any particular equipment. Plant tissues such as suspended cultured cells, embryos, and embryogenic calli are stirred with DNA in a buffer mixture containing silicon carbide fibers. The silicon carbide fibers penetrate plant cell walls and cell membranes and consequently foreign genes can flow into the cells. Transformation efficiency is affected by the size of fibers, stirring conditions, shape of the container, plant materials, and characteristics of the plant cells (thickness of the cell wall). This technique is simple, inexpensive, and applicable to many kinds of plant tissues. However, transformation efficiency is quite low, and careful operation is required when using this technique because of the riskiness of the silicon carbide fibers. This method was successfully applied to soft corn calli (Frame et al. 2000), but for hard corn calli, the transformation frequency was low and has rarely been used.

9.1.2.5 Transformation by Micro-injection

Glass micro-capillary injection pipettes are used to introduce foreign DNA into the nucleus or cytoplasm of cells. The micro-injection method has been used for animal cells, which have large cell sizes, but has not been applied successfully in plant cells because the cell wall, which consists of polysaccharides like lignin and cellulose, hinders the insertion of glass micro-capillary pipettes into the cells. Foreign genes can be introduced into plant protoplasts by micro-injection; however, preparing and manipulating protoplasts is a difficult task, and is costly and time consuming. Nevertheless, a major characteristic of the micro-injection method is that it can be used to introduce chromosomes into plant cells in addition to plasmid DNA. Potentially, this technique could be used in functional studies of plant cells and physiological research of plastids.

9.2 Transformation Protocols Applied to Crops

9.2.1 *Rice (Oryza sativa)*

9.2.1.1 Background

The direct and indirect introduction of DNA into protoplasts mediated by particle bombardment (also called gene gun) and *Agrobacterium* has been used for rice transformation. But *Agrobacterium*-mediated transformation is now the favored method because complicated processes for plant regeneration from protoplasts can be skipped, and low copy numbers and the ease of foreign gene fixation reduce gene silencing. However, the genotype-dependent problem has not been resolved, which means that *Agrobacterium*-mediated transformation is confined to a few specific cultivars. A commonly used protocol for rice *Agrobacterium*-mediated transformation is described in Sect. 9.2.1.2 below.

9.2.1.2 *Agrobacterium*-Mediated Transformation

① Plant material: Brown rice without seedcoats are prepared for transformation. The naked seeds are sterilized once in ethanol for 10 min, then in 2–3 % sodium hypochlorite solution for 10 min three times, and washed with sterilized water more than three times.

② Callus induction: 10–20 sterilized seeds are placed on medium containing 20–25 mL of callus induction medium (2N6+2,4-D 2-mg/L) and cultured at 27 °C in dark for 3–4 weeks.

③ Callus selection and proliferation: Pale yellow-colored calli are selected from various types of calli after the callus induction stage and preprocessed on 2N6 medium for 3 days to stabilize the calli.

④ Preparation of *Agrobacterium* and transformation: *Agrobacterium*-carrying expression vectors are spread on AB or YEP medium and cultured at 27 °C in dark for 3–5 days. Although various strains of *Agrobacterium* could be used, strains LBA4404 and EHA101 are widely used. Glycerol stocks of *Agrobacterium* carrying preferred plasmids are made in 0.5- or 1.5-mL Eppendorf tubes and stored long-term at −80 °C. For transformation, *Agrobacterium* is spread directly onto an AB medium plate. An *Agrobacterium* colony is picked using a sterilized loop or spatula, inoculated in 15-mL AAM (100 mM acetosyringone supplemented) medium contained in a 50-mL falcon tube, and grown in dark for 2 days. The 2 day-grown *Agrobacterium* suspension is diluted to 1.5–3.0 of the OD_{650} value for final suspension and co-cultivation.

⑤ Co-cultivation of *Agrobacterium*: Pale yellow calli 1–2 mm in diameter are placed in a petri-dish with 20 mL of *Agrobacterium* suspension, and shaken well for 30 min. The suspension is removed, and the infected calli are placed onto sterilized filter paper to remove the rest of the suspension. They are then transferred onto 2N6-AS medium, and co-cultured in dark for 3 days.

⑥ Selection of transformed calli: Three days later, the calli are washed three times with cefotaxime solution (250 mg/L) to remove *Agrobacterium*, and completely dried on sterilized filter paper. The calli are placed on callus selection medium (2N6-CP) and resistant calli are obtained after 2–3 weeks of cultivation at 27 °C in dark. A selection marker gene, *hpt* (hygromycin phosphotransferase) or *bar* (bialaphos resistant) is usually used and the applicable concentration for selection is 50 and 6 mg/L, respectively. Well-grown callus colonies are sub-cultured 2–3 times on selection medium, then carefully selected and proliferated.

⑦ Plant regeneration from selected calli: Surviving calli are selected on MSR-CP medium containing 3-mg/L PPT (when the *bar* gene is the selection marker) at 25 °C in light for 4 weeks. In general, a non-transformed callus will take about 4 weeks to generate plants, but a transformed callus will take 1–2 weeks more. Shoots differentiated from calli are transferred on MS medium (the basal medium for tissue-culture of crops) without any plant hormones to induce roots. When the regenerated plants are acclimated in 1 % HYPONEX solution for 7–10 days, they are transplanted into pots for seeds. The plants are analyzed by PCR, southern blot, and northern blot.

9.2.2 Barley (Hordeum vulgare)

9.2.2.1 Background

Barley is used worldwide for food, beer, and forage. Barley, which contains β-glucan and vitamin B1, has been recognized as a healthy food. However, compared with rice, corn, and other monocotyledonous crop plants, transformation of barley is known to be recalcitrant. Barley plants can be regenerated easily from the embryonic disk of immature embryos. Various transformation techniques such as particle bombardment, PEG-mediated, and electroporation-mediated have been used in barley. *Agrobacterium*-mediated transformation was attempted in this crop (Tingay et al. 1997); however, almost no regeneration was achieved from anthers or immature embryos of Korean barley varieties. Barley transformation using the particle bombardment and *Agrobacterium*-mediated methods are described in Sects. 9.2.2.2 and 9.2.2.3 below.

9.2.2.2 Particle Bombardment Method

① Sterilization: Immature seeds (0.5 mm) from mature plants are submerged in 70 % ethanol for 1 min, surface sterilized with 1.25 % NaOCl solution, washed three times with aseptic water for 10 min, and dried on sterilized filter papers.

② Pre-cultivation: Inside a sterile-air hood, the sterilized immature seeds are peeled off, and the immature embryos are cultured on PL medium containing 2.5-mg/L 2,4-D. At this stage, embryonic disks cut from the immature embryos are collected on the center of the medium, 40 embryonic disks per plate, and culture at 25 ± 1 °C in dark for 3 days.

③ Preparation of M10 particle bombardment: 66.7-mg tungsten of 0.7 μm in size is prepared in a 1-mL tube using micro-carrier, mixed with 70 % ethanol to sterilize. Then the supernatant is discarded, and the metal powder is washed two times with aseptic water. Finally, the particles are put into a test tube containing 1 mL of 50 % glycerol and used for 2 weeks by storing in a refrigerator.

④ DNA coating: 45-μL M10 prepared in step ③, 10-μL DNA, 50-μl 2.5 M $CaCl_2$, and 20-μL 100 mM spermidine are added into a 1-mL tube orderly, and vortexed 2–3 min followed by centrifugation. The supernatant is poured off, and the pellet is washed gently, first with 140-μL 70 % ethanol and then with 140-μL 100 % ethanol, without any disturbance. Finally, 240 μL of 100 % ethanol is added, and the solution is vortexed, and 5 μL is dispensed for each bombardment.

⑤ Callus induction: The micro-carrier coated with the gene of interest is shot into pre-cultured immature barley embryos by the particle bombardment method (see step ④). One day later, the bombarded immature embryos are transferred onto PL medium and cultured at 25 ± 1 °C in dark for callus induction.

⑥ Selection: 2–3 days after particle bombardment, the transformed immature embryos are transferred to MS medium with 2–2.5-mg/L 2,4-D (selection marker is treated moderately according to the vector used) at 25 ± 1 °C in dark for 3–4 weeks.

⑦ Regeneration and growth of plants: Calli that grew well on the selection medium show resistance and are proliferated and transferred onto MS medium containing 1-mg/L BAP for shoot induction. To grow complete plantlets, the primary shoots are then transferred to 1/2 MS medium and cultivated at 25 ± 1 °C with a 16-h photoperiod. After 3 weeks, the plantlets are moved to pots to mature.

9.2.2.3 *Agrobacterium*-Mediated Transformation

① Preparation of *Agrobacterium* solution: One day before co-cultivation, 200 μL of 2-day-cultured bacterial suspension is added to a 5-mL falcon tube and suspended at 28 °C, 150 rpm for 1 day. When the OD_{600} value reaches 1.4–1.7, the culture is used as the final co-cultivation solution.

② Preparation of immature embryos: Immature embryos (1.5 mm) are isolated from well-grown barley grains, as is done in the particle bombardment method. The seeds are kept as fresh as possible in a sealed tube or petri-dish. The immature embryos are removed from the immature seeds using sharp scalpels, and placed with the embryonic disks face upward.

③ Co-cultivation: 200–500 μL of *Agrobacterium* solution is poured onto the immature embryos in each plate and the plates are incubated for 20–30 min. After incubation, the immature embryos are cultured at 28 °C in dark for 2–3 days.

④ Selection of callus: The immature embryos are cultured or sub-cultured on BCI medium (Wan and Lemaux 1994) containing selection agent for three times at 2-week intervals, until a callus begins to grow.

⑤ Plant regeneration: Prosperously growing calli are transferred to FHG medium (Olsen 1987) containing 0.5-mg/L BAP, 1.0 mg/L IAA, and 50 mg/L hygromycin, cultured at 25 °C in light, and then sub-cultured every 2 weeks for four times to induce shoots.

⑥ Induction of whole plants: Regenerating shoots induced on the selection medium are moved to $^1/_2$ BCI rooting medium (Shrawat et al. 2007) at 25 °C in light for 2 weeks. Plantlets are transplanted into soil after suitable acclimating to obtain mature plants. Leaf disks are collected from the surviving plants, and checked for insertion and expression of target genes by PCR, southern blot, and northern blot.

9.2.3 Corn (Zea mays)

9.2.3.1 Background

Corn is the most cultivated crop worldwide, greater than rice and wheat, and is an important forage and diet crop. Corn has also been used to produce ethanol as a bioenergy source. Thereby, the development of new corn cultivars using molecular breeding methods is actively in process. In early studies of corn, transformation was done by both *Agrobacterium*-mediated and particle bombardment methods. Later, researchers preferred to use *Agrobacterium*-mediated transformation methods on immature embryos or embryogenic calli because of various advantages. This strategy applied to explants of immature embryos and embryogenic calli of corn are described in Sect. 9.2.3.2 below (Carbone 2013).

9.2.3.2 *Agrobacterium*-Mediated Transformation

① Preparation of immature embryos and embryogenic calli: Hi-II cultivar plants, which have been proved to be the best choice for transformation, are planted in a greenhouse. During the flowering stage, pollen is collected to pollinate the ears at 9–10 am. Nine to 10 days post-pollination, immature embryos of 2 mm in size are used as the material. Freshly isolated immature embryos are inoculated on MS medium containing MS salt, Eriksson vitamins, sucrose, MES, L-proline, myo-inositol, casamino acid, BBL agar, $AgNO_3$, thiamine-HCl, and 2,4-D (pH 5.8) to induce embryogenic calli. The selected embryogenic calli are sub-cultured on the same medium to proliferate type II calli. Immature embryos and embryogenic calli are used for *Agrobacterium*-mediated transformation.

② Inoculation and co-cultivation: Immature embryos or embryogenic calli are inoculated in an *Agrobacterium* suspension (C58CI) for 5 min, then the suspension is removed with pipettes, and the tissues are dried on filter paper. The infected immature embryos or embryogenic embryos are transferred to solid co-cultivation medium and cultured at 24 °C in dark for 2 days.

③ Selection: After co-cultivation, the immature embryos or embryogenic calli are activated on delay medium without selectable agent for 6–7 days. For immature embryos, the embryo axes are removed before the transfer. The embryonic calli are transferred directly to selection medium. Prosperously growing calli are selected while sub-culturing every 2 weeks for four times.

④ Induction of somatic embryos and regeneration of plants: Prosperously growing calli are transferred from the selection medium to 1st regeneration medium for 7 days in dark to induce somatic embryos. Then, the somatic embryos are moved to 2nd regeneration medium in light to obtain whole plantlets. Acclimated plantlets are transplanted into soil and cultivated in a greenhouse to obtain seeds for the next generation of corn.

9.2.4 Wheat (Triticum aestivum)

9.2.4.1 Background

The study of *Agrobacterium*-mediated transformation in wheat started in 1990, and some kanamycin-resistant transgenic wheat plants were obtained. For transformation, mainly immature embryos known for their high totipotency are used. *Agrobacterium*-mediated transformation as applied to wheat is described in Sect. 9.2.4.2 below (Zhou et al. 1995; Cheng et al. 1997; Zhang et al. 2014).

9.2.4.2 *Agrobacterium*-Mediated Transformation

① Preparation of immature embryos and embryogenic calli: Wheat variety 'Bobwhite' (highly recommended for its high regeneration rate) is sowed in a greenhouse. About 14 days post-pollination, immature grains of wheat are sampled, surface-sterilized with 70 % ethanol for more than three times, and peeled off on a clean bench. The immature embryos are used as materials for *Agrobacterium*-mediated transformation. Embryogenic calli are cultured and selected from the immature embryos on CM4C medium for 10–25 days in the dark. Then, the embryogenic calli are chopped into 2-mm pieces and used for co-cultivation.

② Preparation of *Agrobacterium* suspension: The C58 strain harboring a vector carrying *nptII* selectable marker and *β-glucuronidase* (GUS) reporter gene is applied. The strain is inoculated in LB medium with appropriate antibiotics in shaker, and used for co-cultivation when the OD_{660} reaches 1–2.

③ Co-cultivation: *Agrobacterium* suspension is added to the prepared wheat immature embryos or embryogenic calli at 23–24 °C in dark for 3 h. After inoculation, the suspension is removed with pipettes and the infected tissues are transferred to aseptic filter paper to completely remove the suspension. The inoculated immature embryos or embryogenic calli are transferred onto MS medium with 10-g/L glucose and 200-μM acetosyringone or CM4C medium, and co-cultivate at 24–26 °C in the dark for 2–3 days.

④ Selection: After co-cultivation, the immature embryos or embryogenic calli are cultured by resting on CM4C solid medium containing 250-mg/L carbenicillin for 2–5 days. Then, newly growing calli are selected and induced on CM4C medium containing 25-mg/L G418 and 250-mg/L carbenicillin.

⑤ Transformant induction: Carefully selected embryogenic calli are transferred onto 1st regeneration medium of MMS0.2C containing 1.95-g/L MES, 0.2-mg/L 2,4-D, 100-mg/L ascorbic acid, 40-g/L maltose, 2-g/L gelrite, 25-mg/L G418, and 250-mg/L carbenicillin as well as MS salts and vitamins to induce green shoots. Next, young shoots are moved to the 2nd regeneration medium of MMS0C, which is the same as the 1st regeneration medium except that it does not contain 2,4-D. Plants are moved into soil when they are more than 3 cm in height.

⑥ Transformants check: Young leaves are taken from the healthy plants developed on 2nd regeneration medium and judged for transformants by GUS assay.

9.2.5 Onion (*Allium cepa*)

9.2.5.1 Background

Onion is the fifth most widely grown vegetable in the world, and its global yield has gradually increased. The recognition of onion as an important health food plus the development of processed foods, many of which contain onion, have increased the consumption of onion. To our knowledge, it may take more than 10 years to breed a new onion variety, longer than any other crop. Therefore, the development of new breeding technologies for onion is in high demand. Several research teams have used biotechnology approaches to try to improve onion varieties. For example, the development of a regeneration system and insect-resistant transformants for onion has been reported by Colin Eady (1995), while a transgenic onion with an insect-resistant gene was successfully developed by Holland (Zheng et al. 2001). However, more research is still needed to achieve stable gene expression, progeny production, and transformant tests. Recently, *Agrobacterium*-mediated transformation has been used in onion as described in Sect. 9.2.5.2 below (Aswath et al. 2007).

9.2.5.2 *Agrobacterium*-Mediated Transformation

① Preparation of onion materials: Onion seeds are submerged in 70 % ethanol for 30 s, surface-sterilized in 2 % sodium hypochlorite solution, washed with aseptic water more than twice, and dried on filter paper. To isolate the mature embryos more easily, the onion seeds are submerged in aseptic water for about 24 h in a refrigerator before sterilization.

② Callus induction: MS basal medium supplemented with 5-μM 2,4-D is used for callus induction from the mature embryos of onion. After cultivating the mature embryos for 6 weeks, hard and yellowish embryogenic calli are selected, and sub-cultured on the same medium for proliferation. Finely chopped embryogenic calli are used for co-cultivation with *Agrobacterium*.

③ Preparation of *Agrobacterium* suspension: *Agrobacterium* with transformed expression vector is spread on YEP medium and cultured at 27 °C in dark for 3–5 days. Various strains of *A. tumefaciens* like LBA4404 and EHA101 have been used for onion. Even though different selection agents could be applied according to the expression vectors that are used, in this protocol, mannose is used as the selection chemical. To prepare the *Agrobacterium* suspension, an *Agrobacterium* colony is inoculated using sterilized loops in YEP liquid medium supplemented with appropriate antibiotics, and cultured for 2 days in a shaker until the OD_{650} reaches 1.5.

④ Co-cultivation: Embryogenic callus clumps of onion are co-cultivated with the prepared *Agrobacterium* suspension for 30 min. After infection, the cultures are centrifuged briefly, and supernatant is removed completely by pipette and dried on aseptic filter paper. Then, the infected embryogenic calli are cultured on MS basal medium containing 5-μM 2,4-D, 20-g/L sucrose, and 100 μM acetosyringone at 28 °C in dark for 3 days.

⑤ Selection and induction of somatic embryos: After co-cultivation, embryogenic calli are selected on MS medium containing 5-μM 2,4-D, 10-g/L sucrose, 10-g/L mannose, and 300-mg/L cefotaxime, and sub-cultured every 2–3 weeks for proliferation. To induce somatic embryos from well-grown embryogenic calli, the embryogenic calli that show mannose resistance are cultured on the medium with 5-μM kinetin, 1-μM Abscisic acid, and 10-g/L mannose for more than 3 weeks in light.

⑥ Plant regeneration: Soft and green calli that emerge on the surface of the embryogenic calli are transferred onto the next selection medium consisting of MS basal components and 10-g/L mannose but no hormones for further development, and followed by the induction of germination on 1/2 MS medium containing 10-g/L mannose. Acclimated plantlets are moved to soil. Genomic DNA is isolated from the leaves of the surviving putative transgenic plants, and identified by PCR, southern blot, or northern blot.

9.2.6 Soybean (Glycine max)

9.2.6.1 Background

Both *Agrobacterium*-mediated transformation and particle bombardment strategies have been used for soybean. In the early days, mainly particle bombardment was used with somatic embryos, and genetically modified soybeans with herbicide-resistance and insect-resistance were developed using this method. However, the copy number of the inserted genes in these transgenic plants was high and the stability of the inserted genes was disturbed by gene silencing and rearrangement in the progeny (Vaucheret et al. 1998; Dai et al. 2001). Also, mainly somatic embryos were used, and to obtain somatic embryos, immature seeds are needed; therefore, securing the right material all year-round was difficult. Further, genotypes from which somatic embryos can be isolated are limited.

Recently a cotyledon node-based *Agrobacterium*-mediated transformation method has been reported (Fig. 9.1). It is known that cotyledonary nodes, the portion that includes tissues of the growing point, are not appropriate targets for transformation in other crops. If the growing point is included, effective selection is difficult because of the rapid growing rate, and the frequency of chimera plants generation is very high when differentiated cells are present. However, in soybeans, a portion that includes the growing point is used as a target of the transformation, because the extent of responsiveness of other tissues has been found to be very low, and direct organogenesis scarcely occurred (Zhang et al. 1999). To enhance the efficiency of the *Agrobacterium*-mediated transformation method for soybean, various treatments have been added to the primary method. The most infective treatment is the addition of an antioxidant such as L-cysteine, sodium thiosulfate, and DTT. The effect of antioxidant addition has been elucidated in other plants including grape, sugarcane, and rice. It has been reported that the optimum composition and concentration of antioxidant in soybean suppresses the activity of two oxidases, polyphenol oxidase (PPO) and pyranose oxidase (POD), inducing browning of the inside of plant cells, consequently contributing to the formation of a primary leaf and an increase of transformation efficiency (Olhoft and Somers 2001; Olhoft et al. 2003). Another treatment that has been used to increase the transformation rate in soybean

Fig. 9.1 Scheme for genetic transformation of soybean (*Glycine max* (L.) Merrill) cotyledonary nodes

is cutting the explants at an exact point. In soybean transformation experiments that had high efficiency (more than 10 % stable transformation), the explant was cut very precisely at an extremely narrow region between the axillary bud and the hypocotyl (Paz et al. 2006). Hypocotyl-based *Agrobacterium*-mediated transformation has also been developed and this method produced explants more easily than the cotyledonary node-based approach. Specially, it has been reported that the addition of silver nitrate together with the antioxidant enhances the efficiency of the antioxidation effect (Wang and Xu 2008). The cotyledonary node-based *Agrobacterium*-mediated transformation method is described in Sect. 9.2.6.2 below.

9.2.6.2 *Agrobacterium*-Mediated Transformation

Plant Materials Soybean seeds are surface-sterilized in 70 % ethanol for 30 s, then sterilizes in 1 % sodium hypochlorite with shaking for 30 min and washed with sterilized water for 10 min three times. The seeds are sowed on germination medium (Table 9.2) and germinated at 24 °C in an incubation room under 16/8 h of light/dark for 7 days. The size of true leaves similar to that of the cotyledons is usually used for inoculation. The appropriate time for the inoculation step for each variety is determined by observing the germination state after sowing.

***Agrobacterium* Cultivation** *Agrobacterium* is cultured at 25 °C at 250 rpm with shaking until the OD_{600} reaches 1.0. The pellet is harvested and re-suspended with liquid co-culture medium containing an antioxidant like L-cysteine, sodium thiosulfate, or DDT to $OD_{600}=0.5$. (Culture compounds and hormones are added on the day of inoculation.)

***Agrobacterium* Infection and Co-cultivation** Cut 1 cm of the hypocotyl located under the cotyledon of germinated seedlings. Vertically cut the hypocotyl by inserting a scalpel between the two cotyledons, and remove the true leaves. Make wounds with a scalpel between the axillary bud and cotyledonary node 10 times under a microscope. The scalpel is first daubed with a concentrated *Agrobacterium* culture ($OD_{600}=2$). Put about 50 explants into 25-mL co-culture medium (Table 9.2), sonicate for 10 s, vacuum for 30 s, inoculate for 30 min, and then remove the *Agrobacterium* solution on the surface of the explants with sterilized filter papers. Place a sheet of filter paper on co-culture medium, put the adaxial side of the explants face down on the filter paper, and culture at 25 °C for 5 days in dark.

Selection and Regeneration After 5 days of co-cultivation, remove the *Agrobacterium* by washing the explants lightly with 1/2 shoot induction medium (SIM) (Table 9.2). After removing the water, seven explants per plate are set at a right angle with SIM without selectable antibiotics. After 2 weeks, all portions except the shoots are removed. The shoots are then transferred to SIM with selectable antibiotics placing the shoots with the adaxial side down. After 2 weeks, browned shoots/shoot pads are cut out with a scalpel and placed in shoot elongation medium (SEM) (Table 9.2) with selectable antibiotics. Every 2 weeks, the shoots

Table 9.2 Composition of media mentioned in this chapter

GM	CCM	SIM	SEM	RM
B5 or MS salt	B5 or MS salt	B5 or MS salt	B5 or MS salt	B5 or MS salt
B5 vitamins	B5 vitamins	B5 vitamins	B5 vitamins	B5 vitamins
MES 3 mM	MES 20 mM	MES 3 mM	MES 3 mM	MES 3 mM
	BAP 7.5 mM	BAP 7.5 mM	GA 1.5 mM	IBA 5 mM
	GA 0.7 mM	Cefotaxime 500 ppm	Cefotaxime 500 ppm	Cefotaxime 500 ppm
	Acetosyringone 0.2 mM			
	L-cysteine* 3.3 mM			
	Sodium-thiosulfate 1.0 ppm			
	DDT 1.0 mM			
Sucrose 3 %	Sucrose 3 %	Sucrose 3 %	Sucrose 3 %	Sucrose 3 %
Agar 0.8 %	Agar 0.8 %	Agar 0.8 %	Agar 0.8 %	Agar 0.8 %
pH 5.8	pH 5.4	pH 5.6	pH 5.6	pH 5.6

*L-cysteine should be kept in foil because they are sensitive to light
Abbreviates: *GM* germination medium, *CCM* co-culture medium, *SIM* shoot induction medium, *SEM* shoot elongation medium, *RM* rooting medium, *B5* Gamborg' B-5 medium, *MS* Murashige and Skoog medium, *MES* 2-(N-morpholino)ethanesulfonic acid, *BAP* 6-Benzylaminopurine, *GA* Gibberellic acid, *IBA* Indole-3-butyric acid, *DDT* 1,1,1-trichloro-2,2-bis-(4′-chlorophenyl)ethane

are transferred to new SEM with selectable antibiotics and the shoot pad is cut out continuously. After 8–10 weeks of inoculation, the shoots that have elongated to about 4 cm are transferred to rooting medium.

Acclimation Rooting is induced when the shoots are 4–5 cm and have 5–6 leaves (Fig. 9.1). To stimulate root induction, the bases of the shoots are dipped in 1-mg/L IBA solution before transfer to the rooting medium. Transfer shoots with roots to soil and acclimate, afterwards, transfer to pots for seed fructification.

9.2.7 *Potato (Solanum tuberosum)*

9.2.7.1 Background

Potato is known to be relatively easy to tissue-culture and regenerate, and has been widely used for genetic transformation. *Agrobacterium*-mediated transformation is most commonly used for the potato transformation and the process for gene introduction involves callus induction, genesis of shoots from induced callus, and healthy growth of generated shoots. Callus formation was reported to be stimulated by appropriate concentrations of cytokinin and auxin, shoot genesis was accelerated by removal of auxin, and continual development of shoots was facilitated by addition of gibberellin (Gaspar et al. 1996). Regeneration methods have been categorized as one-step, two-step, or three-step methods. One-step methods use one kind of culture medium from callus formation to shoot development; two-step methods use different media for callus formation, and for shoot induction and development stage; and three-step methods use different media for each of the three stages. Regeneration efficiency changes mainly according to which method is applied and optimal hormone combinations and concentrations. Even in the same crops, there are differences according to the varieties that are used.

9.2.7.2 *Agrobacterium*-Mediated Transformation

Plant Materials For transformation, *in vitro* cultured plantlets for the cultivar of interest can be used; however, if no *in vitro* cultured plants are available, transformation can be started from a tuber. Because it is difficult to germinate from the true seeds of potato, potato has to be propagated in various ways. Good quality, virus- or pathogen-free potato should be used. Shoots that have sprouted from dormant potato can be cleanly broken off, washed with running water, surface sterilized with 70 % ethanol for 5 min, and washed with distilled water. The potato shoots are submerged in 10 % sodium hypochlorite solution for 10–15 min (15 min is adequate for healthy shoots), then washed with sterilized water more than three times to remove the sodium hypochlorite. From here, all the work should be done on a clean bench. After completely drying the sample, the stems are proliferated on MS medium. Culturing conditions are 16 h of light period (4,000 lux light intensity) and 23 °C. The required numbers of *in vitro* cultured stems can be secured by sub-cultivation over a certain period. For transformation, leaves, stems, and tubers have been used, but leaf disks are most generally used. For stems, the internodes are used but particular attention must be paid to obtaining internodes as far as possible from nodes so as not to include growing points. It is important to note that there is a high chance of obtaining fake plants when leaves are used. The transformation of many different cultivars has been reported, but different conditions from those described here need to be used for red potato, which is known to be difficult to transform (Dale and Hampson 1995).

***Agrobacterium* Cultivation** The virulence of *Agrobacterium* is different for different crops; therefore, it is advisable to choose the strain with the highest efficiency in the crop of interest. The most commonly used strain is LBA4404 because it is known to have high transformation efficiency in potato (Ooms et al. 1987). Antibiotics are selected according to the strain and vector for transformation. *Agrobacterium* is streaked onto YEP solid medium and cultured at 28 °C for 36–48 h. A single colony is inoculated to YEP liquid medium supplemented with antibiotics (addition of acetosyringone is optional) and cultured until $OD_{600}=0.5$–0.6 for co-cultivation with explants. The OD_{600} should not exceed 1.0.

Pre-treatment of Explants Pre-processing of potato leaves for co-cultivation with *Agrobacterium* can increase the regeneration efficiency. Various medium compositions can be used, but floating the leaves on MS liquid medium with 80-mg/L NH_4NO_3, 14.7-mg/L $CaCl_2$, 5.0-ppm NAA, and 5.0-ppm BAP for a day is the strategy. Floating the potato leaves before placing them on the callus formation medium improved both the callus induction rate and shoot generation regardless of the cultivars.

Co-cultivation When the *Agrobacterium* culture is ready, wounds are made in the *in vitro* cultured leaves for co-cultivation. Excessive wounding like shattering cells when cutting leaves from a stem and large wounded regions should be avoided. Larger wounding regions could produce large callus formations, but exorbitant energy is needed for the wounds to recover, which often makes it difficult to obtain healthy transformants. Enough bacterial suspension is poured to submerge the leaf disks. For cultivars that show high regeneration rates like *S.tuberosum* cv. Desiree, acetosyringone need not be added, but when the regeneration rate is low, acetosyringone could be helpful. In our laboratory, shaking is not applied during the co-cultivation step, but shaking has been applied in other laboratories. After submerging for about 10–15 min, the leaf disks are dried completely on sterilized filter paper. The leaf disks should be treated gently, not with tweezers, to avoid withering of leaf disks.

Regeneration Induction While in most other crops, the regenerated entity is induced directly from a cutting plane, in potato the shoot formation is indirect because a callus is induced from a cutting plane and shoot induction occurs from the callus. So after co-cultivation, the leaf disks are placed first on MS medium containing 2-mg/L 2,4-D for callus formation. To improve the chance for insertion of the target T-DNA gene into the plant genome, antibiotics commonly used as selection pressure are not contained in the medium. After 2 days, the leaf disks are transferred to regeneration medium for shoot induction. Because the regeneration rate differs according to the concentrations of the hormone treatment and the potato varieties, it is better to first determine the optimum regeneration conditions for the variety of interest. It has been reported previously that $AgNO_3$ can increase the regeneration rate. Regeneration medium is generally the MS medium supplemented with 0.01-mg/L NAA, 0.1-mg/L gibberellic acid, and 2.0-mg/L zeatin.

Transformant Selection The precise selection of transformed plants is important and depends on the selectable markers used in the vectors. Until now, the most commonly used vectors contain a kanamycin-resistant gene. For selection of transformants, the regeneration medium is supplemented with 500-mg/L carbenicillin to remove any bacteria that remain on the leaf disks after co-cultivation, and kanamycin for selectable pressure. The optimum concentration of kanamycin that least affects regeneration of the potato shoots and hinders growth of non-transformed shoots is used. Under these conditions, the Desiree variety formed only a few calli with 50-mg/L kanamycin, but transformed shoots could be selected because *S.tuberosum* cv. Sumi, cv. Atlantic, and cv. Norland varieties have less resistance to kanamycin than cv. Desiree, which caused them to stop growing, and also caused withering of the non-transformed cv. Desiree calli or plants. Nevertheless, non-transformed regenerated plants could survive on selection medium by obtaining tolerance to antibiotics during long-term cultivation of over 8 weeks, so an effective way of reducing negative plants is to use 100-mg/L kanamycin when multi shoots are formed by potato calli. The shoot for which the roots are grown completely on medium containing antibiotics is more likely the transformed plant.

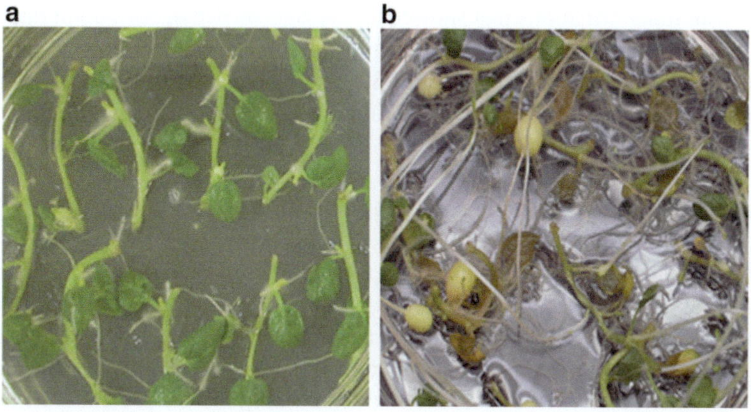

Fig. 9.2 (**a**) Propagation of transgenic potato shoots, (**b**) Tuber induction from the transgenic shoots

Fixation and Maintenance of Transformants Shoots that have grown into independent entities after 10 weeks of cultivation on the medium with antibiotics are proliferated on MS medium with kanamycin and no hormones and analyzed to select confirmed transformed plants (Fig. 9.2). Because vegetatively propagated plants like potato can be maintained by *in vitro* cultivation, it is relatively convenient to fix lines rather than seed propagated plants. However, loss of genes or variations in genes has been reported to occur by continuous cultivation; therefore, consecutive cultivation on selection medium and gene re-analysis should be done periodically.

9.2.8 Tomato (Solanum lycopersicum)

9.2.8.1 Background

Tomato is another important solanaceous crop and the genetics and genomics of tomato have been studied more than other plant crops, so tomato is used widely for cultivar development using transformation techniques and gene functional studies. Tomato transformation is usually done using *Agrobacterium*-mediated transformation method, and young leaf, hypocotyl, or cotyledon have been used as explants. It has been reported that cotyledon produces the highest efficiency. Transformation of tomato using mannose as a selectable marker instead of antibiotic-resistant genes (Sigareva et al. 2004) has been reported. Tomato plastid transformation (Nugent et al. 2005) also has been reported. Micro-Tom is a cherry tomato that has been identified as a model plant in molecular biological, and gene analysis through transformation has been widely studied. Dan et al. (2006) reported mass scale

Micro-Tom transformations that showed an average 56 % transformation rate using kanamycin and glyphosate selection. Transformation efficiency differs according to the genotype, but cotyledon-based transformations have been shown to transform plants with a large variety of genotype (Ellul et al. 2003a; Park et al. 2003).

9.2.8.2 *Agrobacterium*-Mediated Transformation

Plant Materials Seeds are sterilized for transformation generally with 70 % ethanol for 30 s, washed three times with distilled water, submerged in 25 % commercial Clorox and Tween-20, then washed three times with distilled water and germinated on 1/2 MS medium containing 2 % sucrose at 25 °C for about 7 days. When the cotyledon emerges from the seed, it is cut, and cultured on pre-culture medium (1-mg/L BA and 0.1-mg/L NAA) in dark for 1 day.

***Agrobacterium* Cultivation and Co-cultivation** The tomato strains LBA4404, EHA101, and EHA105 have no special differences, but it is easier to remove *Agrobacterium* from strain LBA4404 rather than the EHA strains after co-cultivation. *Agrobacterium* is streaked on YEP solid medium, and a single colony is re-inoculated after cultivation at 28 °C, for 36–48 h. Inoculate to 20 mL of YEP liquid medium containing 50-mg/L rifampicin, 50-mg/L kanamycin, 200-μM acetosyringone, and the suspension is cultured until the OD_{600} reaches 0.7. The cell pellet is harvested and re-suspended to 20-mL of 1/2 MS medium (2.2-g/LMS salt, 2 % sucrose, pH 6.0), and cultured with stirring. The virulence-induced bacterial solution and the cotyledon fragments are then shaken together at 22 °C for 10 min. The cotyledon fragments are placed on sterilized tissues to absorb and remove the bacterial solution. The cotyledon fragments are then placed on co-cultivation medium with the back of the cotyledon facing the medium and cultured in dark for 48 h.

Selection and Shoot Elongation After co-cultivation, the explants are cultured on selection medium, B5 MS salt containing 3 % sucrose, 2.0-mg/L zeatin, 0.2- mg/L IAA, 100-mg/L Km, and 300-mg/L Cb for 1–2 months. Callus formation is observed on the selection medium after about 5 weeks. Calli are isolated from the explants and transferred to shoot elongation medium; in some cases, the explants turned yellow-brown and died. Tomato transformants are usually selected from shoots generated from calli (indirect shoot formation). Shoot genesis can be observed after a week of transferring the calli to shoot medium, B5 MS salt containing 3 % sucrose, 2.0-mg/L zeatin, 0.2-mg/L IAA, 100-mg/L Km, and 300-mg/L Cb, and the shoots begin to grow after 2–3 weeks. When the shoots are about 1–2 cm in height (after about 2–3 months of cultivation), they are transferred to root medium. It is difficult to generate shoots from cherry tomato calli when the calli enters the auxes is stage, finally aging and becoming dark brown. Generally, tomato calli with lots of moisture do not give rise to shoots and consequently become white or dark brown.

Rooting and Acclimation Process The cultured shoots are isolated from calli on rooting medium, 1/2 MS containing 2 % sucrose, 20-mg/L Km, and 300-mg/L Cb

Fig. 9.3 (**a**) Cotyledon of tomato seedlings for experimental materials, (**b**) Shoot induction, (**c**) transformed tomato plant

for about 5–6 weeks. When transferring the shoots, the shoots including some of the callus are cut and placed on root medium. After about 2 weeks on the rooting medium, root induction is observed, and the roots grow enough to acclimate at about 5–6 weeks. When the induced roots reach 5–10 cm in height, the plants can be transferred to Jiffy pots (www.jiffygroup.com) and when the roots extend out of the Jiffypots, the plants are transferred to bigger pots (Fig. 9.3).

9.2.9 Pepper (Capsicum annuum)

9.2.9.1 Background

Pepper is a high value vegetable crop, which ranks seventh in vegetable cultivation area worldwide, and is a commercially important crop that is used as food, dye, and medicine by approximately 5 billion people. Pepper transformation is difficult and lags behind the other major cereal and vegetable crops, and although pepper transformation has been tried in many laboratories, it has usually failed. Integrating the results of the few successful studies indicates that the efficiency of pepper transformation depends on the virulence degree of the *Agrobacterium* strain, the genotypes of the pepper, and selection of the marker in the vector system. Therefore, a methodical and reproducible system for pepper transformation is urgently required.

Generating pepper transformants from explants using the direct shoot formation method is very difficult, so generating shoots from calli is the preferred method. Here, two methods for generating shoots, callus-mediated shoot formation (CMSF) from calli and callus-induced transformation after callus formation based on Lee et al. (2004, 2009) are described below.

9.2.9.2 *Agrobacterium*-Mediated Transformation

Plant Materials Any pepper variety can be used for transformation; however, if the aim is to develop a particular variety for further breeding, the corresponding line should be chosen. Generally, the seeds are sterilized with 95 % ethanol for 30 s, washed three times with distilled water, submerged in 50 % commercial Clorox for about 10 min, then washed three times with distilled water and germinated on 1/2 MS medium at 25 °C in an incubation room. Cotyledons and hypocotyls from 8 to 10 day-old plants are used as explants.

***Agrobacterium* Cultivation** *Agrobacterium* is cultured on YEP liquid medium to which antibiotics and acetosyringone have been added. The *Agrobacterium* is cultured until the OD_{600} reaches 0.3–0.8, and does not exceed $OD_{600}=1.0$. The cells are harvested and re-suspended in 1/2 MS medium (2.2-g/LMS salt, 1.5 % sucrose, pH 5.7) up to OD_{600} 0.3–0.5. The virulence-induced bacterial solution and cotyledon fragments are shaken together at 22 °C for 10–20 min.

Correlation Between *Agrobacterium* Strains and Virulence The virulence of *Agrobacterium* and the degree of host infection differ in different crops; therefore, depending on the strain used for the transformation, the infection efficiency of *Agrobacterium* can change in different pepper varieties. For example, a strain of P915 line showed the highest regeneration rate, and EHA105, which is commonly used for transformation because of its high virulence, also showed a high regeneration rate.

Callus-Mediated Shoot Formation Method A protocol for pepper transformation using the CMSF method is described in Table 9.3. The most important point when using this method is the selection of transgenic plants, which means that large numbers of directly grown shoots are discarded during the process of regeneration and calli that can give shoots are selected carefully.

Callus-Induced Transformation Method The transformation efficiency of the CMSF method, which transforms explants and induces shoots from calli that are naturally formed on a cutting plane of the explants, depends on the induction rate of the naturally occurring calli. Thus inducing calli artificially and generating shoots from them could enhance the success rate of the whole transformation. During screening, the hormones for callus induction and the concentration of auxin should be chosen so that there is not only mass induction of calli but also that the induced calli have the ability to produce shoots.

Table 9.3 Pepper transformation method using callus-mediated shoot formation

Procedure	Medium and buffer	Duration
Germination	1/2MS (MS* 2.2 g L^{-1}, 1.5 % sucrose, 0.8 % agar, pH 5.7)	8–10 days
Pre-culture	MS (MS 4.4 g L^{-1}, 3.0 % sucrose, 0.8 % agar, pH 5.7)	2–36 h
	2.0-ppm zeatin, 0.05-ppm NAA (or 0.1-ppm IAA)	
Co-culture	MS + 2.0-ppm zeatin, 0.05-ppm NAA	38–96 h in dark
	Washing buffer (1/2MS + 500–800-ppm cefotaxime)	
Selection	MS + 2.0-ppm zeatin, 0.05-ppm NAA + 80–100 mgL^{-1} kanamycin, 300-ppm cefotaxime	Callus formation: 4–5 weeks
		Callus development: 2–3 weeks
Shooting	MS + 2.0-mgL^{-1} zeatin, 0.01-mgL^{-1} NAA + 60–100-mgL^{-1} kanamycin, 300-mgL^{-1} cefotaxime	Shoot formation: 1–2 weeks
		Shoot elongation: 6–8 weeks
Rooting	MS + kanamycin 20–30 mgL^{-1}, cefotaxime 200 mgL^{-1}	Root formation: 4–5 weeks
		10 cm height: 2–3 weeks

**MS* Murashige and Skoog medium
Abbreviates: *NAA* 1-Naphthaleneacetic acid, *IAA* Indole-3-butyric acid

9.2.10 Watermelon (Citrullus lanatus)

9.2.10.1 Background

Watermelon, a horticultural crop that originated in tropical and subtropical areas, is an important crop of the Cucurbitaceae (cucumber) family. Watermelon is known to be difficult to transform. A watermelon transformation system was first reported by Choi et al. (1994), and more recently transformed plants generated from cotyledonary fragments through organogenesis have been reported (Ellul et al. 2003b; Gaba et al. 2004). However, a stable transformation system could not be established from those studies because limited selectable markers were used and success was dependent on a specific laboratory or cultivar. Selectable marker genes that have been used for watermelon are *npt*, and rarely *pmi* (phosphormannose isomerase) gene. The *bar* gene, a herbicide-resistance gene, which is known to be a highly effective selectable marker in a large number of crops including Cucurbitaceae, has not been used as a selectable marker in watermelon. For the development of a stable and highly efficient transformation system for crops, it is important to choose the *Agrobacterium* strain with the highest virulence and one in which the *bar* gene can be used as a selectable marker.

9.2.10.2 *Agrobacterium*-Mediated Transformation

Plant Materials Watermelon seeds (*C.lanatus* cv. Daesan) are surface sterilized with 70 % ethanol for 1 min and then with 0.4 % hypochlorite for 15 min. Ten seeds per petri dish are placed onto MS basal medium containing 2 % sucrose and 0.8 %

agar, and germinated at 25 °C in dark for 5 days. After 5 days, cotyledon fragments are excised carefully so as not to include shoot tips and used as explants for co-cultivation with *Agrobacterium.*

Agrobacterium *Agrobacterium* solution with OD_{600}=0.5–0.6 is centrifuged at 4,000 g for 10 min. The pellet is harvested and re-suspended in co-cultivation solution. The prepared cotyledon fragments are submerged in the re-suspended bacterial solution and shaken at regular intervals. After 10 min, sterilized filter paper is placed on the surface of the co-cultivation medium (1/2 MS, 3 % sucrose, 20-mM ME, 3.2-ppm BAP, 0.5-mg/L IBA, 0.5 % phyto agar, 20-mM MES, 200-μM acetosyringone, and 100-ppm L-cysteine, pH 5.4). Then 10–15 fragments per plate are placed on the filter paper and cultured at 24 °C in light.

Selection of Transformed Plants After 2 days of co-cultivation, the explants are transferred to shoot induction medium containing antibiotics and cultured in light, then sub-cultured every 2 weeks. Next, six explants are arranged on one petri dish with the base of the cotyledon fragments submerged in the medium. After 2 weeks of cultivation, all shoots generated from the cotyledon fragments are sub-cultured in new medium, as before. The shoots are cultured for 8–10 weeks at 2-week-intervals. Green healthy shoots are isolated from the base of the cotyledon fragments, transferred to elongation medium, and sub-cultured every 2 weeks until they grow to more than 3 cm in height. Culture usually takes more than 4 weeks. When the *bar* gene is used as the selectable marker, the composition of the shoot induction medium is MS salt, 3 % sucrose, 3-mM MES, 0.8 % phyto agar, 3.2-ppm BA, 0.5-ppm IBA, 50-ppm vancomycin, 50-ppm ticarcillin, 50-ppm cefotaxime, and 5-ppm glufosinate. The shoot elongation and root induction medium contains MS salt, 2 % sucrose, 3-mM MES, 0.1-ppm IBA, 0.05-ppm gibberellic acid ,0.8 % agar, 50-ppm vancomycin, 50-ppm ticarcillin, and 50-ppm cefotaxime, pH 5.8.

Acclimation to Soil When the plants grown in the incubator are 10 cm in height, they are transferred to aseptic soil, covered with plastic, and acclimated in an incubation room for about 7 days. For plants that are beginning to take root, holes are made in the plastic and the plants are acclimated for another 2–3 days in an incubation room. When the new leaf emerges, the plants are transferred to a greenhouse and grown for 2 weeks, then transferred to bigger pots.

9.3 Roadmap for Commercializing Genetically Modified Crops

9.3.1 Introduction

The number of hectares in which biotech crops (genetically modified (GM) crops) are grown exceeded 175 million hectares in 2013 and a record 18 million farmers in both large and small developing countries grew biotech crops (James 2014). The global hectarage of biotech crops has increased more than 100-fold from 1.7 million hectares in 1996 to over 175 million hectares in 2013. This makes biotech crops the fastest adopted crop technology in recent history. This adoption rate speaks for itself in terms of there silience of biotech crops and the benefits they deliver to farmers and consumers. The global value of biotech seed alone was about USD 15.6 billion in 2013. This represents 22 % of the USD 71.5 billion global crop protection market in 2012, and 35 % of the approximately USD 45 billion commercial seed market.

To commercialize a GM crop, many processes must be considered. First, a GM crop that properly expresses a transgene should have commercial merit and marketing value. Second, the GM line should be prepared with other elite inbred lines, so that many different GM F_1 hybrids can be cultivated easily, like non-GM crops. Third, the GM crop should go through human health and environmental risk assessments and pass international biosafety government regulations. Fourth, the GM crop developer must identify the available GM market where farmers are satisfied with the production and expense of cultivating the GM F_1 hybrid. These processes take years and are expensive; therefore, it is important to prepare in advance all plans and steps necessary for the research and development, production, quality assurance, and marketing of a GM crop.

9.3.2 Gene Discovery

There is a limit to what breeders can do to develop a new variety using non-GM breeding methods such as crossing among the same species. GM breeding methods use a selected target gene that can be obtained from another organism, such as a microorganism, insect, another plant, animal, or human to develop new varieties. Finding a target gene that will become a transgene is difficult because the target should be applicable to all crop species and be commercially valuable. In the search for useful genes, generally about 1,000 genes with similar functions are screened to find one commercially useful target gene. Exceptions to this are *Bacillus thuringiensis* and herbicide-resistant genes because, for several decades, farmers have known about their effects and value. Therefore, farmers, scientists, breeders, consumers, and marketing scientists need to interact to identify characters that are necessary for the development of high value-added crops. Once this is decided, it will take at least 2–3 years to find a gene, depending on the targeted character.

9.3.3 Genetic Transformation and Validation

The efficiency of genetic transformation varies depending on the crop and the transformation of some crops has been protracted because of difficulties in *Agrobacterium*-mediated transformation. Therefore, transferring genes into resilient explants requires effort, and a successful and reproducible transformation method needs to be in place. Usually, it takes about 1 year to produce a T_0 plant once the transformation technique is ready.

It is advisable to work with many T_0 plants at the beginning of selection. Usually, 50–100 T_0 plants are sufficient to self-cross and look for the next T_1 generation. Generally, about 100 plants of each T_1 line are screened for the presence or absence of the transgene and the phenotype of interest. Breeders then select the phenotype with the highest level of the desired characteristics and perform self-crosses and back-crosses. Continuous selection in the self-crossed and back-crossed generations can be performed in a large field or greenhouse to identify the best GM line. A GM line should maintain the target gene and have the proper characteristics and horticultural aspects. After selecting the best T_1 plants, validating the final choice(s) requires at least 2–3 years.

9.3.4 Selection of an Event

During the validation process, molecular evaluation among the selected candidate GM crops is conducted as follows: (1) only one copy of the transgene in the GM crop genome; (2) the transgene has been inserted into an intergenic site; (3) no disturbance in the DNA sequence 1-kb upstream and downstream of the insertion site (Fig. 9.4); and (4) no insertion of apart of the transgene (including promoter, terminator, or any region between the left and right borders) into other site in the genome. If the GM crop fulfills these conditions, it is called an event. An event is a prerequisite to conducting risk assessment research and breeding. A careful evaluation at the molecular level is necessary to verify the event line. At least three events are required for experimental repetition to conduct the risk assessment research.

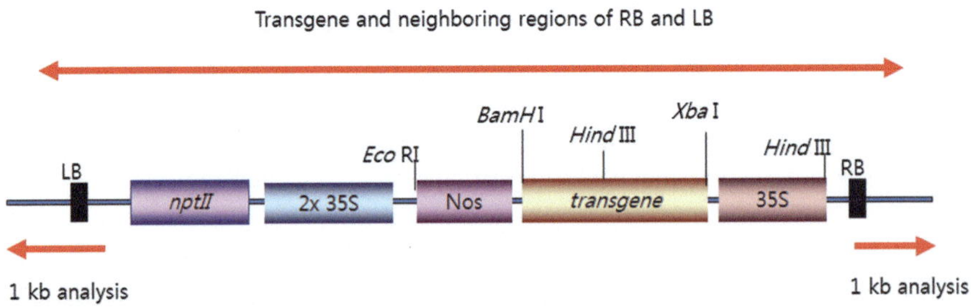

Fig. 9.4 Insertion site of a transgene in the genome of a genetically modified crop. Neighboring regions must be thoroughly investigated because the transgene is inserted randomly. At least 1 kb on both sides of the *left* and *right* borders as well as the internal region of the transgene should be sequenced to identify any DNA sequence disorder. In addition, to be considered an event, no part of the transgene should be repeated in the genome

9.3.5 Developing a GM Line and Establishing Regulatory Data

9.3.5.1 GM Line Development

Transformed crops usually reveal soma-clonal variations that occurred during tissue culture. These variations need to be corrected by self-crossing of several generations. In autogamous crops such as rice, maintaining the T generation by self-crossing and selecting among progeny is a common way to develop a GM line (Fig. 9.5).

However, continuous self-crossing of an allogamous crop causes self-depression and abnormal phenotypes (Fig. 9.6); therefore, continuous backcrossing to the recurrent line is important to stabilize the genome. Two ways of backcrossing have been used. In the first, the selected GM crop is crossed to a non-GM control line (the one that was used for genetic transformation) for the risk assessment. The GM crop should be backcrossed twice, because this should be sufficient to correct the genome for the field test. In the second, a GM breeding line can be constructed by backcrossing several times to the recurrent line of interest and continually self-crossing to stabilize the genome (Fig. 9.7). In this way, the breeder generates a new germplasm with the transgene. It can take up to 5–6 years to develop a GM line depending on the crop.

9.3.5.2 Risk Assessment Study

A human health and environmental risk assessment must be conducted. The criteria for the assessment vary depending on the crop. Table 9.4 lists the items that are necessary to submit to a government agency for a review request. This information

Fig. 9.5 Development of an autogamous crop line. Selection through the T_6–T_8 generations is sufficient to correct a disturbed genome

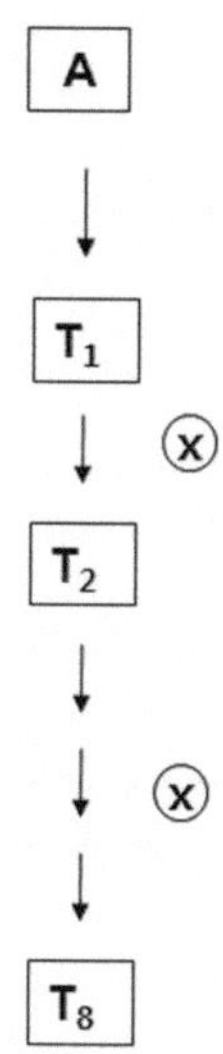

Fig. 9.6 Appearance of an abnormal phenotype caused by self-depression after continuous self-crossing over four generations. The GM (T_4) line did not develop properly because the shoot was weak

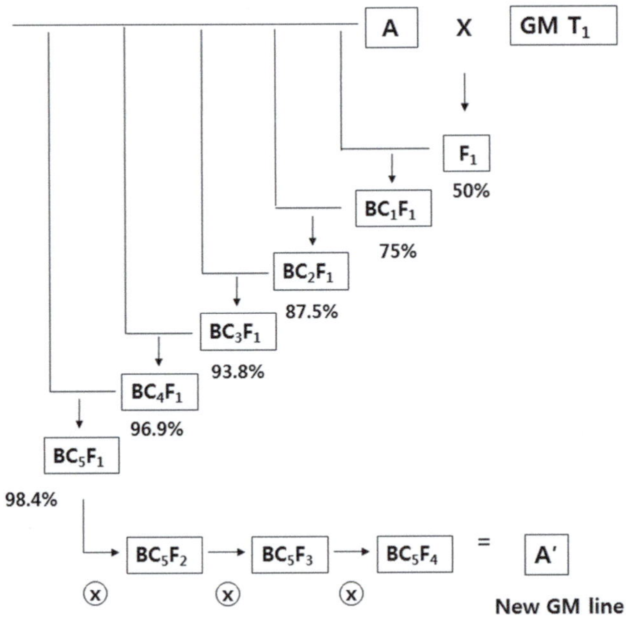

Fig. 9.7 Construction of a GM breeding line by backcrossing and self-crossing to stabilize the genome. The GM T_1 line can be crossed to an elite recurrent line by backcrossing (BC_5) and, consequently, self-crossing (F_4) to obtain a new genetic source (BC_5F_4)

is specific to Korea but the items are similar to what is generally required universally. All these items must be thoroughly worked through in the field and laboratory. In Korea, for example, it takes about 270 days to determine if a GM crop has passed the review. If the GM crop passes the review, the developer can cultivate the GM crop on land; however, the permission to cultivate is restricted to domestic land. If the developer wants to cultivate the GM crop in a foreign country, the GM crop has to go through an inspection review in that country. All countries have their own risk assessment regulations; thus, usually the developer needs to conduct assessment research in each country to obtain permission to cultivate and market a GM crop in that country.

Table 9.4 Items to be submitted to government agencies for risk assessment of GM crop cultivation

1. General data
1.1. Purpose of development
1.2. Benefits of development and uses
2. Data on host organism(s)
2.1. Taxonomic characteristics (e.g., scientific name, common name, and variety/line).
2.2. Distribution in nature
2.3. History of use (including use with the country and other countries, history of cultivation and development through breeding, etc.)
2.4. Biological characteristics (viability and reproduction/propagation capabilities in the natural environment or under test conditions simulating the natural environment)
2.4.1. Mode/cycle of reproduction/propagation and likelihood of cross-hybridization
2.4.2. Interactions with other species in the natural ecosystem and oceanic ecosystem (e.g., natural enemy, pathogenic organisms, competitors, and predators)
2.4.3. Mode of reproduction and reproductive compatibility with other species or relatives
2.4.4. Dissemination (dissemination characteristics of pollen/seed, environmental factors affecting dissemination)
2.5. Capability of producing hazardous materials (including the capability of relatives)
2.6. Reports on toxigenicity and allergenicity of the host and relatives
2.7. Contamination by pathogenic and foreign agents (e.g., viruses)
2.8. Centers of origin and genetic diversity
2.9. Other important physiological characteristics including parasitism
2.10. Research data on pathogenicity or relationship to known pathogens in the case of microorganisms
2.11. Potential to become a weed (become wild)
3. Data on donor organism(s)
3.1. Taxonomic status (e.g., scientific name, common name, and variety/line)
3.2. Distribution status in natural ecosystem and oceanic ecosystem
3.3. History of use by humans
3.4. Biological characteristics
3.5. Toxigenicity
3.6. Data on reported toxigenicity and allergenicity of the donor and relatives
3.7. Research data on pathogenicity or relationship to known pathogens in the case of microorganisms
4. Data on the vector
4.1. Name and source (e.g., GenBank accession number)
4.2. Molecular weight of DNA
4.3. Vector structure
4.4. Restriction map (including genetic elements in the vector, loci, orientation, and sequences)
4.5. Presence of hazardous sequences
4.6 Potential of vector being transferred to other cells or dependency on the host
4.7. Identification and functions of genetic elements in the intermediate host
4.8. Data on intermediate hosts (toxigenicity and allergenicity)

(continued)

Table 9.4 (continued)

5. Data on introduced gene(s)
5.1. Functions (e.g., GenBank accession numbers) and characteristics of introduced gene(s)
5.2. Sources and sequences of components in an introduced gene
5.2.1. Size, name and functions of the introduced gene
5.2.2. Regulators (promoters and terminators)
5.2.3. Selectable marker genes
5.2.4. Other promoters and other factors affecting the DNA functions
5.3. Modifications to the gene for use
5.4. Presence of hazardous sequences
5.5. Sites and orientation of gene sequences in the finalized vector
5.6. Presence of exogenous open reading frames and potential of transcription and expression thereof
6. Data on LMO (living modified organism; i.e., a living GMO) development
6.1. Genetic modification methods (transgenic methods)
6.2. Description on the development process for LMO (e.g., cultivation and breeding)
6.3. Stability of phenotypes of an inserted gene through multiple generations
7. Molecular characterization of LMO
7.1. Data for inserted gene in LMO
7.1.1. Identification results and composition for an inserted gene in LMO
7.1.2. Insertion site of a gene (chromosome or cellular organelle) and adjacent sequences
7.1.3. Number of copies of an inserted gene
7.1.4. Methods to detect and to verify the expression of an inserted gene
7.1.5. Data on stability of an inserted gene
7.1.5.1. Changes in the sequences and size of an introduced gene through multiple generations
7.1.5.2. Changes in expression sites, timing, and level of an introduced gene through multiple generations
7.1.6. Data proving that an inserted gene in the LMO genome does not encode toxins and allergens
7.2. Data on genetic product
7.2.1. Traits/characteristics of genetic product (e.g., proteins, non-coding RNA)
7.2.2. Post-translational modification of an inserted gene (variation after protein formation)
7.2.3. Expression level, timing and site of a marker protein due to an inserted gene or the insertion, and measurement methods and sensitivity thereof
7.2.4. Effects of genetic product on metabolic pathways
7.2.5. Structure and functions of the selectable marker gene and the mechanism of resistance and metabolite(s) thereof
7.2.6. LMO detection and identification methods
8. Comparison between LMO and non-LMO
8.1. Enhanced characteristics and traits after modification
8.2. Difference in viability and propagation between the host and the LMO
8.3. Differences from the host or the species of the host
8.3.1. Viability and reproduction/ propagation capabilities in the natural environment or under test conditions simulating the natural environment
8.3.2. Mode/cycle of reproduction/propagation and likelihood of cross-hybridization

(continued)

Table 9.4 (continued)

8.3.3. Interactions with other species in the natural ecosystem and oceanic ecosystem (e.g., natural enemy, pathogenic organisms, competitors, and predators)
8.3.4. Mode of reproduction and reproductive compatibility with other species or relatives
8.3.5. Dissemination (dissemination characteristics of pollen/seed, environmental factors affecting dissemination)
8.3.6. Production of hazardous materials and residual effects on the ecosystem
9. Detailed data on adverse effects
9.1. Generation of toxic materials (e.g., presence of toxic materials secreted by an organism)
9.2. Potential effects on organisms in the vicinity and the ecosystem
9.3. Potential to become a weed
9.4. Information on the environment to which the LMO is to be introduced
9.4.1. Distance from the production origin of the LMO
9.4.2. Geographic and climatic characteristics and ecological characteristics of flora in the vicinity
9.5. Data for toxicity of the LMO
9.5.1. Potential toxicity of gene product or metabolite of the LMO
9.5.2. Structural similarity between the gene product or metabolite of the LMO and protein toxin.
9.5.3. Average exposure of the LMO through uptake or eating
9.5.4. Biological similarity evaluation between the expressed protein from the LMO and the protein from non-LMO
9.5.5. Single dose toxicity test
9.5.6. Other toxicity test data if necessary
9.6. Data for allergenicity of the LMO
9.6.1. Potential allergenicity of the gene product
9.6.2. Structural similarity between the gene product or metabolite of the LMO and the allergen
9.7. Cross-response evaluation of the patient-specific IgE antibody when the structure of the gene product is similar to the one of allergen.
10. Data on environmental release, monitoring, and disposal of LMO
10.1. In the case of LMO for use as food/feed or for processing (data from production country)
10.1.1. Data on environmental release (duration and time)
10.1.2. Location of release (geographical and geological aspect, relation with biologically important area such as the preservation of the natural environment)
10.1.3. Ecosystem of release (distribution and kinds of species such as allied species)
10.1.4. Method and amount of release
10.1.5. Results and data of release
10.1.6. Information of the release permission (nation, date and number of permission)
10.1.7. Monitoring plan (methods, time, frequency)
10.1.8. LMO inactivation methods
10.1.9. Emergency plan (containment)
10.1.10. Treatment method of LMO waste
10.2. In the case of LMO for cultivation (information of domestic field and containment)
10.2.1. Time
10.2.2. Method

(continued)

Table 9.4 (continued)

10.2.3. Magnitude
10.2.4. Treatment method at the end of test
10.2.5. Results and data
10.2.6. Monitoring plan (method, time, frequency)
10.2.7. LMO inactivation methods
10.2.8. Management plan of emergency status (containment)
10.2.9. Treatment method of LMO waste
11. Approvals and uses in other countries
11.1. Name of country
11.2. Name of organization (agency, institution, firm)
11.3. Organization of risk assessment evaluation
11.4. Permission number
11.5. Permission data
11.6. Status of use
12. Submission of LMO
12.1. Inactivated LMO seed (50 seed) and LMO seed 1 kg
12.2. Inactivated vegetative mass (50 individual)
12.3. Micro-organism case, small volume of inactivated LMO culture
12.4. Information of DNA sequence of the transgene for detecting the LMO
12.5. Information of DNA sequence of the LMO genome data, 5′ and 3′ region of transgene, insertion region including 1 kb forward and backward.
12.6. Other genome information for development of the analysis method.

This list is applicable worldwide

9.3.6 Pre-farming and Performance Testing

To evaluate the performance of a GM crop, breeders usually conduct field tests by crossing the GM line to other lines to evaluate the various F_1 combinations. A performance evaluation is conducted to select commercially valuable F_1 combinations (Fig. 9.8). The selected F_1 hybrid is tested further in farms indifferent regions, and the observations and preferences of the farmers are considered when selecting the final combination. Testing of a hybrid combination may take 2 years or more depending on the variety.

9.3.7 Launching for Commercialization

Once the breeder receives permission from the government agency following the GM line or crop risk assessment and the F_1 hybrid for commercialization has been selected, the breeder determines where the seeds will be produced and cultivated. GM F_1 hybrid crops must pass the regulatory requirements of the country where the breeder aims to sell the F_1 hybrid and a pipeline is constructed to develop a

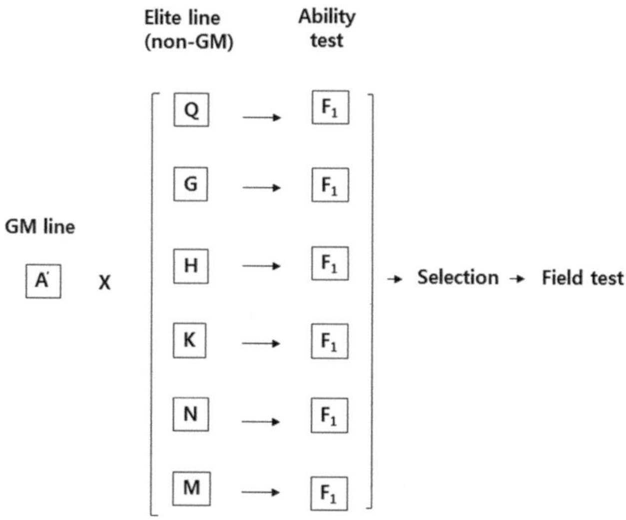

Fig. 9.8 Evaluating the performance of GM crops to select a commercially valuable F_1 hybrid combination. It is common to use a non-GM line as the female MS (male sterile) and the GM line as the male

commercial GM crop. It can take 13–15 years to commercialize a GM crop from the beginning (Fig. 9.9) but if the target gene is available and the GM breeding infrastructure is already in place, it may take 10 years or less. It takes a further 5–6 years to prepare the risk assessment data for exporting the GM crop, depending on the country, and to simultaneously conduct breeding trials in a local market. Therefore, establishing a pipeline to develop commercial GM seeds including a review request and risk assessment data is very important and should be prepared well for time and cost efficiency.

9.4 Development of New Genetic Sources Using GM Technology

9.4.1 Cisgenesis

9.4.1.1 Introduction

To transfer an elite gene or characteristic from a wild-type or open-pollinated crop (donor) to an ordinary line using non-GM breeding processes, the procedure is as follows: (1) male and female plants are crossed to produce the F_1; (2) the F_1 is back-crossed to the recurrent line (i.e., the ordinary line) generally for 5–6 generations (BC_5 or BC_6); and (3) BC_5 or BC_6 are self-crossed for three generations to obtain BC_5F_3 or BC_6F_3. In this way, the gene of interest can be transferred from the donor

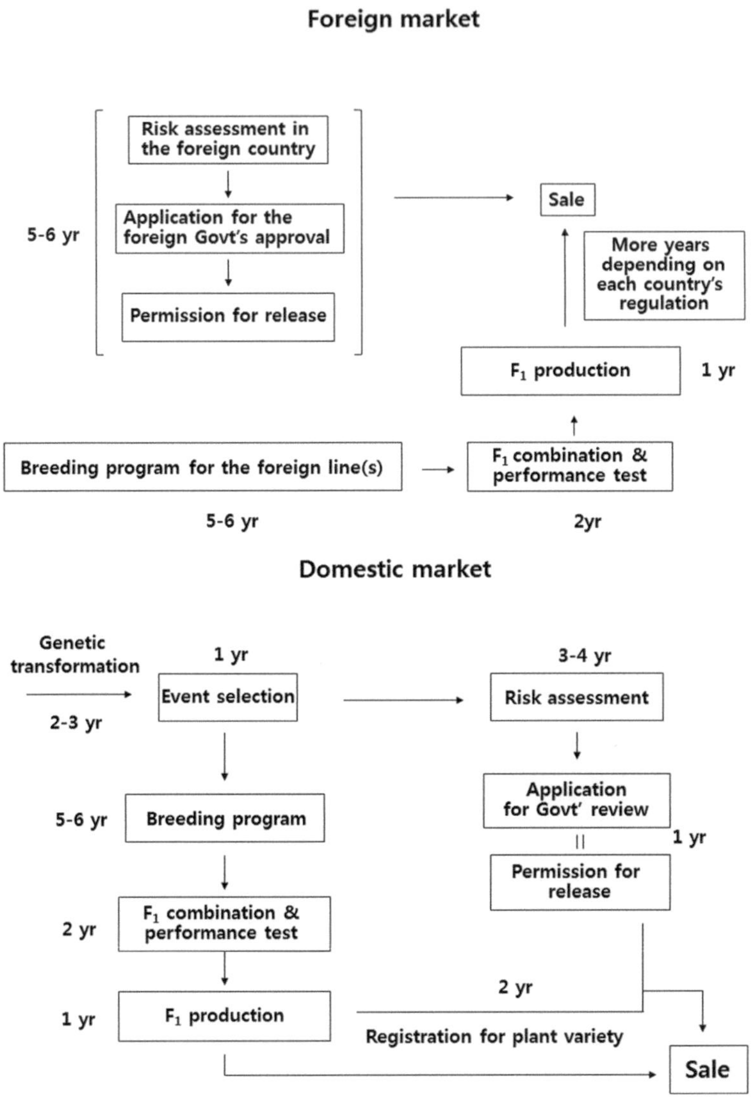

Fig. 9.9 Road map to the commercialization of GM seeds in domestic (Korea) and foreign (outside Korea) markets. The steps and durations may vary depending on each country's regulations but they are similar. The event line can be crossed to a non-GM line in the foreign country to develop a new GM line, and the GM F_1 would fit the local market

crop and the genome can be fixed, often not perfectly but reasonably. This method has traditionally been used worldwide in most crop breeding practices and breeders prefer to use this process even though it takes several years. However, this process may also transfer recessive genes of the donor to the recurrent line by linkage drag from the target gene. Cisgenesis can be used to transfer a gene not by crossing, but by genetic modification. The transgene is not the cDNA but the genomic DNA of the target gene; therefore, the transgene contains its own promoter, exon, and intron. Once the transgene is transferred to the crop and the crop produces seeds and continues for next generations, the vector template, which is designed for deletion, can be selected out from the genome leaving only the whole gene in the transformed crop. Ordinary GM crops contain the cDNA of a target gene instead of the whole gene (Fig. 9.10).

When the genetic relationship between a donor and recipient is distant from each other, it usually takes a long breeding process (several years) using backcrossing to the recipient (recurrent) and self-crossing to fix the genomic background in the recipient crop. When cisgenesis is used, the recipient plant obtains the specific gene directly from the donor plant, therefore the time required to develop a new germplasm line can be reduced. For example, to construct a new tomato line, the average time would be 5–6 years using traditional methods, 7–8 years including the transformation, using ordinary GM methods, and 2–3 years using cisgenesis (Fig. 9.11).

For cisgenesis, the following conditions are required: (1) the whole target gene can be obtained; (2) a recombination-out vector can be constructed; and (3) all vector components, except the whole target gene, can be excluded from the host genome. Although this method uses genomic transformation, it leaves the intact whole gene in the genome and does not include any other vector components; therefore, technically, cisgenesis generates a non-GM plant.

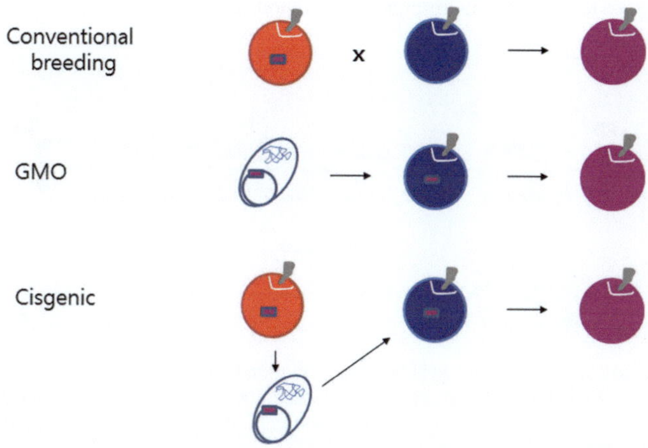

Fig. 9.10 Schematic representation of cisgenesis, genetic modification, and conventional breeding approaches. For conventional breeding, the *purple* gene in the red apple can be transferred to the *blue* apple by crossing; however, it would take a long time to conduct backcrosses and selfcrosses to obtain the *purple* apple. Using genetic modification, the purple gene (cDNA) is transformed in a vector and transferred to the *blue* apple using a genetic transformation, which would generate *purple* apples containing the *purple* gene. Using cisgenesis, the whole *purple* gene (genomic DNA) in the *red* apple is transformed in a vector and transferred to the *blue* apple using a genetic transformation method. During the selection process, the vector components are deleted from the genome leaving the whole *purple* gene in the genome of the *blue* apple, generating a *purple* apple

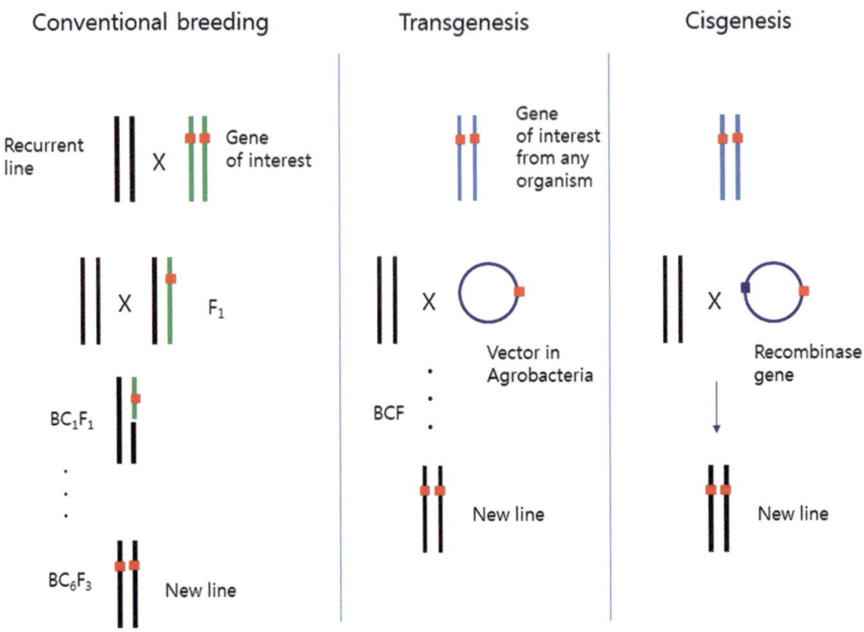

Fig. 9.11 Comparison of conventional breeding, transgenesis, and cisgenesis. Cisgenesis uses whole gene transfer in the same species, while transgenesis transfers a gene from any source such as animals, plants, humans and micro-organisms to another organism

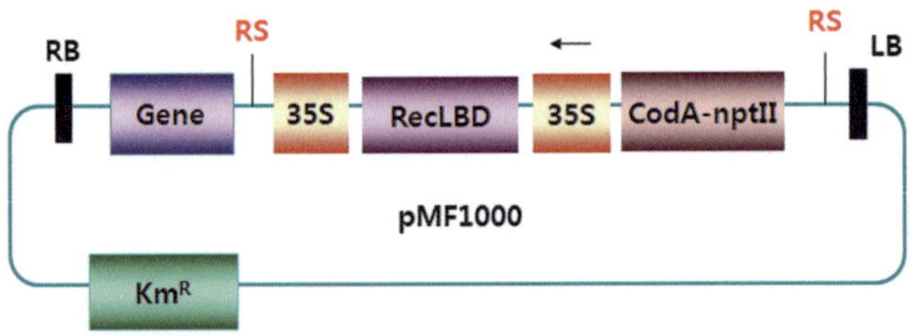

Fig. 9.12 A constructed vector for genetic transformation by cisgenesis. The vector contains there combinase gene (RecLBD) and a target gene. *35S* cucumber mosaic virus promoter, *Km*^{R} kanamycin resistant gene, *CodA* choline oxidase, *nptII* neomycin phosphotransferase type II, *LB* left border, *RB* right border, *RS* recombination site

9.4.1.2 Genetic Transformation for Cisgenesis

The constructed vector (Fig. 9.12) is transformed by *Agrobacterium* and the insert region between the right and left borders are located in the genome. The inserted region contains a gene encoding recombinase. It is advisable to choose an elite line for genetic transformation; however, because the transformation efficiency depends on the genotype, the line that shows the highest regeneration rate should be used.

9.4.1.3 Recombination Mechanism

After the genetic transformation, explants are cultured on MS medium to which dexamethasone has been added to activate the recombinase (Fig. 9.13), which cuts the recombination site leaving only the target gene.

9.4.1.4 Successful Uses of Cisgenesis

A GM apple resistant to scab disease was constructed successfully with the scab resistance gene *HcrVf2* (Vanblaere et al. 2011) that has been used in the recombination vector (Fig. 9.14). The segment in the vector between the left and right borders is inserted into the host genome, and the segment between the recombination sites is removed by recombinase-mediated deletion (Fig. 9.14). This is a useful alternative approach that can help to avoid public controversy about the use of GM in food crops. In addition, Jo et al. (2014) developed a marker-free transformation pipeline to select potato plants functionally expressing a stack of late-blight-resistance genes using cisgenesis.

Therefore, cisgenesis is a promising breeding tool that introduces native genes from the same species to elite lines using GM technology, taking less breeding time and thereby retaining favorable characteristics of established varieties.

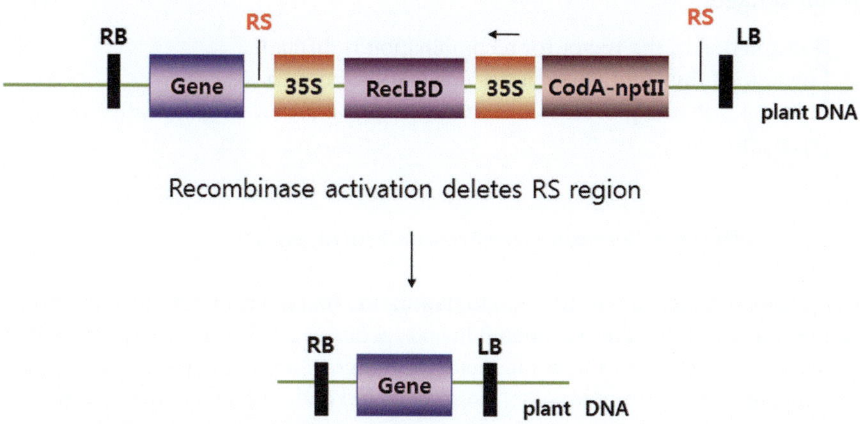

Fig. 9.13 Recombination mechanism is activated after genetic transformation using the constructed vector in Fig. 9.12. Recombinase is activated when dexamethasone is added, and the region between the recombination sites (RSs) with in the right and left boundaries (RB and LB) is deleted, leaving only the target gene in the host genome

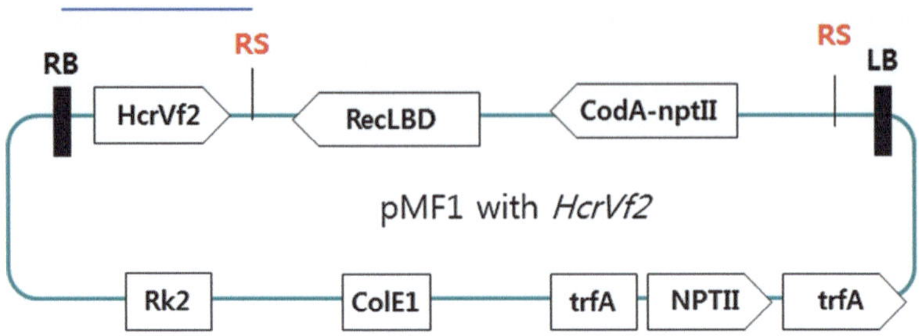

Fig. 9.14 Schematic representation of the pMF1 recombination vector containing the apple scab resistance gene (*HcrVf2*). The gene is controlled by its native regulatory sequences (Referred from Vanblaere et al. 2011). *HcrVf2* apple scab resistance gene from apple cv. Florina. Fusion marker gene *coda-nptII*, hybrid gene for positive (*nptII*) and negative (*codA*) selection. *RecLBD* translational fusion of recombinase R-LBD. Rk2 and ColE1, origins of replication. *trfA* replication gene. *nptIII* kanamycin resistance gene

9.4.1.5 Advantages and Disadvantages of Cisgenesis

Advantages

- Transferring a gene from wild-type in the same species
- Avoiding linkage drag from recessive alleles when backcrosses and self-crosses are conducted
- No need to fix the genome after cisgenesis; saving time for line construction
- Easier stacking of elite alleles.

Disadvantages

- Construction of the vector for recombination is difficult
- Low rate of transformation causes difficulty for cisgenesis
- It is not yet accepted that the modified plant is not a genetically modified organism.

9.4.1.6 Difference Between Cisgenesis and Intragenesis

Cisgenesis transfers a whole gene, including the native promoter, introns, exons, and terminator. Intragenesis (Schouten and Jacobsen 2008) uses the same method to transfer the gene but the components of the gene are not native; instead, they are constructed *in vitro* with other promoters and coding regions. In other words, cisgenesis uses an intact gene directly from the same species, while intragenesis uses a gene from the same species but a promoter and terminator from another gene.

9.4.2 Reverse Breeding

9.4.2.1 Introduction

To develop F_1 hybrids, breeders construct elite lines by crossing lines to develop and select the best horticultural characteristics. Parental lines often do not look particularly good, so it is interesting to consider how breeders select parental lines that contain the vigorous characteristics required in the hybrid. In ordinary breeding practice, the selection of elite offspring (F_1) is not easy and, as a result, reverse breeding was introduced. Reverse breeding is used to generate homozygote lines using F_1 and F_2 that have already been segregated, by blocking recombination during the cell division process (Fig. 9.15). In this way, the original parental lines can be obtained from F_1 and F_2 and new genetic sources with various genetic combinations can be obtained as homozygotes. Therefore, reverse breeding can provide new sources for breeding; therefore, this technology has had a significant impact in breeding programs.

When germ cells are fertilized, chromosome recombination occurs and the cells divide to become daughter cells. If no recombination occurs during the meiotic division, unaltered chromosomes will be transferred to the daughter cells. Pollen cells

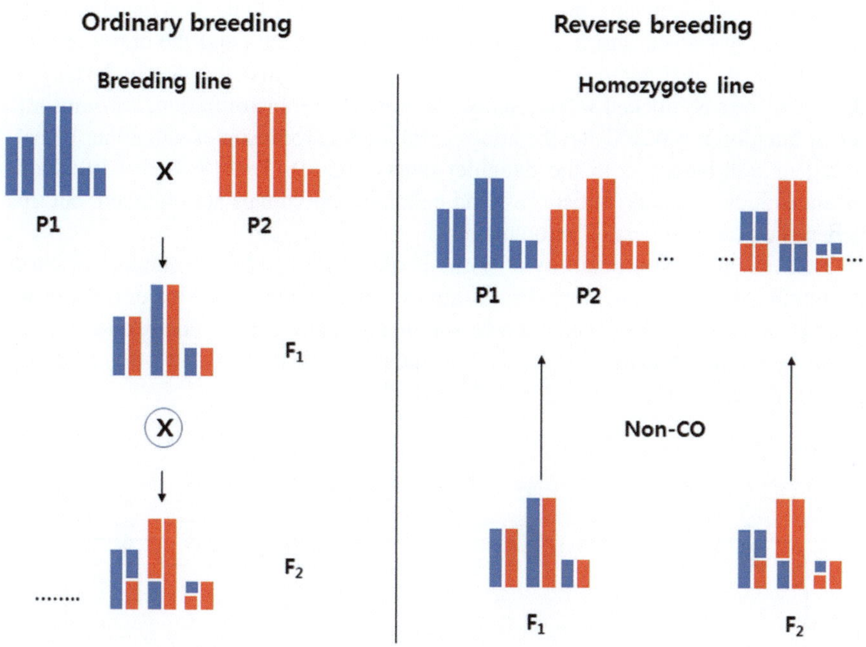

Fig. 9.15 Comparison of F_1 breeding and reverse breeding

can then be obtained and cultured to generate a doubled haploid (DH) line. To ensure that no recombination occurs during meiosis, the specific gene that controls recombination is blocked.

9.4.2.2 Preparation of Reverse Breeding

- Vector construction and genetic transformation: A small region of the disrupted meiotic cDNA1 gene (*DMC1*), including several hundred base pairs of the highly conserved region, are isolated and prepared as a transgene (Fig. 9.16) (Wijnker et al. 2012).
- An RNAi vector containing the transgene leads to gene silencing by knocking out *DMC1*.
- For genetic transformation, it is important to use genotypes that show a high regeneration rate.

9.4.2.3 Selection of Non-recombinant Plants

Normally, chromosome recombination occurs before meiotic division during germ cell development and chromosomes are separated equally to the two daughter cells. Sometimes, however, chromosomal non-disjunction occurs and the chromosomes are not equally distributed. When this occurs, normal germ cells will not develop. If the cross-over is blocked intentionally using genetic transformation, chromosome recombination is blocked and the non-recombinant chromosomes will either be balanced or non-balanced in the daughter cells (Figs. 9.17, 9.18, and 9.19). Non-balanced individuals do not survive and balanced individuals are obtained, but this depends on the numbers of chromosomes.

Theoretically, the probability of normal balancing is $(0.5)^x$ (x = number of chromosomes); therefore, the more chromosomes a plant contains, the lower is the probability of getting balanced non-recombinant individuals. For example, for *Arabidopsis* the probability of normal balancing is $(0.5)^5 = 0.0313$ (3 %), while for pepper, the probability is $(0.5)^{12} = 0.0002$ (0.02 %).

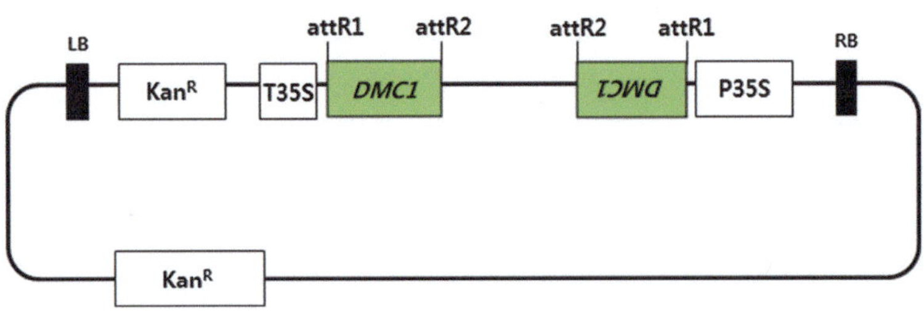

Fig. 9.16 RNAi vector with a part of the *DMC1* gene for vector construction

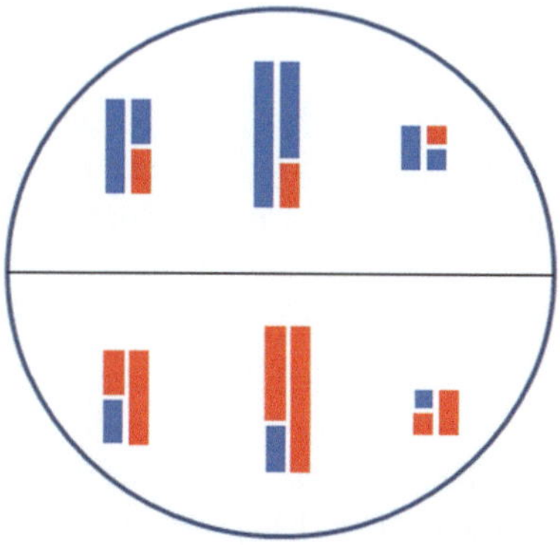

Fig. 9.17 Meiotic cell division with normal recombination of chromosomes

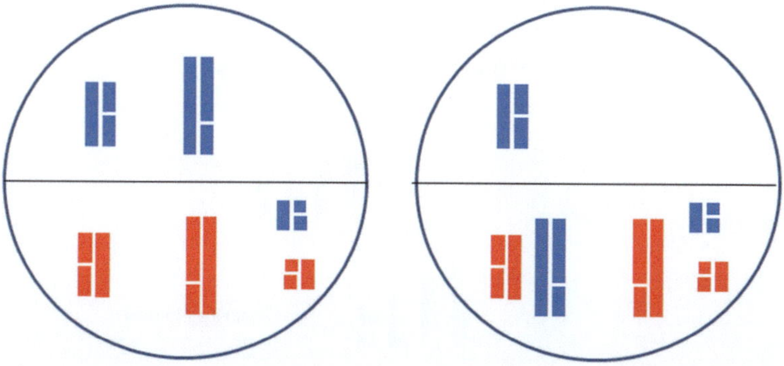

Fig. 9.18 Meiotic cell division with non-recombination and abnormal balancing of chromosomes

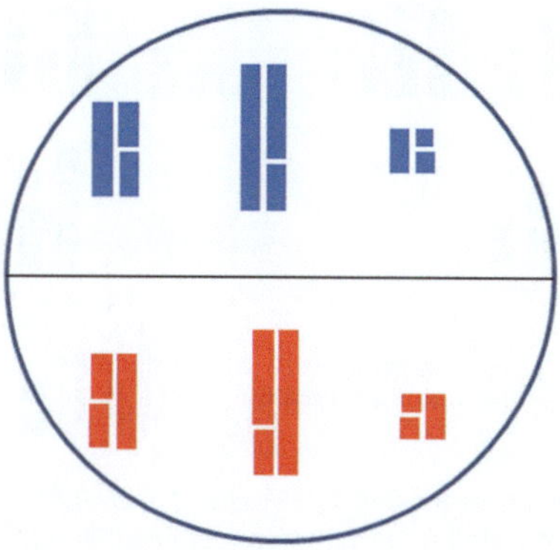

Fig. 9.19 Meiotic cell division with non-recombination and normal balancing of chromosomes

9.4.2.4 Procedure of Reverse Breeding

Transforming F_1 The F_1 hybrid is a product of crossing male and female parental lines and is genetically heterozygote. F_1 hybrids can be transformed with a vector that carries the *DMC1* gene sequence. The transformant selected as the one that blocks chromosome recombination would be a homozygote. The reverse breeding process is described in Fig. 9.20 (modified from Dirks et al. (2009)).

A F_1 heterozygote is the starting hybrid. F_1 seeds are planted to obtain the explants tissues. Explants are transformed with the transgene; the chromosomes of the transformants would not be recombined. After meiotic division, various combinations of chromosomes are generated. The pollen is cultured to obtain the DH line, which is a homozygote. The more DH lines obtained, the better the chance of obtaining chromosome substitution lines. By self-crossing the selected DH lines, new genetic sources can be maintained.

Parental lines for F_1 as well as various combination lines can be found among homozygotes and substitution lines, parental lines for F_1 can be found as well as various combination lines (Wijnker et al. 2012) and new lines can be obtained

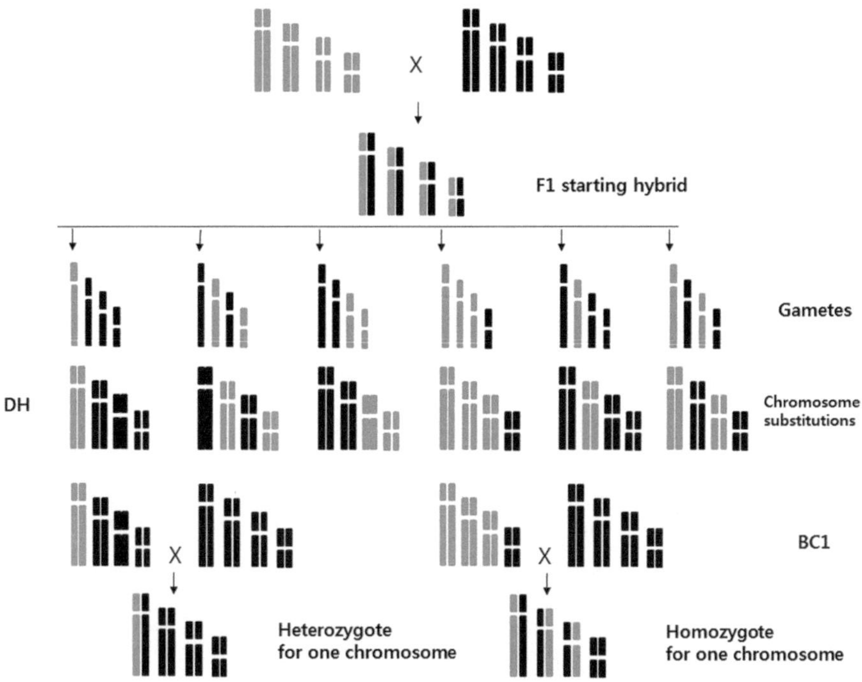

Fig. 9.20 Reverse breeding process using F_1 (Modified from Dirks et al. 2009). After meiotic division, various combinations of chromosomes are obtained: complete *black*, complete *gray*, or one *black* one *gray* as in the F_1 starting hybrid. The doubled haploid (DH) line is a homozygote. The more DH lines obtained the better the chance of generating chromosome substitution lines. By self-crossing the selected DH lines, new genetic sources are maintained

when these various combinations are crossed. When the markers for each chromosome are known, then the chromosomal composition of the new lines can be identified, indicating that the tailor-made breeding for each chromosome could be performed.

Transforming F_2 Chromosomal composition is more segregated in F_2. Each individual plant has its own chromosomal composition based on the recombination. Various compositions would be better as breeding sources (Fig. 9.21).

F_2 is the starting hybrid. The F_2 seeds are planted to obtain the cotyledon or hypocotyl tissues. The explants are transformed with the transgene; the chromosomes of the transformants would not be recombined. After meiotic division, various combinations of chromosomes are generated. The pollen is cultured to obtain the DH line, which is a homozygote. The more DH lines obtained, the better the chance of obtaining chromosome substitution lines. By self-crossing the selected DH lines, new genetic sources can be maintained. Compared with F_1 transformation, F_2 transformation generates more varied homozygotes because the chromosomal composition of the F_2 starting hybrids is more segregated.

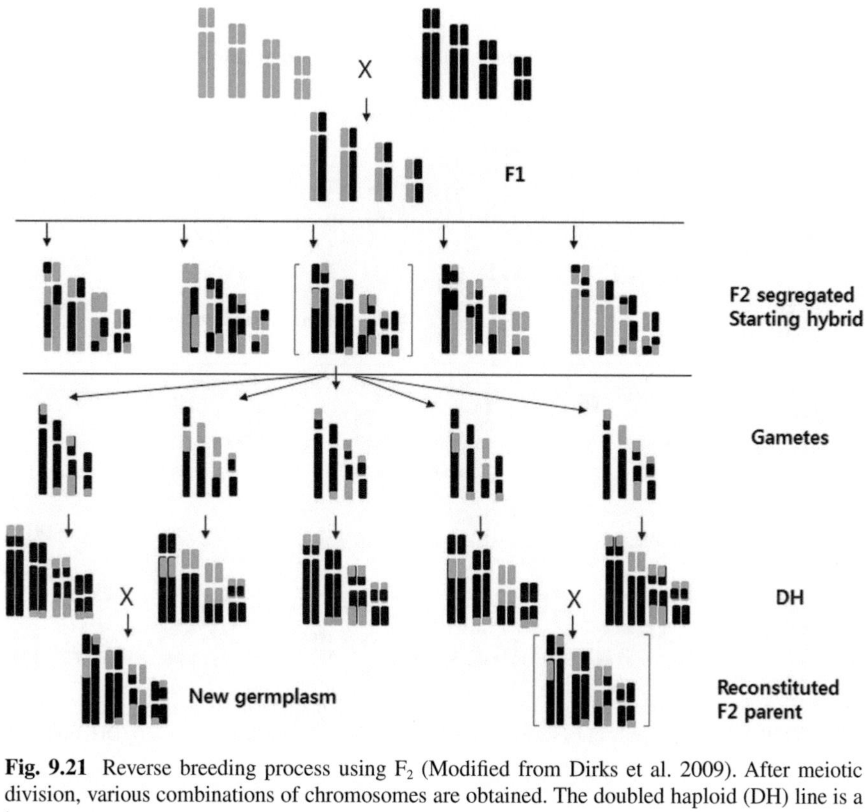

Fig. 9.21 Reverse breeding process using F_2 (Modified from Dirks et al. 2009). After meiotic division, various combinations of chromosomes are obtained. The doubled haploid (DH) line is a homozygote

Selection of Individuals After Reverse Breeding Because the transgene is inserted, it should be selected from the genome to obtain non-GM plants. The creation of non-GM hybrids is the final target for reverse breeding. Figure 9.22 shows that about 50 % of the germ cells in F_2 reverse breeding contain the transgene (red spot). Therefore, when the DH line is obtained, individuals that contain the transgene can be detected by PCR analysis and removed.

When homozygotes, which do not contain the transgene, are selected, the lines are maintained by self-crossing and the selected lines can be crossed to obtain new and various genetic germplasm. This germplasm will provide varied sources because of recombination. Therefore, reverse breeding is an important breeding technology to develop and create genetic sources.

9.4.2.5 Advantages and Disadvantages of Reverse Breeding

Advantages

- Various chromosome combinations can generate balanced parental lines as well as various homozygotes that can become new genetic sources.
- Tailor-made breeding using chromosome substitution lines is possible.
- GM technology can be used to generate new non-GM sources.

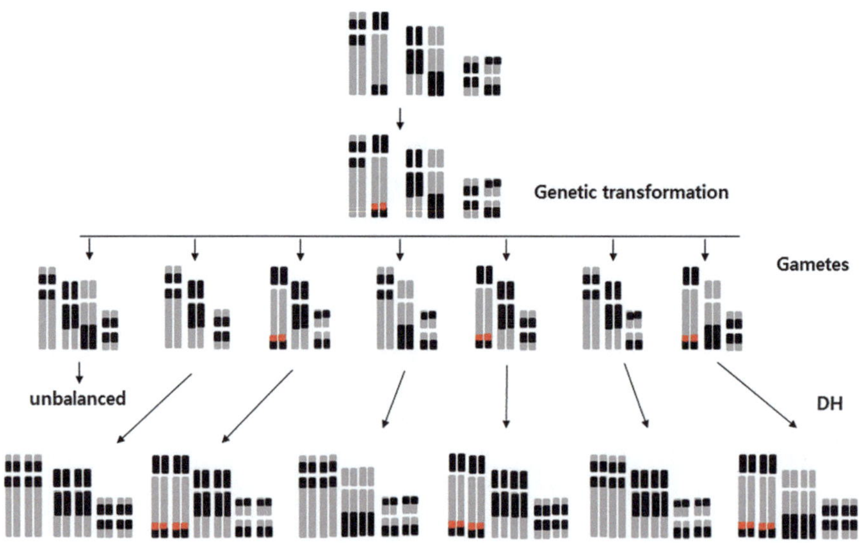

Fig. 9.22 Selection process of F_2 reverse breeding (modified from Wijnker and de Jong 2008). About 50 % of the germ cells in F_2 reverse breeding contain the transgene (*red spot*). DH, doubled haploid

Disadvantages

- When the number of chromosomes is more than 12, it is hard to obtain balanced germ cells.
- It would take a long time to confirm that the selected plant is non-GM.

References

Aswath CR, Mo SY, Kim DH et al (2007) Agrobacterium an biolistic transformation of onion using non-antibiotic selection marker phosphomannose isomerase. Plant Cell Rep 25:92–99

Carbone K (2013) Cultivars: chemical properties, antioxidant activities and health benefits. Nova Science Publishers, Inc. New York, pp 237–251

Cheng M, Fry JE, Pang S et al (1997) Genetic transformation of wheat mediated by *Agrobacterium tumefaciens*. Plant Physiol 115:971–980

Chilton MD, Drummond MH, Merlo DJ (1977) Stable incorporation of plasmid DNA into higher plant cells: the molecular basis of crown gall tumorigenesis. Cell 11:263–271

Choi PS, Soh WY, Kim YS et al (1994) Genetic transformation and plant regeneration of watermelon using *Agrobacterium tumefaciens*. Plant Cell Rep 13:344–348

Dai S, Zheng P, Marmey P et al (2001) Comparative analysis of transgenic rice plants obtained by Agrobacterium-mediated transformation and particle bombardment. Mol Breed 7:25–33

Dale PJ, Hampson KK (1995) An assessment of morphogenic and transformation efficiency in a range of varieties of potato (*Solanum tuberosum* L.). Euphytica 85:101–108

Dan Y, Yan H, Munykwa T et al (2006) MicroTom-a high throughput model transformation system for functional genomics. Plant Cell Rep 25:432–441

Dirks R, Van Dun K, De Snoo CB et al (2009) Reverse breeding: a novel breeding approach based on engineered meiosis. Plant Biotechnol J 7:837–845

Eady CC (1995) Review towards the transformation of onions (*Allium cepa* L.). NZ J Crop Hortic Sci 23:239–250

Ellul P, Garcia-Sogo B, Pineda B et al (2003a) The ploidy level of transgenic plants in *Agrobacterium* mediated transformation of tomato cotyledons (*Lycopersicon esculentum* L. Mill.) is genotype and procedure dependent. Theor Appl Genet 106(2):231–238

Ellul P, Rios G, Atares A et al (2003b) The expression of the Saccharomyces cerevisiae HAL1 gene increase salt tolerance in transgenic watermelon (*Citrullus lanatus* Thunb.) Matsum. & Makai. Theor Appl Genet 107:462–469

Filichkin SA, Gelvin SB (1993) Formation of a putative relaxation intermediate during T-DNA processing directed by the *Agrobacterium tumefaciens* VirD1, D2 endonuclease. Mol Microbiol 8:915–926

Frame BR, Zhang H, Cocciolone S (2000) Production of transgenic maize from bombarded Type II callus: effect of gold particle size and callus morphology on transformation efficiency. In Vitro Cell Dev Biol Plant 36:21–29

Gaba V, Zelcer A, Gal-On A (2004) Cucurbit biotechnology-the importance of virus resistance. In Vitro Cell Dev Biol Plant 40:346–358

Gahakwa D, Maqbool SB, Fu X et al (2000) Transgenic rice as a system to study the stability of transgene expression: multiple heterologous transgenes show similar behavior in diverse genetic backgrounds. Theor Appl Genet 101:388–399

Gaspar T, Kevers C, Penel C et al (1996) Plant hormones and plant growth regulators in plant tissue culture. In Vitro Cell Dev Biol Plant 32:272–289

James C (2014) Global status of commercialized biotech/GM crops: 2014. *ISAAA* Brief No. 49. Ithaca, NY

Jo KR, Kim CJ, Kim SJ et al (2014) Development of late blight resistant potatoes by cisgene stacking. BMC Biotechnol 14:50–60

Lee YH, Kim HS, Kim JY et al (2004) A new selection method for pepper transformation: callus-mediated shoot formation. Plant Cell Rep 23:50–58

Lee YH, Jung M, Shin SH et al (2009) Transgenic peppers that are highly tolerant to a new CMV pathotype. Plant Cell Rep 28:223–232

Mysore KS, Nam J, Gelvin SB (2000) An Arabidopsis histone H2A mutant is deficient in *Agrobacterium* T-DNA integration. Proc Natl Acad Sci 97:948–953

Nandakumar R, Babu S, Wang ZY (2011) DNA delivery systems. In: DanY, Ow DW (eds), Historical technology developments in plant transformation technology revolution. Bentham Science Publishers, USA, pp 3–24

Nugent GD, ten Have M, van der Gulik A et al (2005) Plastid transformants of tomato selected using mutations affecting ribosome structure. Plant Cell Rep 24:341–349

Olhoft PM, Somers DA (2001) L-Cysteine increases Agrobacterium-mediated T-DNA delivery into soybean cotyledonary node cells. Plant Cell Rep 20:706–711

Olhoft PM, Flagel LE, Donovan CM et al (2003) Efficient soybean transformation using hygromycin B selection in the cotyledonary-node method. Planta 5:723–735

Olsen FL (1987) Induction of microspore embryogenesis in cultured anthers of *Hordeum vulgare*: the effects of ammonium nitrate, glutamic acid and asparagines as nitrogen sources. Carlsberg Res Commun 52:393–404

Ooms G, Burrell MM, Karp A et al (1987) Genetic transformation in two potato cultivars with T-DNA from disarmed *Agrobacterium*. Theor Appl Genet 73:744–750

Park SH, Morris JL, Park JE et al (2003) Efficient and genotype-independent Agrobacterium-mediated tomato transformation. J Plant Physiol 160:1253–1257

Paz MM, Martinez JC, Kalvig AB et al (2006) Improved cotyledonary node method using an alternative explants derived from mature seed for efficient Agrobacterium-mediated soybean transformation. Plant Cell Rep 25:206–213

Schouten HJ, Jacobsen E (2008) Cisgenesis and intragenesis, sisters in innovative plant breeding. Trends Plant Sci 13:260–261

Shrawat AK, Becker D, Lorz H (2007) *Agrobacterium* mediated genetic transformation of barley (*Hordeum vulgare* L.). Plant Sci 172:281–290

Sigareva M, Spivey R, Wilits MG et al (2004) An efficiency mannose selection protocol for tomato that has no adverse effect on the ploidy level of transgenic plants. Plant Cell Rep 23:234–245

Tingay S, McElroy D, Kalla R (1997) *Agrobacterium* tumefaciens-mediated barley transformation. Plant J 11:1369–1376

Tzfira T, Citovsky V (2000) From host recognition to T-DNA integration: the function of bacterial and plant genes in the *Agrobacterium*–plant cell interaction. Mol Plant Pathol 1:201–212

Vanblaere T, Szankowski I, Schaart J et al (2011) The development of a cisgenic apple plant. J Biotechnol 154:304–311

Vaucheret H, Beclin C, Elmayan T et al (1998) Transgene-induced gene silencing in plants. Plant J 16:651–659

Wan Y, Lemaux PG (1994) Generation of large numbers of independently transformed fertile barley plants. Plant Physiol 104:37–48

Wang G, Xu Y (2008) Hypocotyl-based Agrobacterium-mediated transformation of soybean (Glycine max) and application for RNA interference. Plant Cell Rep 27:1177–1184

Wijnker E, de Jong H (2008) Managing meiotic recombination in plant breeding. Trends Plant Sci 13:640–646

Wijnker E, van Dun K, de Snoo CB et al (2012) Reverse breeding in Arabidopsis thaliana generates homozygous parental lines from a heterozygous plant. Nat Genet 44:467–470

Zhang ZY, Xing AQ, Staswick P et al (1999) The use of glufosinate as a selective agent in Agrobacterium-mediated transformation of soybean. Plant Cell Tissue Organ Cult 56:37–46

Zhang W, Wang K, Lin ZS (2014) Production and identification of haploid dwarf male sterile wheat plants induced by corn inducer. Bot Stud 55:1–8

Zhao P, Wang K, Zhang W (2014) Review and inspiration of plant proteins involved in the transformation processing of T-DNA initiated by *Agrobacterium*. Sctientica Agric Sin 47(13):2504–2518

Zheng SJ, Khrustaleva L, Henken B et al (2001) *Agrobacterium tumefaciens*-mediated transformation of *Allium cepa*L.: the production of transgenic onion and shallots. Mol Breed 7:101–115

Zhou H, Arrowsmith J, Fromm M et al (1995) Glyphosate-tolerant CP4 and GOX genes as a selectable marker in wheat transformation. Plant Cell Rep 15:159–163

Zhu Y, Nam J, Carpita NC (2003) *Agrobacterium*-mediated root transformation is inhibited by mutation of an Arabidopsis cellulose synthase-like gene. Plant Physiol 133:1000–1010

Index

H.-J. Koh et al. (eds.), *Current Technologies in Plant Molecular Breeding*, DOI 10.1007/978-94-017-9996-6

S

T

U

V

MIX
Papier aus verantwortungsvollen Quellen
Paper from responsible sources
FSC® C105338

If you have any concerns about our products,
you can contact us on
ProductSafety@springernature.com

In case Publisher is established outside the EU,
the EU authorized representative is:
Springer Nature Customer Service Center GmbH
Europaplatz 3, 69115 Heidelberg, Germany

Printed by Libri Plureos GmbH
in Hamburg, Germany